Oxidative Stress Biomarkers and Antioxidant Protocols

METHODS IN MOLECULAR BIOLOGY™

John M. Walker, Series Editor

METHODS IN MOLECULAR BIOLOGY™

Oxidative Stress Biomarkers and Antioxidant Protocols

Edited by

Donald Armstrong

University of Florida, Gainesville, FL

Humana Press ✳ **Totowa, New Jersey**

Cover illustration: The false color image on the cover was provided by Professor Susumu Terakawa and illustrates glucose-induced superoxide production in the islet of Langerhans from rat pancreas maintained in culture. Superoxide was monitored in the dark by MCLA-dependent chemiluminescence as described in Chapter 24 of the companion volume to this book, titled "*Ultrastructural and Molecular Biology Protocols*," Vol. 196 of the Methods in Molecular Biology™ Series.

Production Editor: Kim Hoather-Potter.

Cover design by Patricia F. Cleary.

For additional copies, pricing for bulk purchases, and/or information about other Humana titles, contact Humana at the above address or at any of the following numbers: Tel: 973-256-1699; Fax: 973-256-8341; E-mail: humana@humanapr.com, or visit our Website at www.humanapress.com

Library of Congress Cataloging in Publication Data available.

Preface

The first protocols book, *Free Radical and Antioxidant Protocols (1)* was published in late 1998. Sections were divided into three parts, covering selected biochemical techniques for measuring oxidative stress, antioxidant (AOX) activity, and combined applications. In choosing the 40 methods to be included in that book, I realized there were considerably more of equal value than that which we could have presented in a single volume. To produce a comprehensive resource, this book and a third are being compiled to expand coverage of the field.

A summary of papers *(2)* published on this important subject emphasizes the continuing rapid growth in oxidative stress investigations relating to our understanding of biochemical reactions, their relevance to pathophysiological mechanisms, how disease may arise, and how therapeutic intervention may be achieved *(3)*. Although there is some overlap between the categories, the analysis shown below illustrates where current studies are concentrated and are almost evenly distributed between free radicals and AOX. Over the last 4 yr, there has been a 55% increase in the number of papers published in the area.

Table 1
Recent Citations of Oxidative Stress Biomarkers

	1997	1998	1999	2000
Free radical mechanisms	60	72	75	92
Free radicals in disease	78	88	109	111
AOX mechanisms	87	91	150	160
AOX in disease	94	122	155	204
Applications for treatment	0	2	5	8
TOTAL	**319**	**375**	**494**	**575**

Oxidative Stress Biomarkers and Antioxidant Protocols has added 33 more high-tech methods written by 73 authors from prestigious universities/institutes around the world, which together with our previous volume 108, provide a wide range of procedures for evaluating perturbations in cell function resulting from increased oxidative stress. Although primarily a reference for research, these two books also provide easy-to-follow directions that make them readily adaptable for academic use as a laboratory manual for graduate students in the basic sciences.

Of particular interest is the final chapter, which describes how the grouping of data from more than two biomarkers can be used to derive an appropriate statistical measure of change in the biological systems under study. The ability to more accurately interpret oxidative stress results in terms of either free radicals or AOX by using data from each to characterize laboratory or clinical observations, greatly enhances the value of this specific biostatistical approach.

I thank the Department of Small Animal Clinical Services, University of Florida College of Veterinary Medicine, and the Department of Clinical Laboratory Science, University at Buffalo for administrative support and facilities to produce this book. Professor John Walker, the *Methods in Molecular Biology*™ Series Editor, was helpful in the review process. Linda Rose and Chris Armstrong provided essential secretarial assistance and Aqeela Afzal compiled the literature search data shown in Table 1. I am indebted to authors in this volume and colleagues who alerted me to other technologies that were ultimately included to broaden its scope.

Donald Armstrong

References

1. Armstrong, D. (ed.) (1998) *Free Radical and Antioxidant Protocols, Methods in Molecular Biology, vol. 108.* Humana Press Inc., Totowa, NJ.
2. Internet Grateful Med (2001).
3. Armstrong, D. (1994) *Free Radicals in Diagnostic Medicine, Advances in Experimental Medicine and Biology, vol. 366.* Plenum Press, NY.

Contents

Contents

Contributors

AQEELA AFZAL • *Department of Small Animal Clinical Sciences, University of Florida, College of Veterinary Medicine, Gainesville, FL*

KURSHID ALAM • *Department of Biochemistry, Faculty of Medicine, Aligarh Muslim University, Aligarh, India*

RASHID ALI • *Department of Biochemistry, Faculty of Medicine, Aligarh Muslim University, Aligarh, India*

MOHAMMED AFZAL • *Department of Biological Sciences, Faculty of Science, Kuwait University, Safat, Kuwait*

DONALD ARMSTRONG • *Department of Biotechnology and Clinical Laboratory Sciences, State University of New York, Buffalo, NY; Free Radical and Antioxidant Laboratory and Department of Small Animal Clinical Sciences, University of Florida, College of Veterinary Medicine, Gainesville, FL*

UMA ARORA • *Institute of Toxicology, GSF-National Research Center for Environment and Health, Neuherberg, Germany*

MARIA GIULIA BATTELLI • *Department of Experimental Pathology, University of Bologna, Bologna, Italy*

JOHN W. BAYNES • *Department of Chemistry and Biochemistry, University of South Carolina, Columbia, SC*

NICHOLAS BODOR • *Department of Pharmaceutics, Center for Drug Discovery, University of Florida, College of Pharmacy, Gainesville, FL*

DOUGLAS BORCHMAN • *Department of Ophthalmology and Visual Sciences, University of Louisville School of Medicine, Louisville, KY*

ANNA BRAJTER-TOTH • *Department of Chemistry, University of Florida, Gainesville, FL*

ROBERTO BRAVO • *Department of Chemistry, University of Florida, Gainesville, FL*

RICHARD W. BROWNE • *Department of Biotechnology and Clinical Laboratory Science, State University of New York, Buffalo, NY*

PETER BUCHWALD • *Department of Pharmaceutics, Center for Drug Discovery, University of Florida, College of Pharmacy, Gainesville, FL*

I. HENDRIKJE BUSS • *Department of Pathology, Free Radical Research Group, Christchurch School of Medicine and Health Sciences, Christchurch, New Zealand*

GERARDO D. CASTRO • *Centro de Investigaciones Toxicologicas, (CEITOX)-CITEFA/CONICET, Buenos Aires, Argentina*

JOSÉ A. CASTRO • *Centro de Investigaciones Toxicologicas, (CEITOX)-CITEFA/CONICET, Buenos Aires, Argentina*

HAW-WEN CHEN • *Department of Nutrition, Chung Shan Medical University, Taiwan, Republic of China*

ANDREW R. COLLINS • *Rowett Research Institute, Aberdeen, UK*

MICHAEL J. DAVIES • *The Heart Research Institute, Sydney, Australia*

ROGER T. DEAN • *The Heart Research Institute, Sydney, Australia*

YASHWANT G. DESHPANDE • *Division of General Surgery, Department of Surgery, St. Louis University School of Medicine and Health, St. Louis, MO*

ALEXANDRINE DURING • *Phytonutrients Laboratory, Beltsville Human Nutrition Research Center, USDA, Beltsville, MD*

MÁRIA DUŠINSKÁ • *Institute of Preventive and Clinical Medicine, Bratislava, Slovak Republic*

GORDON A. FRANCIS • *Lipid and Lipoprotein Research Group, Faculty of Medicine and Oral Health Sciences, University of Alberta, Edmonton, Canada*

SHANLIN FU • *The Heart Research Institute, Sydney, Australia*

YOSHIHIRO HIGUCHI • *Department of Pharmacology, Kanazawa University School of Medicine, Kanazawa, Japan*

ANDREW JONES • *Department of Small Animal Clinical Sciences, University of Florida, College of Veterinary Medicine, Gainesville, FL*

DONALD L. KAMINSKI • *Division of General Surgery, Department of Surgery, St. Louis University, St. Louis, MO*

YOJI KATO • *School of Humanities for Environmental Policy and Technology, Himeji Institute of Technology, Himeji, Japan*

DOMINIK KLEIN • *Institute of Toxicology, GSF-National Research Center for Environment and Health, Neuherberg, Germany*

LIOUBOV G. KOROTCHKINA • *Department of Biochemistry, School of Medicine and Biomedical Sciences, State University of New York, Buffalo, NY*

BRUCE S. KRISTAL • *Departments of Biochemistry and Neuroscience, Burke Medical Research Institute, Weill Medical College of Cornell University and Dementia Research Service, White Plains, NY*

LEONARD A. LEVIN • *Department of Ophthalmology and Visual Science, University of Wisconsin Medical School, Madison, WI*

CHONG-KUEI LII • *Department of Nutrition, Chung Shan Medical University, Taiwan, Republic of China*

MITSUHARU MASUDA • *Unit of Endogenous Cancer Risk Factors, International Agency for Research on Cancer, Lyon, France*

WAYNE R. MATSON • *ESA Inc., Chelmsford, MA*

SHINICHI MIYAIRI • *Laboratory of Pharmaceutical Chemistry, College of Pharmacy, Nihon University, Chiba, Japan*

BÉNÉDICTE MORIN • *The Heart Research Institute, Sydney, Australia*

JASON D. MORROW • *Departments of Medicine and Pharmacology, Vanderbilt University School of Medicine, Nashville, TN*

AUDRIC S. MOSES • *Department of Medicine, University of Alberta, Edmonton, Canada*

SILVIA MUSIANI • *Department of Experimental Pathology, University of Bologna, Bologna, Italy*

AKIRA NAGANUMA • *Laboratory of Molecular and Biochemical Toxicology, Tohoku University, Sendai, Japan*

AKIHIKO NAGAO • *National Food Research Institute, Ministry of Agriculture, Forestry, and Fisheries, Ibaraki, Japan*

ROSNAH MD. NOR • *Food Technology and Nutritional Unit, Malaysian Palm Oil Board (MPOB), Kuala Lumpur, Malaysia*

HIROSHI OHSHIMA • *Unit of Endogenous Cancer Risk Factors, International Agency for Research on Cancer, Lyon, France*

TOSHIHIKO OSAWA • *Nagoya University Graduate School of Bioagricultural Sciences, Nagoya, Japan*

NINDER PANESAR • *Division of General Surgery, Department of Surgery, St. Louis University School of Medicine and Health, St. Louis, MO*

ELIZABETH J. PARKS • *Department of Food Science and Nutrition, University of Minnesota, Twin Cities, St. Paul, MN*

MULCHAND S. PATEL • *Department of Biochemistry, School of Medicine and Biomedical Sciences, State University of New York, Buffalo, NY*

BRIGITTE PIGNATELLI • *Unit of Endogenous Cancer Risk Factors, International Agency for Research on Cancer, Lyon, France*

NARSING A. RAO • *Department of Ophthalmology, Doheny Eye Institute, University of Southern California, Los Angeles, CA*

L. JACKSON ROBERTS, II • *Departments of Medicine and Pharmacology, Vanderbilt University, Nashville, TN*

MITUAKI SANO • *Laboratory of Health Science, School of Pharmaceutical Science, Shizuoka Sangyo University, Shizuoka, Japan*

SHIN SATO • *Department of Food and Nutrition, Hakodate Junior College, Hokkaido, Japan*

ENRIQUE F. SCHISTERMAN • *UCLA School of Public Health, Cedars Sinai Medical Center, Los Angeles, CA*

J. NIKKI SHAW • *Department of Chemistry and Biochemistry, University of South Carolina, Columbia, SC*

SANTOSH SINHA • *Department of Ophthalmology and Visual Sciences, University of Louisville, Louisville, KY*

JAMES CECIL SMITH, JR. • *Phytonutrients Laboratory, Beltsville Human Nutrition Research Center, USDA, Beltsville, MD*

JOSE SOUZA • *Departments of Biochemistry and Biophysics, Stokes Research Institute, Children's Hospital of Pennsylvania, The University of Pennsylvania Medical Center, Philadelphia, PA*

DAWN M. STICKLE • *Department of Chemistry, University of Florida, Gainesville, FL*

KARL H. SUMMER • *Institute of Toxicology, GSF-National Research Center for Environment and Health, Neuherberg, Germany*

KALYANA SUNDRAM • *Food Technology and Nutritional Unit, Malaysian Palm Oil Board (MPOB), Kuala Lumpur, Malaysia*

CSABA SZABÓ • *Inotek Corporation, Beverly, MA*

SUSAN R. THORPE • *Department of Chemistry and Biochemistry, University of South Carolina, Columbia, SC*

ISAO TOMITA • *Laboratory of Health Science, School of Pharmaceutical Science, Shizuoka Sangyo University, Shizuoka, Japan*

KEIZO UMEGAKI • *Laboratory of Health Science, School of Pharmaceutical Science, Shizuoka Sangyo University, Shizuoka, Japan*

KAREN VIGNEAU-CALLAHAN • *ESA Inc., Chelmsford, MA*

LÁSZLÓ VIRÁG • *Department of Medical Chemistry, Faculty of Medicine, University of Debrecen, Debrecen, Hungary*

HONGJIE WANG • *The Heart Research Institute, Sydney, Australia*

CHRISTINE C. WINTERBOURN • *Department of Pathology, Free Radical Research Group, Christchurch School of Medicine and Health Sciences, Christchurch, New Zealand*

GUEY-SHUANG WU • *Doheny Eye Institute, University of Southern California, Los Angeles, CA*

YORIHIRO YAMAMOTO • *Department of Chemistry and Biotechnology, Graduate School of Engineering, The University of Tokyo, Tokyo, Japan*

SATOSHI YAMASHITA • *Research Center for Advanced Science and Technology, University of Tokyo, Tokyo, Japan*

VLADAMIR YERMILOV • *Unit of Endogenous Cancer Risk Factors, International Agency for Research on Cancer, Lyon, France*

I

Techniques for Free Radical Derived Biomarkers

1

Human Xanthine Oxidoreductase Determination by a Competitive ELISA

Maria Giulia Battelli and Silvia Musiani

1. Introduction

Xanthine oxidoreductase (XO) catalyses the oxidation of hypoxanthine to xanthine and of the latter to uric acid. The enzyme is present in traces in most of human tissues, including plasma, being more abundant in milk, liver, and intestine *(1)*.

Both in experimental animals and in humans (reviewed in **ref.** *2*), an increased plasma level of XO has been associated with various pathological conditions in which the enzyme may leak out from impaired cells *(3)*, and may be released from damaged tissues into circulation. During the oxidation of substrates the enzyme generate superoxide anion and hydrogen peroxide, which have cytotoxic effects *(4)*. The active oxygen species produced by the activity of XO may also amplify the damage and could cause tissue injury even at other sites *(5)*.

Consistently with the considerable amount of XO activity found in the liver, this organ is the main source of serum enzyme in experimental pathology *(6)* and possibly also in humans. Although the physiopathological meaning of an increased level of serum XO has not been clarified, the elevation of XO in human serum observed in association with hepatic damage (*see* for example **refs.** *7,8*) suggests a clinical value for its determination in the differential diagnosis of liver diseases.

In normal human serum the level of XO is very low, and can be detected only with particularly sensitive methods, which in most of the cases detect the protein from its enzymatic activity. Only the enzyme immunoassay method *(2)*

From: *Methods in Molecular Biology, vol. 186: Oxidative Stress Biomarkers and Antioxidant Protocols*
Edited by: D. Armstrong © Humana Press Inc., Totowa, NJ

has the advantage of measuring both active and inactive XO protein and is also convenient for routine execution in clinical laboratories.

We describe here the purification of XO from human milk and the determination of its enzymatic activity, the preparation of rabbit polyclonal anti-serum and its purification, and the determination of human serum XO by the competitive immunoenzymatic test.

2. Materials

2.1. Equipment

1. Stirrer.
2. Conductivity and pH meter.
3. Centrifuge (*see* **Note 1**).
4. Fraction collector, equipped with peristaltic pump, UV detector, and recorder.
5. UV and VIS spectrophotometer (*see* **Note 2**).
6. Microplate reader equipped with a 405 nm filter.
7. Chromatographic columns: 2.6×40 cm; 2.6×40 cm; 5×20 cm and 1×10 cm.
8. Chromatographic resins: Hypatite C (Clarkson, Williamsport, PA); CF11 cellulose (Whatman, Maidstone, Kent, UK); Sephadex G25 coarse and CNBr-activated Sepharose 4B (Pharmacia LKB Biotech, Uppsala, Sweden); DE52 (Whatman) (*see* **Note 3**).
9. Dialysis tubing 20/32", siliconized Vacutainer tubes, microtiter plates.
10. PC with software for statistical analysis.

2.2. Reagents

1. Ammonium sulfate, 4-nitrophenylphosphate disodium salt exahydrate, isobutanol.
2. Complete and incomplete Freund's adjuvant (Gibco, Grand Island, NY), bovine serum albumin (BSA) and antirabbit IgG-alkaline phosphatase conjugate (Sigma Chemical Co., St. Louis, MO).
3. Solutions: 5×10^{-1} *M* NaOH; 10^{-1} *M* Titriplex; 10^{-1} *M* Na salicylate; 10^{-1} *M* 2-mercaptoethanol; 3×10^{-4} *M* xanthine sodium salt; DE52 solution: a mixture of Titriplex, Na salicylate and 2-mercaptoethanol, 10^{-4} *M* each, adjusted to pH 9.0 with KOH, store at 0°C; 10^{-1} *M* KH$_2$PO$_4$ in DE52 solution, store at 0°C; substrate solution: 4-nitrophenylphosphate disodium salt exahydrate 1 mg/mL in diethanolamine buffer (*see* below in **Subheading 2.2., item 4**).
4. Buffers, store all buffers at 0°C:
   a. Hypatite C buffer: 2×10^{-1} *M* Na phosphate buffer, pH 6.0, containing 10^{-4} *M* Titriplex, 10^{-4} *M* Na salicylate, and 10^{-4} *M* 2-mercaptoethanol.
   b. Phosphate-buffered saline (PBS): 5×10^{-3} *M* Na phosphate buffer, pH 7.5, containing 1.4×10^{-2} *M* NaCl.
   c. PBS-Tween: 0.05% (v/v) Tween 20 in PBS.

    d. Determination buffer: 1 M Tris-HCl (hydroxymethyl) aminomethane buffer, pH 8.1.

    e. Coupling buffer: 10^{-1} M Na carbonate buffer, pH 8.3, containing 5×10^{-1} M NaCl.

    f. Adhesion buffer: 5×10^{-2} M Na carbonate buffer, pH 9.6.

    g. Affinity buffer: 5×10^{-3} M Na phosphate buffer, pH 7.5, containing 5×10^{-1} M NaCl.

    h. Diethanolamine buffer: 1 M diethanolamine buffer containing 5×10^{-4} M $MgCl_2$ and 3.1×10^{-3} M NaN_3, pH 9.8.

2.3. Animals and Human Materials

1. New Zealand rabbits weighing 2–3 kg.
2. Pooled human milk, store in 0.5-L aliquots at –20°C.
3. Pooled human serum (HS) from healthy donors, store in 1.5-mL aliquots at –20°C.

3. Methods

3.1. Preparation of Chromatographic Columns

1. Activate 160 mL Hypatite C with 0.5 L NaOH solution by slowly stirring for 60 min at 45°C. Wash the resin with distilled water to pH 7.0. Wet 18 g CF11 cellulose with distilled water until swollen. Mix the two resins and pack into a 2.6×40 cm column. Equilibrate with Hypatite C buffer and check the effluent by the conductivity and pH meter.
2. Wet 20 g Sephadex G25 coarse with water until swollen. Pack into a 2.6×40 cm column and equilibrate with DE52 solution.
3. Wash 300 mL DE52 with DE52 solution until the equilibrium is reached, then pack into a 5×20 cm column.
4. To couple CNBr-activated Sepharose 4B resin to HS protein follow manufacturer's instructions using 1 g resin. Dialyze twice 0.5 mL HS against 0.5 L coupling buffer, then remove the precipitate by centrifugation at 12,000g for 2 min. Allow the coupling reaction to proceed for 2 h at room temperature, then block excess remaining groups and wash the resin from uncoupled ligand following manufacturer's instructions. Pack into a 1×10 cm column and equilibrate with affinity buffer.

3.2. Determination of Enzyme Activity

XO activity was determined spectrophotometrically for 5 min at 25°C by measuring the A_{292}, which indicates the formation of uric acid from xanthine. In 1 mL final volume add 0.1 mL determination buffer, 0.2 mL xanthine solution (avoid in reference cuvet), and 0.1 mL sample. A unit of enzyme activity is defined as the formation of 1 µmole/min uric acid, utilizing the extinction coefficient $E^{M}_{292nm,1cm} = 11.000$ to calculate the amount of uric acid produced.

3.3. Purification of Human Milk Xanthine Oxidoreductase

All XO purification steps are carried out at 4°C.

1. Thaw 0.5 L human milk and add Titriplex, Na salicylate, and 2-mercaptoethanol solutions, 0.5 mL each. Mix 84 mL iso-butanol previously cooled at –20°C to milk, then add 97 g solid ammonium sulfate, a small amount at a time, keeping on stirring for at least 1.5 h.
2. Centrifuge the mixture at 18,000g for 30 min, then collect the aqueous phase, under the fat phase. Add to the aqueous phase 110 g/L solid ammonium sulfate, few at a time, keeping on stirring for at least 1.5 h.
3. Centrifuge as in **Subheading 3.3., step 2**, then collect the floating precipitate, dissolve with Hypatite C buffer (approx 75 mL) and dialyze three times against 1.5 L Hypatite C buffer for at least 3 h each time.
4. Remove the precipitate remaining after the dialysis by centrifugation at 30,000g for 30 min, then adsorb the clear supernatant on the Hypatite C column connected to the fraction collector at a pump speed of 60 mL/h. Wash the column with Hypatite C buffer (approx 1 L).
5. Discard the pick of unabsorbed proteins without enzymatic activity. When the plot is back to the baseline, elute the column with 5% (w/v) ammonium sulfate in Hypatite C buffer and determine XO activity in the fractions. Pool the active fractions and add 361 mg/mL ammonium sulfate stirring for 1.5 h.
6. Collect the precipitate by centrifugation as above in **Subheading 3.3., step 4**, then dissolve in the minimum required volume of DE52 solution and gel filter on the Sephadex G25 column connected to the fraction collector with DE52 solution at 60 mL/h.
7. Pool the protein-containing fractions and immediately apply them to the DE52 column connected to the fraction collector. Elute at 60 mL/h, without any washing, with a 0–0.1 M linear gradient of KH_2PO_4 in DE52 solution (total volume 1.6 L).
8. Determine XO activity in the fractions and pool the active ones. Concentrate by precipitation with 361 mg/mL ammonium sulfate and centrifugation as above in **Subheading 3.3., step 4**. Dissolve the precipitate in the minimum required volume of PBS and dialyze against PBS as in **Subheading 3.3., step 3**.
9. Determine protein concentration of purified human milk XO, divide into 0.5 mL aliquots and store at –80°C. When required, thaw an aliquot, divide it into 25-µL aliquots and store at –20°C.

3.4. Production and Purification of Antiserum

Use rabbits to produce the antibodies following the national guidelines for the care and use of laboratory animals.

1. Collect preimmune blood of each anesthetized animal from an ear incision in siliconized Vacutainer tubes. Allow clot formation for 1 h at room temperature,

then 12–18 h at 4°C. Harvest clot-free serum and centrifuge at 1,500g for 5 min at 4°C to remove residual cells, then divide into 1.5-mL aliquots and store at –80°C.

2. To each rabbit inject subcutaneously 650 µg human milk XO in complete Freund's adjuvant, dividing the dose into three different sites on the hind leg. Boost each animal with 350 µg of purified enzyme in incomplete Freund's adjuvant at 3 and 6 wk after the primary injection.

3. Collect the blood at 5 d intervals after the second injection, separate serum, and store as in **Subheading 3.4., step 1**.

4. Clear rabbit antiserum of antihuman antibodies, nonspecific for XO, by affinity chromatography on the column packed with CNBr-activated Sepharose 4B coupled to HS proteins, connected to the fraction collector. Thaw 1.5 mL rabbit antiserum, remove the precipitate by centrifugation at 12,000g for 2 min and apply to the column. Elute at 20 mL/h with affinity buffer and pool the protein containing fractions.

5. Determine protein concentration of purified antiserum and store in 2-mL aliquots at –20°C. When required, thaw an aliquot, divide it into smaller aliquots depending on titration results (*see* below in **Subheading 3.4., step 6**) and store at –20°C.

6. Perform a checkerboard titration of purified rabbit antiserum on the standard competition curve, including control wells (*see* below in **Subheading 3.5.**), to determine the concentrations of primary antibody to be used in enzyme-linked immunosorbent assay (ELISA) (*see* **Note 4**).

3.5. Determination of Xanthine Oxidoreductase by ELISA

1. Antigen adhesion: coat microtiter plates with XO (5 µg/mL) in adhesion buffer (100 µL/well) overnight at 4°C. Include control wells (A) with only adhesion buffer, (B) with BSA (5 µg/mL) in adhesion buffer, (C) and (D) with XO (5 µg/mL) in adhesion buffer.

2. Washing: wash with 200 µL/well PBS-Tween. Repeat the washing once.

3. Saturation: incubate plates 1 h at 37°C with 200 µL/well 1% BSA in PBS-Tween to reduce nonspecific binding. Wash twice as above in **Subheading 3.5., step 2**.

4. Primary antibody and competition between bound and unbound XO.
   a. Standard curve: to each well, add in sequence (i) 50 µL HS, (ii) 25 µL human milk XO in scalar concentrations between 0.4 and 250 µg/mL PBS-Tween, (iii) 25 µL purified rabbit antiserum in PBS-Tween (approx 20 µg protein, depending on the titration results; *see* **Subheading 3.4., step 6**).
   b. Controls: to each well, add 50 µL HS and 25 µL PBS-Tween, then add in (A) 25 µL PBS-Tween; in (B) and (D) 25 µL purified rabbit antiserum in PBS-Tween; in (C) 25 µL rabbit preimmune serum in PBS-Tween at the same protein concentration as purified rabbit antiserum.
   c. Samples: to each well, add in sequence (i) 50 µL serum sample from human subjects (*see* **Note 5**), (ii) 25 µL PBS-Tween and (iii) 25 µL purified rabbit antiserum in PBS-Tween.

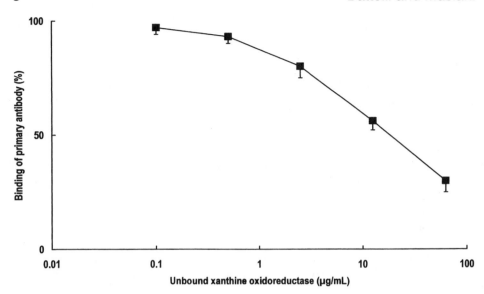

Fig. 1. The values are means ± S.D. of 10 experiments with triplicate samples of ELISA standard competition curve. The A_{405} readings were normalized by expressing them as percentage of positive (D) control results (*see* **Subheading 3.5., step 8**).

5. Incubate plates at 37°C for 3 h, then wash five times as in **Subheading 3.5., step 2**.
6. Secondary antibody: add 100 μL/well antirabbit IgG-alkaline phosphatase conjugate appropriately diluted in PBS-Tween following the manufacturer's instructions.
7. Incubate plates at 37°C for 1 h, then wash five times as in **Subheading 3.5., step 2**.
8. Alkaline phosphatase substrate: add 100 μL/well of a freshly prepared substrate solution and incubate at 37°C for 1–3 h for color development until an absorbance value of approx 1.3 at 405 nm is reached in positive (D) control. Stop the reaction with 100 μL/well 10^{-1} *M* NaOH and measure the A_{405} by the microplate reader, subtracting from sample values the absorbance derived from unspecific binding of negative (B) control.
9. Use linear-regression analysis to obtain the standard competition curve and calculate XO concentration in HS samples.

3.6. Results

3.6.1. Standard Competition Curve of ELISA

The results of the standard competition curve are shown in **Fig. 1**. The correlation coefficient calculated by linear-regression analysis was $R = 0.9$. The sensitivity of the test allows the determination of unbound XO in the

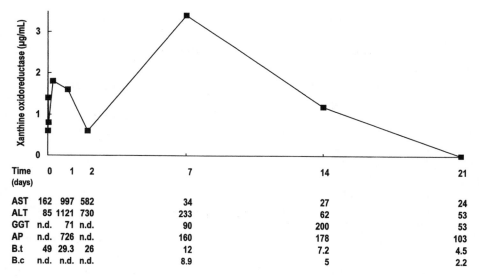

Time (days)	0	1	2	7	14	21
AST	162	997	582	34	27	24
ALT	85	1121	730	233	62	53
GGT	n.d.	71	n.d.	90	200	53
AP	n.d.	726	n.d.	160	178	103
B.t	49	29.3	26	12	7.2	4.5
B.c	n.d.	n.d.	n.d.	8.9	5	2.2

Fig. 2. The concentration of serum XO by competitive ELISA and routine markers of liver injury were determined at the indicated time after liver transplantation. Normal values are reported in parenthesis: AST, Aspartate amino transferase (<50 UI/L); ALT, Alanine amino transferase (<50 UI/L); GGT, γ-Glutamyl transpeptidase (<50 UI/L); AP, Alkaline phosphatase (<240 UI/L); B.t, Total Bilirubin (<1.1 mg/dL); B.c, Conjugated Bilirubin (<0.25 mg/dL). The patient serum was kindly provided by Dr. L. Bolondi, Department of Medicine and Gastroenterology, S. Orsola Hospital, Bologna, Italy.

indicated range (0.01–6.25 µg/well) corresponding to a concentration in HS of 0.2–125 µg/mL XO.

XO concentration in the serum of healthy subjects was lower than 1 µg/mL *(2)*.

3.6.2. Determination of XO by ELISA after Liver Transplantation

The serum level of XO was determined in a patient affected by primary liver cirrhosis at different times after liver transplantation (*see* **Fig. 2**). The values of other markers of hepatic damage are also reported. The elevation of transaminases indicates necrosis of hepatocytes. Alkaline phosphatase and γ-glutamyltranspeptidase are markers of cholestasis. The increment of total bilirubin in serum may be correlated to cholestasis or to reduction of liver function, and is associated to conjugated bilirubin values since bilirubin conjugation requires conserved liver function.

XO shows a biphasic slope and, although the report of a single case does not allow any conclusive consideration, nevertheless, the first peak is coincident with postischemic reperfusion injury of the graft, while the second peak seems to anticipate the pattern of γ-glutamyltranspeptidase.

4. Notes

1. Although a single middle-range centrifuge equipped with adequate rotors could satisfy all the requirements for described procedures, we use a Beckman J2-21 centrifuge for XO purification, a Beckman GS-6R centrifuge to separate serum from rabbit and human blood clot, and an Eppendorf 5415 centrifuge to remove precipitate from small volume samples.
2. The availability of a spectrophotometer with timed automatic change of cell is not essential but helpful to perform XO determination.
3. Hypatite C/cellulose column can be regenerated only once by elution with 1 *M* NaCl until the stabilization of the base line on the recorder, followed by equilibration with Hypatite C buffer. Other chromatographic resins can be used many times if cleaning and storage are performed following manufacturer's instructions.
4. Perform the titration of purified rabbit antiserum on the standard competition curve after each purification procedure, since the concentration of primary antibody to be used in ELISA may vary from one preparation to another.
5. Human serum samples are obtained by centrifugation from clotted blood and stored in 200-µL aliquots at –80°C before XO determination by competitive ELISA. Serum samples from healthy subjects should be included with each determination.

Acknowledgments

This work was supported by the Ministero dell'Istruzione, dell'Università e della Ricerca, Rome, by the University of Bologna, funds for selected research topics, by the Pallotti's Legacy for Cancer Research, and by the Associazione Italiana per la Ricerca sul Cancro, Milan (S. Musiani was supported by a fellowship from FIRC).

References

1. Linder, N., Rapola, J., and Raivio, K. O. (1999) Cellular expression of xanthine oxidoreductase protein in normal human tissues. *Lab. Invest.* **79,** 967–974.
2. Battelli, M. G., Abbondanza, A., Musiani, S., Buonamici, L., Strocchi, P., Tazzari, P. L., et al. (1999) Determination of xanthine oxidase in human serum by a competitive enzyme-linked immunosorbent assay (ELISA). *Clin. Chim. Acta* **281,** 147–158.
3. Battelli, M. G., Abbondanza, A., and Stirpe, F. (1992) Effects of hypoxia and ethanol on the xanthine oxidase of isolated hepatocytes: conversion from D to O form and leakage from cells. *Chem. Biol. Interactions* **83,** 73–84.
4. de Groot, H. and Littauer, A. (1989) Hypoxia, reactive oxygen, and cell injury. *Free Rad. Biol. Med.* **6,** 541–551.

5. Weinbroum, A. A., Hochhauser, E., Rudick, V., Kluger, Y., Karchevsky, E., Graf, E., and Vidne, B. A. (1999) Multiple organ dysfunction after remote circulatory arrest: common pathway of radical oxygen species? *J. Trauma* **47,** 691–698.
6. Battelli, M. G., Buonamici, L., Polito, L., Bolognesi, A., and Stirpe, F. (1996) Hepatotoxicity of ricin, saporin or a saporin immunotoxin: xanthine oxidase activity in rat liver and blood serum. *Virchows Arch.* **427,** 529–535.
7. Yamamoto, T., Moriwaki, Y., Takahashi, S., Tsutsumi, Z., Yamakita, J., Nasako, Y., et al. (1996) Determination of human plasma xanthine oxidase activity by high-performance liquid chromatography. *J. Chromatogr. B. Biomed. Appl.* **681,** 395–400.
8. Pesonen, E. J., Linder, N., Raivio, K. O., Sarnesto, A., Lapatto, R., Hockerstedt, K., et al. (1998) Circulating xanthine oxidase and neutrophil activation during human liver transplantation. *Gastroenterology* **114,** 1009–1015.

2

Simultaneous Determination of Polyunsaturated Fatty Acids and Corresponding Monohydroperoxy and Monohydroxy Peroxidation Products by HPLC

Richard W. Browne and Donald Armstrong

1. Introduction

Lipid peroxidation (LPO) is a prominent manifestation of free radical (FR) activity in biological systems. The primary target of FR attack on lipids is the 1,4-pentadiene structure of a polyunsaturated fatty acid (PUFA), which are either free or esterified to cholesterol or glycerol. Initiation occurs when a FR abstracts a methylene hydrogen from PUFA. In this reaction the FR is quenched and a PUFA centered alkoxyl radical (L•) is formed. L• then undergoes a spontaneous rearrangement of its double bonds forming a conjugated diene. Reaction of L• with molecular oxygen produces a PUFA-centered peroxyl radical (LOO•). Propagation occurs when either L• or LOO• act as initiating FR and attack a neighboring PUFA in a tightly packed lipid bilayer structure of a membrane or within a lipoprotein. The product of this reaction is a new L•, which can further propagate the reaction and form a lipid hydroperoxide (LHP) *(1)*. Termination occurs when an antioxidant (AOX) molecule capable of absorbing the intermediate free radicals, or free-radical scavengers, interrupts this chain reaction.

The hydroperoxide moiety of LHP can be reduced by divalent metal ions or glutathione-dependent peroxidases (phospholipid glutathione peroxidase or, following hydrolysis to free fatty acids, glutathione peroxidase) to an alcohol, yielding a hydroxy derivative (LOH). LHP and LOH represent the primary stable end products of lipid peroxidation *(2)*. Since biological samples are comprised many different LPO products can vary in carbon chain length and degree of unsaturation as well as regioisomerism of the position of the hydroperoxy

From: *Methods in Molecular Biology, vol. 186: Oxidative Stress Biomarkers and Antioxidant Protocols*
Edited by: D. Armstrong © Humana Press Inc., Totowa, NJ

or hydroxy group relative to the carbon chain *(3)*. Furthermore, the native unoxidized PUFA composition of a system inherently effects the possible LPO products that are generated. Because of this, simultaneous determination of both the substrate and its derivative oxidation products has been suggested *(4,5)*.

We have previously described a reverse-phase high-performance liquid chromatography (RP-HPLC) technique capable of separating regioisomeric species of LHP and LOH derived from the four major PUFA found in human plasma; linoleic, arachidonic, linolenic, and docosahexaenoic acid *(6)*. Following total lipid extraction, alkaline hydrolysis and reextraction of the liberated fatty acids, two separate systems with different mobile-phase conditions and analytical columns were used, one for LOH and LHP and the second for the native unoxidized PUFA *(7)*. We report here on an alteration of this methodology allowing simultaneous determination of LHP, LOH, and PUFA on a single chromatographic separation.

This present methodology sacrifices a small amount of resolution of LHP and LOH for inclusive determination of PUFAs in a single isocratic run. Use of diode-array detection allows determination of the PUFA at 215 nm and the conjugated diene of LHP and LOH at 236nm. This method is useful for the determination of total LHP and LOH relative to their precursor PUFA within 20 min after injection.

2. Materials

2.1. Instruments and Equipment

2.1.1. Analytical HPLC System (Shimadzu Scientific Instruments, Columbia, MD)

1. Shimadzu LC-6A Pump.
2. Shimadzu SIL-7A Autosampler/injector.
3. Shimadzu SPD-M6A UV/VIS Photodiode Array.
4. PONY/IBM 486Dx PC.
5. Shimadzu CLASS-VP Chromatography Software.

2.1.2. HPLC Columns

1. Supelco LC-18 analytical, 4.6 × 250 mm, 5 micron particle size, 100 Å pore (Supelco, Bellafonte, NJ).
2. Supelcoguard C-18 (4.6 × 20 mm, 5 micron particle size, 100 Å pore).
3. Supelcoguard Guard Column Cartridge.

2.2. Reagents and Solvents

1. Unless otherwise indicated, reagents were obtained from Sigma Chemical Co. (St. Louis, MO).

2. All organic solvents were HPLC grade and are obtained from J.T. Baker Chemical Co. (Phillipsburg, NJ). Solvents were filtered through 0.22 micron, nylon, filter membranes immediately prior to use.

2.3. Standards

1. Linoleic (18:2ω6), linolenic (18:3ω3), arachidonic (20:4ω6), and docosahexaenoic acids (22:6ω6) and 5-hydroxy eicosatetraenoic acid methylester were purchased from Sigma in their highest purity.
2. Conjugated linoleic acid (CLA, iso-linoleic acid) was purchased from Cayman Chemical Co. (Ann Arbor, MI). Calibration solutions were prepared by mass and dissolved in HPLC-grade ethanol and stored under argon at –80°C prior to HPLC analysis.
3. LHP standards were obtained as ethanolic solutions from Cayman (*see* **Note 1**).

3. Methods

3.1. LHP Standards

Calibration solutions are prepared in ethanol using a 160UV scanning spectrophotometer (Shimadzu Scientific Instruments, Kyoto, Japan) with the molar extinction coefficients provided by Caymen Chem. Co. **Table 1** lists the standards along with their shorthand nomenclature and molar extinction coefficients (*see* **Note 2**). Following individual standard peak identification, a hydroperoxy HPLC mixture is used on a daily basis to adjust for retention time fluctuations.

3.2. LOH Standards

LOH standards are prepared from LHP standards by methanolic sodium borohydried reduction as previously described *(6)*. Following chloroform reextraction the LOH standards are dissolved in ethanol and calibration solutions prepared and stored as described for LHP. Following individual standard peak identification a hydroxy HPLC mixture is used on a daily basis to adjust for retention-time fluctuations (*see* **Note 3**).

3.3. HPLC Conditions

1. HPLC mobile phase consisted of 0.1% acetic acid/acetonitrile/tetrahydrofuran (41:41:18 v/v/v), which was premixed, filtered, and degassed under vacuum sonication. The mobile phase is continuously sparged with helium during analysis.
2. System flow rate is 1.3 mL/min and pressure of 175 Kg/cm^3.
3. The diode array monitored the column effluent from 200–300 nm with specific analysis channels of 236 and 215 nm with 8 nm bandwidth for the LHP/LOH and PUFA, respectively.

Table 1
Nomenclature, Retention Times and Molar Absorptivities
of Polyunsaturated Fatty Acid Hydroperoxides
and Hydroxy Derivative Standards

Retention time (min)	Name	Shorthand name	Absorbance maximum	Molar extinction coefficient
7.9	13-hydroxy-octadecatrieneoic acid	13-HOTE	235	23,000
8.9	13-hydroperoxy octadecatrieneoic acid	13HpOTE	235	23,000
9.8	13-hydroxy-octadecadieneoic acid	13-HODE	234	23,000
10.5	9- hydroxy-octadecadieneoic acid	9-HODE	234	23,000
11.3	13-hydroperoxy octadecadieneoic acid	13-HpODE	234	23,000
12.1	9-hydroperoxy octadecadieneoic acid	9-HpODE	234	23,000
12.2	15-hydroxy-eicosatetraenoic acid	15-HETE	236	27,000
13.1	12-hydroxy-eicosatetraenoic acid	12-HETE	237	27,000
13.5	12-hydroperoxy-eicosatetraenoic acid	12-HpETE	237	27,000
13.8	8-hydroxy-eicosatetraenoic acid	8-HETE	237	27,000
14.8	8-hydroperoxy-eicosatetraenoic acid	8-HpETE	237	27,000
15.3	5-hydroxy-eicosatetraenoic acid	5-HETE	236	27,000
16.4	5-hydroperoxy-eicosatetraenoic acid	5-HpETE	236	27,000
21.2	5-hydroxy-eicosatetraenoic acid methyl ester	Internal Standard (I.S.)	236	27,000
30.5	octadecatrieneoic acid (linoleic)	18:3	215	?
40.0	Docosahexaenoic acid	22:6	215	?
44.9	Eicosatetraenoic acid (arachidonic acid)	20:4	214	?
48.5	octadecadieneoic acid (linolenic acid)	18:2	209	?

3.4. Sample Extraction

1. EDTA plasma was collected into evacuated blood-collection tubes.
2. Hexane/isopropanol (HIP) total lipid extracts are prepared by adding 1.0 mL isopropanol to 0.5 mL ethylenediaminetetraacetic acid (EDTA) plasma.
3. Two mL of hexane is added, the vial perfused with nitrogen, capped, vortexed for 1 min, centrifuged for 3 min at 3,000g and the upper-hexane phase collected by aspiration (*see* **Note 4**). The extraction is repeated three times and the hexane layers are pooled and evaporated to dryness under nitrogen.
4. Alkaline hydrolysis of total dried lipid extracts are performed by dissolving in 0.95 mL of degassed, absolute ethanol. Fifty mL of 10 *M* sodium hydroxide (NaOH) is added, the sample perfused with nitrogen, capped, heated at 60°C for 20 min, and neutralized with 30 µL glacial acetic acid.
5. One hundred m_mL of 1.0 nmol/L 5-HETE-ME is added as internal standard. The ethanol is evaporated under nitrogen, the sample dissolved in 1.0 mL water, extracted twice with 2.0 mL of n-heptane, the upper phase collected and pooled, evaporated under nitrogen, and the residue dissolved in 250 m_mL of ethanol.

3.5. HPLC Analysis

1. Immediately prior to injection 250 µL of water is added to samples (*see* **Note 5**). One hundred fifty m_mL of the sample is injected into the HPLC system, eluted isocratically with mobile-phase conditions described in **Subheading 3.3., step 1**, over 60 min and monitored at 200–300 nm by the photodiode array (*see* **Note 6**).
2. Quantification is based on an external calibration curve using ethanolic standards prepared on a Shimadzu 160 UV scanning spectrophotometer applying the max and absorptivity coefficients provided by the manufacturer. Serial dilutions are made in ethanol/water 50:50 (v/v) and standard curves generated by triplicate injections of each calibrator. Sample concentrations are interpolated from standard curves and corrected for recovery of the 5-HETE-ME internal standard (*see* **Note 3**).

3.6. Results

Figure 1 shows two displays of the chromatographic data from a mixed preparation of standards. **Figure 1A** shows the two dimensional graph of a slice through the diode array three dimentional data at 236 nm identifying LHP and LOH peaks. **Figure 1B** shows the slice at 215 nm identifying the native PUFA. **Figure 2** shows chromatographic data obtained from a human plasma sample prepared as described in **Subheading 3.4.**

4. Notes

1. LHP and LOH standards can be purchased or synthesized. The synthesis of standards is described in detail in volume 108 of this series (*8*).
2. If LOH standards are synthesized by methanolic sodium borohydried reduction, rather than purchased in purified form, it is necessary to perform calibrations of

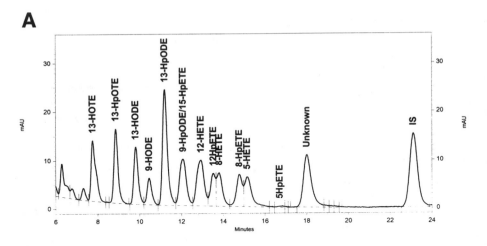

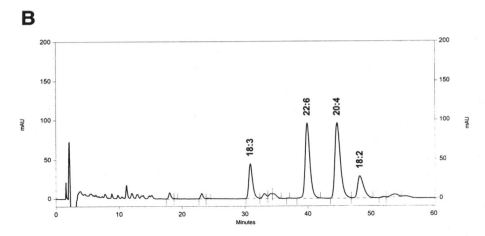

Fig. 1. Simultaneous chromatograms of LHP and LOH standards at 236 nm (**A**) and native unoxidized PUFA standards at 25 nm (**B**). *See* **Subheading 3.3.** for HPLC conditions and **Table 1** for nomenclature of peaks.

LHP and LOH separately since trace amounts of the sodium borohydried in the LOH preparation may reduce LHP upon mixing.
3. It is critical that all solvents, especially those used for extraction, are thoroughly degassed to remove dissolved oxygen and prevent lipid oxidation during processing. We routinely accomplish this by placing solvents in an ultrasonic water bath and applying a vacuum followed by 15 min of helium sparging. Screw-cap extraction vials are perfused with nitrogen or argon and immediately capped prior to vortexing or incubations.

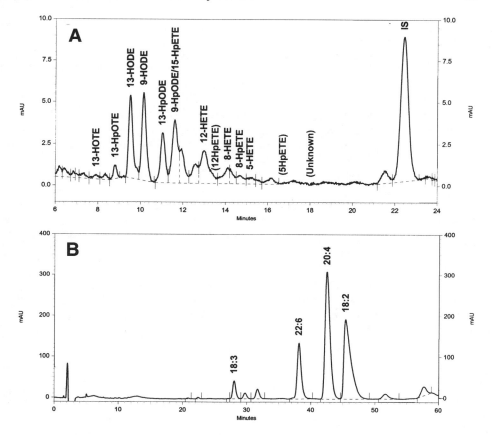

Fig. 2. Simultaneous chromatograms of LHP and LOH at 236 nm (**A**) and native unoxidized PUFA at 215 nm (**B**) isolated from human plasma by total lipid extraction, and saponification.

4. Samples need to be injected in a solution that is at least 50% water in order to ensure good mass transfer of the sample to the stationary phase. Samples injected in pure solvent such as ethanol give extremely broad peaks and poor resolution.
5. It should be noted that a photodiode array is not necessary for this methodology and a simple two-channel UV detector could be used. Integration of peak areas is performed at 236 nm with an 8 nm bandwidth for LOH and LHP. This wavelength and bandwidth are chosen to encompass the wavelength of maximum absorbance (I_{max}) of the ODEs at 234 nm and ETE at 237 nm.

References

1. Porter, N. (1990) Autooxidation of polyunsaturated fatty acids: initiation, propagation, and product distribution, in *Membrane Lipid Oxidation* (Vigo-Pelfrey, C., ed.), CRC Press, Boca Raton, pp. 33–62.

2. Porter, N., Wolfe, R., and Weenan, H. (1979) The free radical oxidation of polyunsaturated licithins. *Lipids* **15,** 163–167.
3. Teng, J. I. and Smith, L. L. (1985) High performance liquid chromatography of linoleic acid hydroperoxids and their corresponding alcohol derivatives. *J. Chromatog.* **350,** 445–451.
4. DiPeierro, D., Tavazzi, B., Lazzarino, G., Galvano, M., Bartolini, M., and Giardina, B. (1997) Separation of representative lipid compounds of biological membranes and lipid derivatives from peroxidized polyunsaturated fatty acids by reverse phase high-performance liquid chromatography. *Free Rad. Res.* **26(4),** 307–317.
5. Banni, S., Contini, M. S., Angioni, E., Deiana, M., Dessi, M. A., Melis, M. P., et al. (1996) A novel approach to study linoleic acid autooxidation: importance of simultaneous detection of the substrate and its derivative oxidation products. *Free Rad. Res.* **25(1),** 43–53.
6. Browne, R. W. and Armstrong, D. (2000) HPLC analysis of lipid derived polyun-saturated fatty acid peroxidation products in oxidatively modified human plasma. *Clin. Chem.* **46(6),** 829–836.
7. Browne, R. and Armstrong, D. (1998) Separation of hydroxy and hydroperoxy polyunsaturated fatty acids by high pressure liquid chromatography, in *Methods in Molecular Biology, Free Radicals and Antioxidant Protocols* (Armstrong, D., ed.), Humana Press, Totowa, NJ, pp. 147–155.

3

Determination of Products of Lipid Oxidation by Infrared Spectroscopy

Douglas Borchman and Santosh Sinha

1. Introduction

Lipid peroxidation is initialed as activated oxygen reacts with the double bonds on the lipid hydrocarbon chains (1). Depending on the type of lipid, type of oxidant, and severity of the oxidation, a variety of lipid-peroxidation products are formed (1). The major products of lipid peroxidation are moieties containing hydroxyls, hydroperoxyls, aldehydes, ketones, caroxylic acids, and *trans* double bonds. Infrared spectroscopy is a sensitive technique that can detect all of these groups and is uniquely sensitive in detecting hydroxyl and hydroperoxyl groups (2–8). The detection of lipid hydoxyl and hydroperoxyl groups is especially useful for quantifying the oxidation of mono unsaturated lipids, such as those found in the ocular lens, where secondary products of lipid oxidation such as malondialdeyde are not readily formed (9,11).

Retinal lipids are highly unsaturated, containing as many as 6 C = C bonds per hydrocarbon chain, and are very sensitive to lipid oxidation. In this study, lipids from bovine retinal-rod disk membranes were extracted (12), layered onto a silver chloride window, and allowed to oxidize in atmospheric oxygen. Products of lipid oxidation were measured vs time using infrared spectroscopy.

2. Materials

2.1. Equipment

1. Infrared spectrometer with digital output interfaced with a computer.
2. Centrifuge.
3. Microprobe sonicator.

From: *Methods in Molecular Biology, vol. 186: Oxidative Stress Biomarkers and Antioxidant Protocols*
Edited by: D. Armstrong © Humana Press Inc., Totowa, NJ

4. Lyophylizer.
5. Dry air source for spectrometer chamber (*see* **Note 1**).
6. Software for analysis of infrared spectra such as Grams 386 software (version 2.04, Galactic Industries Corporation, Salem, NH).
7. Vortex Mixer.
8. Parafilm (American National Can, Greenwich CT).
9. Scintilation vial caps (*see* **Note 2**).

2.2. Reagents

1. The following solvents are spectroscopic grade obtained from Fisher Scientific (Pittsburgh, PA): methanol, n-hexane, isopropanol.
2. Argon (prepurified).

3. Methods
3.1. Sample Processing

In this study rod outer-segment disk membranes were prepared by differential centrifugation *(12)* (*see* **Note 3**).

3.2. Sample Extraction

Bubble all reagents with argon for 10 min prior to use. Place the samples, such as those described above in glass test tubes filled with argon gas. Sonicate the samples in at least 6 volumes of methanol for 20 min in a bath-type sonicator, vortex, and then centrifuge at 7000 rpm (*see* **Note 4**). Decant the clear supernatants and evaporate the methanol under a stream of argon gas. Solubilize the thin lipid film on the bottom of the tube with 2 mL of hexane/isopropanol (2 : 1) and sonicate, vortex, and centrifuge as before. Decant the clear supernatants and evaporate the hexane/isopropanol under a stream of argon gas. Solubilize the thin lipid film on the bottom of the tube with 300 µL methanol to be used for spectroscopic analysis.

3.3. Infrared Spectroscopy Procedure
3.3.1. Preparation of Sample for Infrared Spectroscopy

Layer the lipid sample prepared above onto a AgCl window by placing a small drop of lipid in methanol onto the center of the window and evaporate the methanol under a gentle stream of argon (*see* **Note 2**). Lyophilize the window with the dry lipid film for 12 h to remove methanol and trace amounts of water. Measure infrared spectra of the dried lipid films immediately after removing the windows from the lyophilizer to quantify lipid oxidation as was done for human *(9)* and guinea pig *(11)* lens membranes and as descibed in **Subheading 3.4.**

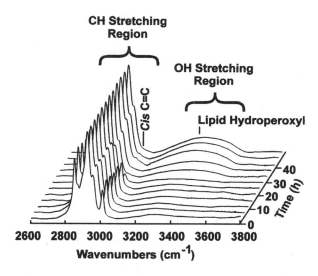

Fig. 1. Infrared CH and OH stretching region of anhydrous films of bovine rod outer-segment lipids exposed to air and light for 0, 0.5, 1, 1.5, 2, 4, 8,12, 16, 24, and 48 h, bottom to top, respectively.

3.3.2. Analysis of Infrared Spectra

In our study 300 interferograms were recorded, co-added, and apodized with a Happ-Genzel function prior to Fourier transformation, yielding an effective spectral resolution of 1.0 cm^{-1}. Fourier self-deconvolution, second derivative, subtraction, and curve-fit analysis were carried out using Grams 386 software (version 2.04, Galactic Industries Corporation, Salem, NH) as described in **Subheading 3.4.**

3.4. Results

3.4.1. Analysis of Hydroxyl and Hydroperoxyl Bands

The CH and OH infrared stretching region for a dried film of rod outer-segment lipid exposed to atmospheric oxygen for a period of 40 h is shown in **Fig. 1**. The intensity of the OH stretching bands (3600–3100 cm^{-1}), reflects the degree of lipid oxidation *(5)* and the amount of hydroxyl containing lipids such as sphingolipids and cholesterol *(see* **Note 5**). To quantify the increase in OH band intensity as a result of lipid oxidation, the areas of the OH and CH stretching bands are measured using the integration program provided by Grams 386 software (version 2.04, Galactic Industries Corporation, Salem, NH). The baselines for the integration were taken near 3600, 3050, and 3050,

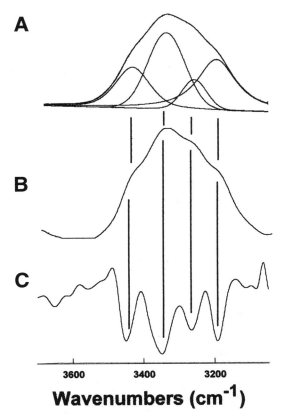

Fig. 2. **(A)** (Upper trace) Infrared CH and OH stretching region of an anhydrous films of bovine rod outer segment lipids exposed to air and light for 20 h. (Lower trace) curve-fit spectrum of upper trace using peak information from **(B)** Fourier self-deconvolution spectrum and **(C)** second derivative spectrum of **(A)**.

2750 cm^{-1} for the CH and OH stretching bands, respectively. The ratio of the intensity of the OH stretching band to the intensity of the CH stretching band region may be used to quantify the change in the number of lipid hydroxyl groups.

The band intensity at 3440 cm^{-1} is sensitive to changes in the number of lipid hydroperoxyl groups formed by oxidation *(4)*. To quantify the amount of lipid hydroperoxyl groups, the number and position of the major bands that compose the OH stretching region must first be determined using Fourier self-deconvolution and second derivative analysis. Note that 4 bands are detected at 3500, 3412, 3345 and 3250 cm^{-1} as seen from the Fourier self deconvolution (**Fig. 2B**) and 2nd derivative spectra (**Fig. 2C**).

Using a conservative number of bands, 4 in this instance, the areas of each of the minor bands in the OH stretching region are determined using the curve fit algorithm from Grams 386 software (version 2.04, Galactic Industries Corporation, Salem, NH) (**Fig. 2A**). The area of the minor band at 3440 cm^{-1} relative to the area of the CH stretching band is indicative of the number of lipid-hydroperoxyl groups. We curve-fit the original spectrum (**Fig. 2A**) although often the Fourier self-deconvoluted spectrum (**Fig. 2B**) is curve-fit. Caution should be taken when interpreting the data obtained using the curve-fit algorithm and difference spectra (the spectrum of the control sample minus the spectrum of the suspected oxidized spectrum) should be used to confirm that changes in the minor bands detected using the curve-fit spectrum are real.

3.4.2. Analysis of cis Double Bonds

The *cis* double bond band is located at 3010 cm^{-1} (**Fig. 1**). When lipids are oxidized, *cis* double bonds of the hydrophobic chains rearrange to form *trans* double bonds (*see* **Fig. 1** in **ref. 9**) and the intensity of the *cis* C = C band decreases (**Fig. 1**) *(4,9)*. Curve fitting as was done for the lipid-hydroxyl region is not necessary for measuring the intensity of the *cis* C = C band and the area of the *cis* double-bond band may simply be measured using the integration program as was done for the total lipid-hydroxyl region.

3.4.3. Analysis of Infrared Carbonyl and Aldehyde Bands

Note the difference spectrum (**Fig. 3C**) shows with oxidation; the lipid-carbonyl groups at 1720 cm^{-1} and aldehyde groups at 1679 cm^{-1} increase. The carbonyl stretching band near 1720 cm^{-1} arises from the acyl-linked hydrocarbon chains of lipids with a glycerol backbone such as phosphatidylcholine or phosphatidylethanolamine and from products of lipid oxidation *(4,9,11)*. The aldehyde bands sometime appear near 1600 cm^{-1} *(3)*, depending upon the salt of the carboxylic acid. To measure the area of these bands, Fourier self-deconvolution and second derivative analysis near the carbonyl region is necessary to determine the position and number of bands in this region. The areas of the bands can then be determined using the curve-fitting algorithm.

3.4.4. Analysis of Infrared trans C = C Band

Note the difference spectrum (**Fig. 3C**) shows with oxidation; the lipid *trans* C = C at 970 cm^{-1} increases. *Trans* double bonds do not occur naturally in lipid hydrocarbon chains, thus, the ratio of the intensity of the *trans* double bond to the intensity of the *cis* double bond may be used as an index of lipid oxidation *(4,9,11)*.

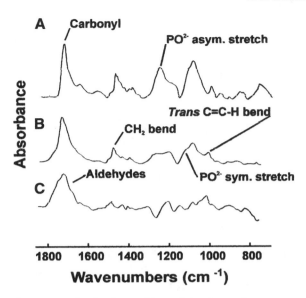

Fig. 3. Infrared spectra of anhydrous films of bovine rod outer-segment lipids. **(A)** Initial infrared spectrum at 0 h (no oxidation) and **(B)** infrared spectrum of lipids exposed to air and light for 48 h (oxidized). **(C)** infrared spectrum **(B)** minus **(A)**. Positive features in the difference spectrum **(C)** indicate increases in band intensity due to oxidation.

4. Notes

1. Purging of the infrared spectrometer with dry air is essential to avoid water vapor interference with the infrared OH, carbonyl and amide regions. We use a Kaeser KLDW series (Edina, MN) dryer that removes CO_2, water trace hydrocarbons and oil from the air used to purge our instrument.

2. It is helpful to place the AgCl windows into plastic scintilation vial type caps so that the sample number can be marked on the caps. Parafilm can be stretched across the caps to contain the samples in the event the lyophilization flask is bumped or overturned. Holes should be poked into the parafilm. To avoid contamination and loss of sample onto the parafilm, care should be taken to avoid contacting the lipid film with the parafilm.

3. Oxidation may be stopped by freezing and storing the membrane or tissue sample in liquid nitrogen. If membranes are to be prepared, reagents should be purged with argon and the membrane preparation procedure performed under an atmosphere of argon where possible. Argon is heavier than air and settles to the bottom of tubes and flasks providing a barrier to oxygen in the air.

4. Depending on the sample, more rigorous sonication or homogenization in MeOH may be necessary to ensure complete extraction. Whole tissue frozen in liquid nitrogen may be pulverized using a mortar and pestle cooled with liquid nitrogen.

A microprobe sonicator such as a Branson Ultrasonics Corporation (Danbury, CT) SONIFER cell disrupter set at an output of 6 should disperse the sample. The sample should be purged with nitrogen, and placed in an ice bath during sonication. Care should be taken not to heat the sample. This can be accomplished by sonicating for only 15 s, then allowing the sample to cool in the ice bath for 1–2 min. Repeat the sonication/cooling 3–6 times to ensure complete dispersion of the sample.

5. In general, infrared spectra of lipid extracted from different membranes will be similar to those in **Figs. 1** and **3**. However, if there is the possibility of lipid compositional changes other than oxidation, lipid composition should be measured to assess the impact of these changes on the spectral characteristics of the samples. For instance, in the study of lens lipids it was found that sphingolipid increased with age and glycerolipid decreased with age. These lipids contribute to the OH, carbonyl, and amide regions of the infrared spectrum and their contribution was accounted for *(11)*.

Acknowledgments

Supported by Public Health Service (Bethesda, MD) grant EYO7975 (D.B), The Kentucky Lions Eye Foundation (Louisville, KY) and an unrestricted grant from Research to Prevent Blindness, New York.

References

1. Vigo-Pelfrey, C. (ed.) (1989) *Membrane Lipid Oxidation (vols. 1–3)* CRC Press, Boca Raton, FL.
2. Privett, O. S., Lundbergm W. O., Khan, N. A., Tolberg, W. E., and Wheeler, D. H. (1993) Structure of hydroperoxides obtained from autoxidized methyl linoleate. *J. Am. Oil Chem. Soc.* **30,** 61–66.
3. Ismail, A. A., van de Voort, F. R., Emo, G., and Sedmanm, J. (1993) Rapid quantitative determination of free fatty acids in fats and oils by Fourier transform infrared spectroscopy. *J. Am. Oil Chem. Soc.* **70,** 335–341.
4. van de Voort, F. R., Ismail, A. A., Sedman, J., and Emo, G. (1994) Monitoring the oxidation of edible oils by Fourier transform infrared spectroscopy. *J. Am. Oil Chem. Soc.* **71,** 243–253.
5. Dugan, L. R., Beadle, B. W., and Henick, A. S. (1949) An infrared absorption study of autoxidized methyl linoteate. *J. Am. Oil Chem. Soc.* **26,** 681–685.
6. Andrews, J. S. and Leonard-Martun, T. (1981) Total lipid and membrane lipid analysis of normal animal and human lenses. *Invest. Ophthalmol.* **21,** 39–45.
7. van de Voort, F. R., Sedman, J., Emo, G., and Ismail, A. A. (1992) A rapid FTIR quality control method for fat and moisture determination in butter. *Food Res. Int.* **25,** 193–198.
8. van de Voort, F. R., Sedman, J., Emo, G., and Ismail, A. A. (1992) Rapid and direct iodine value and saponification number determination of fats and oils by attenuated total reflectance/fourier infrared spectroscopy. *J. Am. Oil Chem. Soc.* **69,** 61–66.

9. Borchman, D. and Yappert, M. C. (1998) Age related lipid oxidation in human lenses. *Invest. Ophthalmol. Vis. Sci.* **39,** 1053–1058.

10. Ahuja, R. P., Borchman, D., Dean, W. L., and Paterson, C. A., Zeng, J., Zhang, S., Ferguson-Yankey, S., and Yappert, M. C. (1999) Ca^{2+}-ATPase kinetics and membrane oxidation relationships in lens epithelial microsomes. *Free Rad. Biol. Med.* **27,** 177–185.

11. Borchman, D., Gibblin, F. J., Yappert, M. C., Leverenz, V. R., Reddy, V. N., Lin, L., Tang, D. (2000) Impact of aging and hyperbaric oxygen in vivo on guinea pig lens lipid and nuclear light scatter. *Invest. Opthalmol. Vis. Sci.* **41,** 3061–3073.

12. Lamba, O. P., Borchman, D., and O'Brien, P. J. (1994) Fourier transform infrared study of the rod outer segment disk and plasma membranes of vertebrate retina. *Biochemistry* **33,** 1704–1712.

4

Detection of Docosahexaenoic Acid Hydroperoxides in Retina by Gas Chromatography/Mass Spectrometry

Guey-Shuang Wu and Narsing A. Rao

1. Introduction

In the retina, as well as in other organs, inflammation-mediated membrane lipid peroxidation is regarded as the major source of unsaturated fatty acid hydroperoxides. It is generally accepted that inflammatory infiltrates, such as polymorphonuclear leukocytes and macropahges, simultaneously generate superoxide and nitric oxide (NO) *(1)*. Although both radicals display low chemical reactivity, their combination product, peroxynitrite, is a potent oxidant capable of oxidizing cellular macromolecules, including membrane lipids *(2)*. Photoreceptor membranes in the retina contain an unusually high concentration of docosahexaenoic acid (22:6, omega 3), amounting to nearly 50% of the total fatty acid pool *(3)*. By virtue of its structure, 22:6 is particularly susceptible to lipid peroxidation induced by peroxynitrite *(2,4)*. Peroxidation of photoreceptor membranes in the retina has been recognized as a key component in various types of photoreceptor degeneration and pathogenesis *(5)*. Interest in the pathological consequences of photoreceptor membrane lipid peroxidation has led to the development of several analytical approaches designed to detect peroxidized lipids in the retina.

Gas chromatography/mass spectrometry (GC/MS) has emerged as the method of choice for detecting peroxidized 22:6 in tissues. Docosahexaenoic acid, with its five skipped dienes in the molecule, is capable of forming 10 positional isomers of hydroperoxides. These isomers are not discernible by thin-layer chromatography (TLC) or high-pressure liquid chromatography

From: *Methods in Molecular Biology, vol. 186: Oxidative Stress Biomarkers and Antioxidant Protocols*
Edited by: D. Armstrong © Humana Press Inc., Totowa, NJ

(HPLC). GC/MS requires only a small amount of tissue to detect and quantitate individual isomers of 22 : 6 hydroperoxide *(6)*.

In this chapter, we describe a GC/MS method, that allows detection of the in vivo formation of 22 : 6 hydroperoxide in inflamed rat retina. Although, in recent years, the direct detection of in vitro generation of phospholipid hydroperoxides has been reported *(7)*, the current method, which hydrolyzes fatty acid hydroperoxide from phospholipid for detection, remains a good, unambiguous and reproducible method for detecting 22 : 6 hydroperoxides in vivo.

2. Materials
2.1. Equipment

1. Vortex mixer.
2. Organic solvent evaporator (such as Reacti-Therm evaporator; Pierce, Rockford, IL).
3. Table centrifuge (such as Beckman GPR centrifuge).
4. Microcentrifuge (such as Eppendorf centrifuge 5415C).
5. Hot plate/stirrer.
6. Tissue homogenizer (such as Polytron; Brinkmann).
7. Dual cavity UV-VIS spectrophotometer (such as Shimatzu model UV-160).
8. Gas chromatography/mass spectrometry instrumentation.
   a. For best results, the gas chromatograph should be equipped with a Hewlett Packard fused silica capillary column (12.5 m × 0.20 mm i.d.) coated with cross-linked methyl silicone gum (0.33 µ thickness).
   b. The gas-chromatograph unit is then coupled with a Hewlett Packard mass selective detector.
   c. The data acquisition is performed by a Hewlett Packard 9000 series computer operated through a Hewlett Packard 59970 MS chemstation.

2.2. Reagents

1. Folch reagent (chlorform/methanol 2 : 1).
2. Butylated hydroxytoluene.
3. Ethanol (photrex grade, J.T. Baker, Philipsburg, NJ).
4. Methanol.
5. Sodium borohydride.
6. M-trifluoromethylphenyl-trimethylammonium hydroxide (METH-PREP II, Alltech, Deerfield, IL).
7. N,O-bis (trimethylsilyl) trifluoroacetamide (Pierce, Rockford, IL).

3. Methods
3.1. Sample Preparation

1. The globes are dissected to obtain retina and choroid. In animals with severe inflammation, isolation of retina alone is difficult, and inclusion of both retina

and choroid appears to make the wet weights considerably more reproducible. Total lipids are extracted from these retinas, by Folch's method *(8)*. The retina and choroid from six rat eyes are combined as one sample and homogenized with a Polytron (Brinkmann, Westbury, NY) in 6 mL of chloroform/methanol (2:1) containing 1 mg of butylated hydroxytoluene per 100 mL of solvent. Following homogenization, the resulting mixture is vortexed for 1 min to ensure a complete extraction. Exactly 1.2 mL of water is added to extract the water-soluble material, and the mixture is centrifuged at 3500 rpm for 20 min at 4°C to separate the layers. The chloroform/methanol layer is collected and the solvent evaporated under nitrogen using a Reacti-Therm evaporator. Approximately, 5–6 mg of total lipids can be obtained at this step (*see* **Note 1**).

3.2. Preparation of Lipids

1. At this point, the level of conjugated dienes, indicative of the extent of lipid peroxidation, should be measured to assure the occurrence of in vivo peroxidation. The total lipids are dissolved in 3.0 mL of ethanol ("photrex" grade, J. T. Baker, Phillipsburg, NJ) for UV absorption measurement. A dual-cavity spectrophotometer is used to record the absorption range of 200–400 nm. Absorbance is measured at 233 nm and a molar extinction coefficient of 25,200 is used to estimate the amount of conjugated dienes produced in the sample *(5)*.
2. The total crude lipids dissolved in 1.5 mL of methanol are reduced by stirring, initially with 50 mg of $NaBH_4$ for 15 min at room temperature, and then with an additional 15 mg of $NaBH_4$ for 15 more min. The excess amount of $NaBH_4$ is decomposed by adding 1.5 mL of water. The final reaction mixture is extracted with 2.0 mL of chloroform and the layer separated by centrifugation. Chloroform is evaporated from the organic layer to leave the total lipids as residue (*see* **Note 2**).
3. The transesterification to release fatty acid moieties from the phospholipids and to methylate carboxyl groups of fatty acid is carried out by dissolving the crude lipids in 0.2 mL of methanol and reacting the solution with 0.3 mL of METH-PREP II (0.2 M m-trifluoromethyl-phenyl-trimethylammonium hydroxide in methanol; Alltech, Deerfield, Il) *(9)*. The reaction mixture is stirred under nitrogen at room temperature for 30 min. After the reaction, 0.8 mL of water is added and the mixture is extracted twice with 2.0 mL of ethyl acetate. The organic layer is separated by centrifugation and evaporated under nitrogen to obtain hydroxy methyl esters. This procedure has been shown to give a nearly quantitative conversion of phospholipids and triglycerides to fatty acid methyl esters *(9)* (*see* **Notes 3–5**).

3.3. Sample Analysis

1. Before running the GC/MS, hydroxy fatty acid methyl esters are derivatized by adding 20–30 µL of N,O-bis (trimethylsilyl) trifluoroacetamide (Pierce, Rockford, Il) to convert hydroxyl groups to trimethylsilyl ethers.
2. Electron impact GC/MS is carried out on a Hewlett Packard gas chromatograph (Model 5890A) coupled with a Hewlett Packard mass selective detector (Model

5970) The data acquisition is performed by Hewlett Packard 9000 series comput-
ers operated through a MS chemstation.
3. The injection port temperature is maintained at 250°C, and the column temperature is programmed from 180–250°C with a rising rate of 4°C/min. The electron
voltage for the detector is set at 70 eV.

3.4. Results

1. In the peroxidation of 22:6, 10 possible positional isomers of hydroperoxide
can be produced from the five sets of 1,4-pentadiene structures in the molecule.
The mass spectrometry of trimethylsilyl ether, methyl ester derivatives of
isomeric hydroxy docosahexaenoic acid methyl ester (HDHE) has previously
been characterized by two studies *(9,10)*. The most intense peaks in mass spectra
of HDHEs result from cleavage alpha to the trimethylsilyl ether group. The base
peaks, therefore, confirm the presence, as well as the position, of the hydroxy
group in these compounds. However, among the two possible modes of alpha
cleavage for all of the HDHEs, usually only one intense cleavage peak will be
observed from either the methyl or the carboxyl side *(10,11)*.
2. The identification of HDHEs in the inflamed retinas can be carried out by GC/MS
using both total ion current scanning and selected ion monitoring methods. In the
total ion scanning method, the HDHEs are seen as a cluster of peaks; retention
times range from 16.0–16.5 minutes under these chromatographic conditions,
and the positional isomers of HDHE are not totally discernible. However, with
the selected ion monitoring mode, a single ion is scanned at one time, and
characterization, as well as quantitation, of individual HDHE is possible.
3. As shown in **Fig. 1**, five hydroperoxide-derived HDHEs, including 10-, 11-, 13-,
14-, and 17-HDHE, can be positively identified and quantitated from the inflamed
retinas. The most intense peaks for these HDHEs are all derived from alpha cleavage
to either side of the trimethylsilyl ether group. However, for all five HDHEs
identified, only one ion is observed from either the methyl or carboxyl side.
4. The characteristic ions for these HDHEs are shown in **Fig. 1**. For other isomers,
including 4-, 7-, 8-, 16-, and 20-HDHEs, the alpha cleavage ions are detected only
in low intensities (*see* **Note 6**). The confirmation of mass spectra is also made
by comparison with MS spectra from an authentic mixture of HDHEs formed by
the autoxidatuion of pure 22:6. The mass spectra, selected ion chromatograms
and retention times for the authentic HDHEs are usually in good agreement with
in vivo peroxidation products. In the bulk phase autoxidation of 22:6, however,
the relative concentration of positional isomers are generally different from
those of in vivo oxidation; normally, the 8- and 16-HDHEs are present in higher
concentrations than the others.

4. Notes

1. This protocol can be adapted with equal success for retina and choroid obtained
from two eyes. For two eyes, the quantities of reagents should be proportionally
reduced.

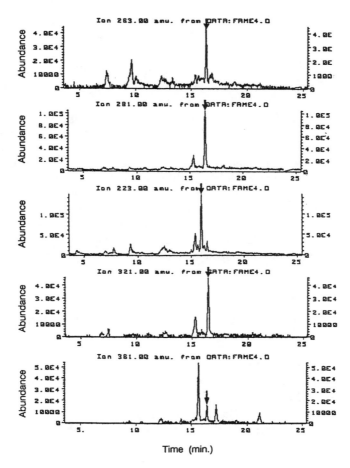

Fig. 1. A selected ion chromatogram obtained from inflamed retina is shown. The lipids from the retina were extracted, reduced and transesterified. The hydroxy fatty acid methyl esters were derivatized by N,O-bis (trimethylsilyl) trifluoroacetamide before analysis. The trimethylsilyl ether, methyl ester derivatives of 10-, 11-, 13-, 14-, and 17-HDHEs are represented by ions m/z 263, 281, 223, 321 and 361, respectively. Ion profiles of HDHEs are presented as relative peak areas. The analyses were repeated three times and a representative chromatogram is shown.

2. Reduction of lipids is carried out with methanolic NaBH₄. The reaction is complete in 20 min on both neutral and polar lipids. In the course of this reduction, the hydroperoxidic group is reduced to the corresponding hydroxy group and can be located through the GC/MS spectra of the corresponding trimethylsilyl group, whereas the other structural features derived from lipid peroxidation, such epoxidic rings that may be present, are not hydrolyzed.

3. Whenever possible, all of the procedures are performed under nitrogen, with the presence of butylated hydroxytoluene, to protect the 22:6 from undue further oxidation.

4. Virtually all of the unsaturated fatty acids in the biological system are bound by ester linkages in phospholipids or triglycerides. Phospholipid and triglyceride peroxides are primary products of lipid peroxidation and have rarely been measured. Most of the procedures involved in detecting lipid peroxidation in vivo involves dissociation of the fatty acid moieties from the phospholipid or triglyceride backbone.

5. This method does not require prior silicic acid chromatography to isolate hydroxy fatty acid methyl esters before the analysis, as has been reported by Hughes (10).

6. In this particular study, the detection and relative intensities of the HEHDs are important and no attempt is made to determine the absolute quantities of these hydroxy fatty acids derived from hydroperoxide. However, using this method, the absolute quantity can also be obtained, provided a suitable deuterated internal standard is used. Many of deuterated internal standards are now commercially available.

Acknowledgment

Supported in part by grant EY12363, National Institute of Health.

References

1. Rodenas, J., Mitjavila, M. T., and Carbonell, T. (1995) Simultaneous generation of nitric oxide and superoxide by inflammatory cells in rats. *Free Radic. Biol. Med.* **18,** 869–875.

2. Beckman, J. S., Chen, J., Ischiropoulos, H., and Crow, J. P. (1994) Oxidative chemistry of peroxynitrite. *Methods Enzymol.* **233,** 229–240.

3. Fliesler, S. J. and Anderson, R. E. (1983) Chemistry and metabolism of lipids in the vertebrate retina. *Prog. Lipid Res.* **22,** 79–131.

4. Mead, J. F. (1984) Free radical mechanism in lipid peroxidation and prostaglandins, in *Free Radicals in Molecular Biology, Aging and Disease* (Armstrong, D., Sohal, R. S., Cutler, R. G., and Slater, T. F., eds.), Raven Press, New York, pp. 53–66.

5. Rao, N. A. (1990) Role of oxygen free radicals in retinal damage associated with experimental uveitis. *Trans. Am. Ophthalmol. Soc.* **88,** 797–845.

6. Wu, G. S., Sevanian, A., and Rao, N. A. (1992) Detection of retinal lipid hydroperoxides in experimental uveitis. *Free Radic. Biol. Med.* **12,** 19–27.

7. Spickett, C. M., Pitt, A. R., and Brown, A. J. (1998) Direct observation of lipid hydroperoxides in phospholipid vesicles by electrospray mass spectrometry. *Free Radic. Biol. Med.* **25,** 613–620.

8. Folch, J., Lees, M., and Sloane Stanley G. H. (1957) A simple method for the isolation and purification of total lipids from animal tissues. *J. Biol. Chem.* **226,** 497–509.

9. Van Kuijk, F. J. G. M., Thomas, D. W., Stephens, R. J., and Dratz, E. A. (1985) Gas chromatography-mass spectrometry method for determination of phospholipid

peroxides: I Transesterification to form methyl esters. *Free Radic. Biol. Med.* **1,** 215–225.
10. Hughes, H., Smith, C. V., Tsokos-Kuhn, J. O., and Mitchell, J. R. (1986) Quantitation of lipid peroxidation products by gas chromatography-mass spectrometry. *Anal. Biochem.* **152,** 107–112.
11. VanRollins, M. and Murphy, R. C. (1984) Autoxidation of docosahexaenoic acid: analysis of ten isomers of hydroxydocosahexaenoate. *J. Lipid Res.* **25,** 507–517.

5

Detection of Lipid Hydroperoxide-Derived Protein Modification with Polyclonal Antibodies

Yoji Kato and Toshihiko Osawa

1. Introduction

The oxidation of lipids can cause the formation of lipid hydroperoxide followed by the degradation of the peroxide. The formed lipid-decomposition products such as aldehydes can easily react with biomolecules such as proteins *(1)*. The aldehyde-derived adduct formation has been examined by many researchers using chemical and immunochemical methods *(1–3)*. However, on the other hand, the lipid hydroperoxide itself might react with protein as already suggested *(4)*. We hypothesize that the lipid hydroperoxide-derived modification occurs and specific products derived from the reaction should be formed. In this chapter, we describe the methodology of the preparation of polyclonal antibodies (PAbs) to lipid hydroperoxide-modified proteins. We also show the preparation of the polyclonal antibody to N^{ε}-(hexanoyl)lysine, which is identified as a novel lipid-lysine adduct from the reaction mixture of linoleic acid hydroperoxide and lysine *(5)*.

2. Materials

2.1. Equipment

1. Microplate reader: SpectraMax 250 (Molecular Device Corp.).
2. Electrophoresis system.
   a. Electrophoresis mini-gel (vertical slab) (8 × 8 cm) (Nihon Eido Co., Ltd., Japan).
   b. Electric power supply: Model NC-1010 (Nihon Eido).
3. Electrotransfer system.
   a. Semi-dry blotting: Horiz Blot (ATTO Co., Japan).
   b. Electric power supply: PowerStation 1000XP (AE-8750; ATTO Co.).

From: *Methods in Molecular Biology, vol. 186: Oxidative Stress Biomarkers and Antioxidant Protocols*
Edited by: D. Armstrong © Humana Press Inc., Totowa, NJ

2.2. Reagents

1. Keyhole limpet hemocyanin (KLH), 1-ethyl-3-(3-dimethylaminopropyl) carbo-diimide (EDC), and N-hydroxysulfosuccinimide (sulfo-NHS) were purchased from Pierce Chemical Co., Rockville, IL.
2. Soybean lipoxygenase (SLO), lipid-free bovine serum albumin (BSA), and arachidonic acid were obtained from Sigma Chemical Co., St. Louis, MO.
3. All reagents are the highest grade available.

3. Methods
3.1. Preparation of Lipid Hydroperoxide

1. 13S-Hydroperoxy-9Z, 11E-octadecadienoic acid (13-HPODE) is prepared as follows: 13-HPODE is made from linoleic acid using soybean lipoxygenase (SLO). The linoleic acid (0.7 mg/mL) is dissolved in 0.1 M borate buffer, pH 9.0. SLO (7600 U/mL) is added to the solution under O_2 with stirring at room temperature. After 90 min, the pH is adjusted to 4.0 with 0.1 M HCl for the termination of the oxidation. The formed peroxide is extracted with a twofold volume of chloroform/methanol (1/1). The extract is evaporated, dissolved in chloroform, and then applied to a TLC plate. The chromatography is performed by development with *n*-hexane/diethyl ether (4/6). The band of 13-HPODE is visualized by a UV lamp and scraped off. The 13-HPODE is dissolved in ethanol and stored under Argon gas at –70°C or –20°C before use. The identification is performed on the basis of ^{1}H-nuclear magnetic resonance spectometry and HPLC coelution with commercially obtained authentic 13-HPODE.
2. 15-Hydroperoxyeicosatetraenoic acid (15-HPETE) is made from arachidonic acid by SLO as described above with some modifications. Briefly, the arachidonic acid (0.76 mg/mL) in the 0.1 M borate buffer is oxidized by SLO (6500 U/mL) under O_2. The extraction and purification are done as described earlier.

3.2. Preparation of Immunogen

1. The 13-HPODE- and 15-HPETE-modified proteins are prepared as follows: The obtained 5 mM 13-HPODE (or 15-HPETE) is incubated with 5 mg/mL of KLH in a phosphate buffer, pH 7.4, at 37°C for 3 d. The protein is dialyzed against PBS at 4°C for 2–3 d with several exchanges of phosphate-buffered saline (PBS). The concentration of the proteins is adjusted with PBS at 1 mg/mL and the proteins are stored at –70°C until immunization.
2. The conjugation of hexanoic acid with protein is performed as follows: Hexanoic acid (2.3 mg), EDC (4.5 mg), and sulfo-NHS (5 mg) are dissolved in 400 µL of dimethylformamide, and the reaction mixture is incubated for 24 h at room temperature. To the solution, 0.95 mL of KLH (10 mg in 0.1 M phosphate buffer, pH 7.4) is added and further incubated for 4 h at room temperature. The obtained hexanoyl proteins are dialyzed against PBS for 3 d at 4°C.

3.3. Immunization and Antibody Generation

These modified KLHs are emulsified with an equal volume of complete Freund's adjuvant to a final concentration of 0.5 mg/mL, and 1 mL of the solution is then intramuscularly injected into a New Zealand White rabbit. After 4 wk, 1 mL of the KLH emulsified with an equal volume of incomplete adjuvant (0.5 mg/mL) is injected as a booster every 2 wk until an adequate antibody generation has occurred. 13-HPODE-modified BSA, 15-HPETE-modified BSA, and hexanoyl BSA are prepared instead of KLH as described in **Subheading 3.2.** and then used for the evaluation of the antibody generation specific to the modified proteins by enzyme-linked immunosorbent assay (ELISA) or Western blotting to be described later.

3.4. ELISA

1. Indirect noncompetitive ELISA is performed as already described *(6,7)*. Briefly, 50 µL of antigen in PBS (typically 0.05 mg/mL) is dispensed into a well of an ELISA plate (Nunc Maxsorp) and kept at 4°C overnight. After blocking with Block Ace (Dainihon Seiyaku, Osaka, Japan), 100 µL of antiserum (1/5000 in PBS containing 0.5% BSA) is added to the well. The binding of the antibody on the coated antigen is evaluated by using goat anti-rabbit IgG antibody-peroxidase conjugates. The color development is done by adding *o*-phenylenediamine and hydrogen peroxide. The O. D. 492 nm is measured by SPECTRA MAX 250 microplate reader (Molecular Devices Corporation).
2. The cross-reactivity of the low molecular-weight compounds with antibody is investigated by indirect competitive ELISA *(6,7)*. As a coating agent, 50 µL of the modified BSA (0.5 µg/mL) is pipetted into the wells and kept at 4°C overnight. At the same time, 50 µL of antiserum (diluted in PBS containing 1% BSA) and 50 µL of sample are mixed in a tube and reacted at 4°C overnight. The plate is washed and 90 µL of the reacted solution is pipetted into a well. The binding of the residual antibody on the coated modified BSAs is estimated as described earlier.

3.5. Sodium Dodecyl Sulfate-Polyacrylamide Gel Electrophoresis (SDS-PAGE) and Western Blotting

1. SDS-PAGE (10% acrylamide) is prepared and run using a stacking gel of 4.5%. Protein (10 µg) is applied to each lane of the gel. The gel is stained with Coomassie Brilliant Blue. For the immunoblot analysis, the migrated protein in the gel is electroblotted to a polyvinylidene difluoride (PVDF) membrane using ATTO Holiz Blot, incubated with 4% Block Ace for blocking, washed three times with Tris-buffered saline (TBS) containing 0.1% Tween 20 (TTBS) for 10 min, and treated overnight with TTBS-diluted antiserum (1/1000) at 4°C.

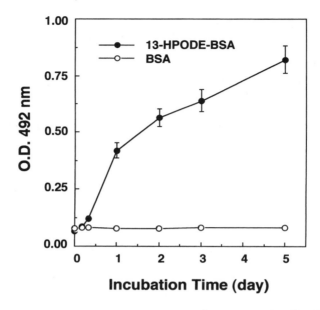

Fig. 1. Time-dependent formation of immunoreactivity during incubation of BSA with 13-HPODE. 13-HPODE (5 mM) and lipid-free BSA (0.5 mg/mL) were incubated in 0.1 M phosphate buffer, pH 7.4, at 37°C for various intervals. The modified protein was precipitated by the addition of ice-cold ethanol to the reaction mixture. The precipitate was dissolved in PBS and then used for ELISA using the anti-13-HPODE-KLH antibody as described in **Subheading 3.4.**

2. Occasionally, for antibodies to the 13-HPODE- and 15-HPETE-modified proteins, the ester bonds on the blotted membrane are hydrolyzed with 0.25 M NaOH (by immersing) for 1 h with gentle shaking prior to the blocking.
3. After treatment with antiserum, the membrane is exposed to the peroxidase-conjugated anti-rabbit IgG antibody (1/2000 in TTBS) and soaked for 1 h at room temperature. The binding of the second antibody to the anti-13-HPODE-KLH (15-HPETE-KLH) antibody is visualized using an ECL reagent (Amersham-Pharmacia).

3.6. Results

3.6.1. Generation of Antibody to 13-HPODE-modified Protein

The obtained antibody to 13-HPODE-KLH recognizes the 13-HPODE-modified BSA (**Fig. 1**) but not the aldehyde-modified BSAs (**Fig. 2**). Based on the characterization *(6)*, it is assumed that the antibody reacts with the COOH terminal moiety as a part of the epitopes. Therefore, to confirm this

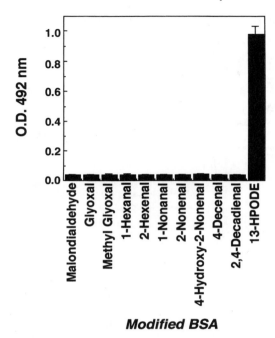

Modified BSA

Fig. 2. Effect of aldehyde-modified proteins on immunoreactivity against anti-13-HPODE-KLH antibody. Aldehydes were incubated with lipid-free BSA in phosphate buffer at 37°C for 24 h. As a control, 13-HPODE was also reacted with lipid-free BSA in a phosphate buffer at 37°C for 3 d. These modified proteins were dialyzed against PBS at 4°C for 3 d and used for ELISA.

fact, the effect of ester bond hydrolysis of the oxidized methyl linoleic acid- and oxidized PC-modified proteins on the formation of the antigenicity was examined. As shown in **Fig. 3**, the generation of the antigenicity of the esterified fatty acid-modified proteins by alkaline treatment is observed, suggesting that the COOH terminal is required for recognition. Also this experiment reveals that the effective alkaline concentration for the ester bond hydrolysis is 0.25 *M*.

3.6.2. Generation of Antibody to 15-HPETE-Modified Protein

The formation of the antibody to 15-HPETE-KLH is also observed (**Fig. 4**). The character of the antibody is similar to that of the antibody to 13-HPODE-KLH (*7*).

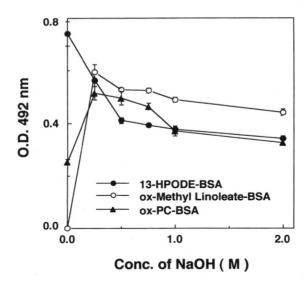

Fig. 3. Effect of alkaline treatment of oxidized esterified fatty acid-modified proteins on the immunoreactivity against anti-13-HPODE-KLH antibody. The methyl linoleate and di-linoleoyl phosphatidylcholine (PC) were thermally oxidized and the peroxidized lipids were incubated with lipid-free BSA in phosphate buffer as described in **Fig. 2** caption. The obtained modified proteins were dispensed into a well, and various concentrations of alkali were added to the well. The plate was incubated for 1 h at 37°C and washed. After blocking of the well, the evaluation of the immunoreactivity was performed as described in **Subheading 3.4.**

3.6.3. Generation of Antibody to N^ε-(Hexanoyl)lysine

We have recently identified N^ε-(hexanoyl)lysine (HEL) as one of the lipid hydroperoxide-derived oxidative modified products *(5)*. We also prepare the polyclonal antibody to the novel lipid-lysine adduct using synthetic hexanoylated KLH as an immunogen. The formation of the antibody recognizing hexanoylated BSA is observed (**Fig. 5**). The antibody to the hexanoylated KLH recognized the synthetic HEL itself. The antibody to HEL can be directly used for estimation of the oxidative modification by the oxidized esterified fatty acid without an alkaline treatment (ester bond hydrolysis) because HEL is the CH_3-terminal derived adduct. Using the anti-HEL antibodies, we have detected the immunoreactive materials in human atherosclerotic plaques *(5)*.

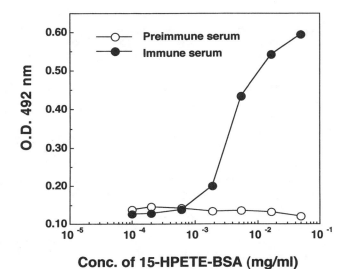

Fig. 4. The generation of the antibody to 15-HPETE-modified protein. One month after the immunization of 15-HPETE-KLH into a rabbit, the serum was collected and then used for ELISA. The reactivity of the immunized serum was compared with the serum before immunization (preimmune serum). The immunized serum was reacted with 15-HPETE-BSA in a dose-dependent manner.

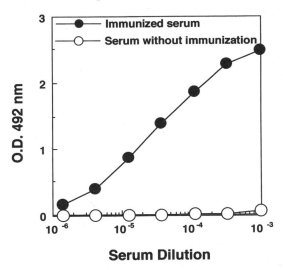

Fig. 5. The generation of the antibody to hexanoylated protein. One month after the immunization of hexanoylated KLH into a rabbit, the serum was collected and then used for ELISA. The reactivity of the immunized serum was compared with the normal rabbit serum.

References

1. Esterbauer, H., Schaur, R. J., and Zollner, H. (1991) Chemistry and biochemistry of 4-hydroxynonenal, malonaldehyde and related aldehydes. *Free Rad. Biol. Med.* **11,** 81–128.
2. Haberland, M. E., Fong, D., and Cheng, L. (1988) Malondialdehyde-altered protein occurs in atheroma of Watanabe heritable hyperlipidemic rabbits. *Science* **241,** 215–218.
3. Palinski, W., Rosenfeld, E., Ylä-Herttuala, S., Gurner, G. C., Socher, S. S., Butler, S. W., et al. (1989) Low density lipoprotein undergoes oxidative modification in vivo. *Proc. Natl. Acad. Sci. USA* **86,** 1372–1376.
4. Fruebis, J., Parthasarathy, S., and Steinberg, D. (1992) Evidence for a concerted reaction between lipid hydroperoxides and polypeptides. *Proc. Natl. Acad. Sci. USA* **89,** 10,588–10,592.
5. Kato, Y., Mori, Y., Makino, M., Morimitsu, Y., Hiroi, S., Ishikawa, T., and Osawa, T. (1999) Formation of N^ε-(hexanonyl)lysine in protein exposed to lipid hydroperoxide. *J. Biol. Chem.* **274,** 20,406–20,414.
6. Kato, Y., Makino, M., and Osawa, T. (1997) Characterization of a specific polyclonal antibody against 13-hydroperoxyoctadecadienoic acid-modified protein: formation of lipid hydroperoxide-modified apo B-100 in oxidized LDL. *J. Lipid Res.* **38,** 1334–1346.
7. Kato, Y. and Osawa, T. (1998) Detection of oxidized phospholipid-protein adducts using anti-15-hydroperoxyeicosatetraenoic acid-modified protein antibody: contribution of esterified fatty acid-protein adduct to oxidative modification of LDL. *Arch. Biochem. Biophys.* **351,** 106–114.

6

Techniques for Determining the Metabolic Pathways of Eicosanoids and for Evaluating the Rate-Controlling Enzymes

Ninder Panesar, Yashwant G. Deshpande, and Donald L. Kaminski

1. Introduction

The term eicosanoid is derived from the chemical name for arachidonic acid, eicosatetraenoic acid. The term eicosanoid includes the prostaglandins (PG), thromboxanes, and leukotrienes *(1)*. Multiple biosynthetic pathways exist for eicosanoid formation *(2)* and specific eicosanoids seem to be produced by specific cell types in response to various inflammatory stimuli *(3)*. The multiple synthetic pathways suggest that multiple sites of inhibition may exist. Phospholipase A_2 (PLA_2) acts on cell membranes to release arachidonic acid *(4)*. While numerous enzymes have demonstrated PLA_2 activity, two groups of enzymes are primarily responsible for the arachidonic acid generated during inflammatory responses. One enzyme with extensive involvement in arachidonic acid formation in response to inflammatory stimuli in many cell types is 14 kDa secretory PLA_2 ($sPLA_2$). $sPLA_2$ requires millimolar concentrations of calcium ion for activity and binds to cell-surface proteoglycans to catalyze the release of fatty acids from the sn-2 position of cell-membrane phospholipids *(5)*. Group IV cytosolic 85 kDa PLA_2 ($cPLA_2$) is also an important enzyme associated with the production of arachidonic acid and subsequent eicosanoid formation during inflammation *(6)*. $cPLA_2$ is also calcium-dependent and selectively hydrolyzes phospholipids with arachidonyl residues at the sn-2 position *(7)*.

Once arachidonic acid is formed, three pathways exist for its metabolism, including the cytochrome p-450, the lipoxygenase and the cyclooxygenase (COX) pathways *(8,9)*. COX enzymes are bifunctional enzymes that serially

From: *Methods in Molecular Biology, vol. 186: Oxidative Stress Biomarkers and Antioxidant Protocols*
Edited by: D. Armstrong © Humana Press Inc., Totowa, NJ

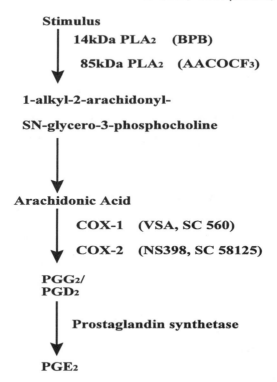

Fig. 1. Schematic diagram of the metabolic pathway producing PGE_2. PLA_2 indicates phospholipase enzymes and COX indicates cyclooxygenase enzymes.

metabolize arachidonic acid to PGG_2 and then to PGH_2 *(10)*. This is the rate-limiting step for the subsequent formation of the biologically active prostanoids including PGD_2, $PGF_{2\alpha}$, prostacyclin, thromboxanes, and PGE_2 *(12)*. There are two COX enzymes. COX-1 is a constitutive enzyme that can be induced by various inflammatory stimuli in some cell types. COX-2 is frequently not identified in unstimulated cells but is inducible and forms prostanoids in response to various inflammatory stimuli in many cell types *(13)*. The metabolic pathways leading to prostanoid formation are illustrated in **Fig. 1**. It is the intention of the present chapter to describe how, in inflammation, the role of the two PLA_2 enzymes and the two COX enzymes in prostanoid formation is determined. The methodology presupposes that the investigator has a cell line that they wish to study and have standard skills, equipment, and materials for maintenance of the cell line.

2. Materials

2.1. Immunoblotting PLA₂ Enzymes

2.1.1. Equipment

1. Mini Protean II cell (Bio-Rad, Rocheville Center, NY).
2. Spectrophotometer (Beckman DU650, Fullerton, CA).
3. Vortexer (Fisher Scientific, Springfield, NJ).
4. Power supply (Bio-Rad).
5. Table top shaker (Fisher).
6. Kodak film developer (Kodak, Rochester, NY).
7. Cassettes for exposing membranes to film (Bio-Rad).
8. Bag sealer.
9. Hybond C extra 0.45 u nitrocellulose membrane (Amersham, Arlington Heights, IL).

2.1.2. Reagents

1. Acrylamide (Bio-Rad).
2. Tris base (Bio-Rad).
3. Bis-Acrylamide (Sigma, St. Louis, MO).
4. SDS (Sigma).
5. Ammonium per sulphate (Sigma).
6. TEMED (Sigma).
7. Glycerol (Sigma).
8. Glycine (Sigma).
9. 2-mercaptoethanol (Sigma).
10. Bromophenol blue (Sigma).
11. Antibodies, primary and secondary (cPLA$_2$, Santa Cruz, CA; sPLA$_2$, Cayman, Ann Arbor, MI).
12. Electrophoresis standard (Cayman).
13. Carnation milk.
14. Tween 20 (Sigma).
15. Enhanced chemiluminescence (Amersham).
16. Triton X 100 (Sigma).

2.2. Quantitating Arachidonic Acid Formation

2.2.1. Equipment

1. 24-well tissue-culture plates (Fisher).
2. Cell scrapers (Fisher).
3. CO_2 incubator (Fisher).
4. Liquid scintillation counter (Beckman).
5. Free-flow laminar hood (Fisher).
6. Scintillation vials (Amersham).

2.2.2. Reagents

1. The sPLA$_2$ inhibitor employed is p-bromophenylacyl bromide (BPB, Sigma) and the maximal inhibitory dose is 40 μM. The cPLA$_2$ inhibitors employed are arachidonyl trifluoromethyl ketone (AACOCF3) and methyl arachidonyl fluorophosphonate (MAMP), which are both available from Cayman, and the maximal inhibitory dose identified in dose-response studies for both cPLA$_2$ inhibitors is 10 μM.
2. Phosphate-buffered saline (PBS, Sigma).
3. Cell-culture media (Sigma).
4. Aqueous scintillation cocktail (Fisher).
5. ^{3}H arachidonic acid (New England Nuclear Life Science Products, Boston, MA).

2.3. Immunoblotting COX Enzymes

2.3.1. Equipment

1. Mini Protean II cell (Bio-Rad, Rocheville Center, NY).

2.3.2. Reagents

1. Antibodies for COX-1 and COX-2, primary and secondary (Oxford Biomedical Research, Oxford, MI).

2.4. Quantitating Prostanoids

2.4.1. Equipment

1. Organic solvent evaporator (Fisher).
2. Nitrogen gas (Air Gas, Mid-America Inc., Bowling Green, KY).
3. Centrifuge (Fisher).
4. Microplate reader (Bio-Rad Model 3550).
5. Cell culture plates (Falcon Products, Newark, NJ).
6. Sep-Pak centrifuge rack and Sep-Pak C18 columns (Waters).
7. Gilson micropipets and Eppendorf pipets (Waters).

2.4.2. Reagents

1. The COX-1 inhibitor employed is valerylsalicyclic acid (Cayman, Ann Arbor, MI) with a maximal inhibitory dose found to be 50 μM. The COX-2 inhibitors are (1-[4-methylsulfonyl(phenyl)-3-trifluoro-methyl-5([4-fluoro] phenyl)pyrazole (SC 58125, Searle, St. Louis, MO) and 2-cyclohexyloxy-4-nitrophenyl methanesulfonamide (NS 398, Cayman) with the maximal inhibitory dose being 100 μM for both COX-2 inhibitors.
2. High-performance liquid chromatography (HPLC) grade methanol, ethanol, ethyl acetate, cyclohexane, petroleum ether, methyl formate, dimethyl sulfoxide (Fisher).
3. Lipopolysaccharide (Escherichia coli LPS, CO 11: B4, Sigma).
4. Calcium ionophore (A23187, Sigma).

5. Interleukin 1β (IL-1β; Merck Dupont, Glenolden, PA).
6. Prostaglandin Elisa immunoassay kits (Cayman).
7. Bradford micro-protein determination kits (Sigma).

3. Methods

3.1. Immunoblotting PLA$_2$ Enzymes

1. For immunoblotting of PLA$_2$ protein, 1×10^6 cells are plated on a 12-well culture plate to yield the requisite amount of cell protein. The cells are washed with PBS twice and then scraped, collected in 15 mL tubes and pelleted at 250 rpm for 5 min at room temperature. The pellets are resuspended in 5 mL of lysis buffer (20 mM Tris-HCl, pH 7.5, 0.9% NaCl, 0.1% Triton X-100, 20 mM β glycerophosphate, 10 µg/mL aprotinin, and 10 µg/mL leupeptin). The samples are left at room temperature for 10 min and then transferred to microcentrifuge tubes and spun at the highest speed for 1 min. For determination of cell protein, cell specimens are solubilized with 0.1 N NaOH for 1 h at 37°C and then sonicated for 10 s. The protein concentrations are determined utilizing the method of Bradford with bovine serum albumin (BSA) as the standard *(14)*.
2. Protein samples (20 µg), PLA$_2$ standards and molecular-weight markers are loaded onto a 10% sodium dodecyl sulfate polyacrylmide gel (SDS-PAGE). The gels are run at 100 V for 2 h until the dye front has reached the bottom of the gel. Proteins are transferred onto a nitrocellulose membrane at 100 V for 1 h in 25 mM Tris base, 192 mM glycine, and 15% methanol using the Bio-Rad trans blot apparatus.
3. The membranes are transferred into (TBS-T) containing 100 mM Tris-HCl, pH 7.6, 150 mM NaCl, and 0.5% Tween 20, rinsed for 5 min to remove any gel particles, and transferred to fresh TBS-T + 10% milk to block the proteins on the membranes and the membranes are incubated overnight at 4°C. The membranes are subsequently incubated separately with the cPLA$_2$ antibody, diluted 1:1000 and the sPLA$_2$ antibody diluted 1:1000 at room temperature for 3 h with gentle shaking. The membranes are then washed with TBS-T + 0.5% milk three times for 5 min, 10 min with TBS-T + 0.5 M NaCl + 0.5 M NaCl + 0.5% milk and washed twice with TBS-T + 0.5% milk for 5 min to remove any unbound antibodies. The membranes are then treated for 30 min with peroxidase-conjugated mouse anti-goat IgG or peroxidase conjugated goat anti-rabbit IgG for the sPLA$_2$ and cPLA$_2$ labeling, respectively. The membranes are then exposed with a chemiluminescent detection system for 1 min, blotted, and exposed for 1 min to Kodak X-O X-ray film.
4. The described technique has been employed to identify PLA$_2$ enzymes in various cell types. The sPLA$_2$ enzyme characteristically migrates to approx the 14 kDa range while the cPLA$_2$ protein migrates to approx the 110 kDA range *(15)*. The cPLA$_2$ antibody is a rabbit polyclonal antibody (PAb) produced by immunization with the amino terminal domain, amino acids 1-216 of human cPLA$_2$ standard. It uniformly identifies the human cPLA$_2$ standard. It does not cross-react with other phospholipases. The sPLA$_2$ antibody is a mouse-produced antibody with the immunogen being purified human 14 kDa sPLA$_2$. It does not cross-react with

other PLA_2 enzymes isolated from pancreas or snake venom and the antibody will detect 0.1 ng $sPLA_2$ protein. The antibody uniformly cross-reacts with the human 14 kDA $sPLA_2$ standard *(16)*. Changes in PLA_2 enzyme expression can be estimated visually or quantitated by densitometry.

3.2. Arachidonic Acid Formation

1. Cells are grown to the required degree of confluence on 24-well cell-culture plates. Cells are then labeled with 0.5 μCi/ml of ^{3}H-labeled arachidonic acid for 20 h. The label is incorporated into membranes phospholipids and will be released when PLA_2 enzyme activity is stimulated. The release of the label can be inhibited by specific PLA_2 inhibitors. The cells should be exposed to an inhibitor for at least 1 h before being stimulated by a cytokine or mitogen. Before adding the experimental substance the wells are washed three times at 5-min intervals with Hank's buffered saline solution (HBSS) containing 0.1 mg/mL of BSA in order to remove any unincorporated ^{3}H arachidonic acid.
2. The cells are then exposed to the stimulant and the specific PLA^2 inhibitor for the appropriate time. At the end of the treatment periods, the supernatants are removed, cleared of cells by centrifugation at 10,000*g* for 5 min, and the supernatants assayed for radioactivity by liquid-scintillation counting.
3. While the methodology is an accurate reflection of PLA_2 activity, some of the radioactive label may be in the form of prostanoids. If quantitation of arachidonic acid only is intended, a method to extract the lipid will need to be employed before counting *(17)*. Arachidonic acid levels can be expressed in relationship to the number of cells plated or the amount of cell protein in each well and can be quantitated as ^{3}H arachidonic acid counts expressed/mg cell protein (*see* **Note 1**).

3.3. Immunoblotting COX Enzymes

1. For immunoblotting of COX protein 1×10^6 cells are plated on a 12-well culture plate to yield the requisite amount of cell protein. The cells are washed with PBS twice and then scraped and collected in 15-mL tubes and pelleted at 2500 rpm for 5 min at room temperature. The pellets are resuspended in 5 mL of lysis buffer (20 m*M* Tris-HCl, pH 7.5, 0.9% NaCl, 0.1% Triton X-100, 20 m*M* β glycerophosphate, 10 μg/mL aprotinin, and 10 μg/mL leupeptin). These samples are left at room temperature for 10 min and then transferred to microfuge tubes and spun at highest speed for 1 min. The protein concentrations are determined by utilizing the method of Bradford *(14)*.
2. Protein samples (15 μg), COX standards and molecular-weight markers are loaded onto a 10% SDS polyacryalmide gel and separated as described in **Subheading 3.1., step 1**.
3. The membranes are transferred into TBS-T containing 100 m*M* Tris HCl, pH 7.6, 150 m*M* NaCl, and 0.5% Tween 20, rinsed for 5 min to remove any gel particles, and transferred to fresh TBS-T + 10% milk to block the proteins on

Table 1
The Effect of Cyclooxygenase Inhibitors on Prostaglandin E2 Formation by Intestinal Epithelial Cells

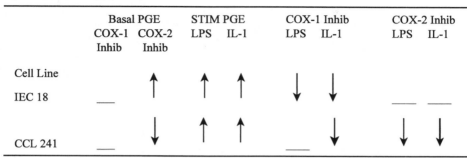

	Basal PGE COX-1 Inhib	Basal PGE COX-2 Inhib	STIM PGE LPS	STIM PGE IL-1	COX-1 Inhib LPS	COX-1 Inhib IL-1	COX-2 Inhib LPS	COX-2 Inhib IL-1
Cell Line IEC 18	—	↑	↑	↑	↓	↓	—	—
CCL 241	—	↓	↑	↑	—	↓	↓	↓

In the above table the ↓ sign indicates that PGE formation was decreased, ____ indicates that there was no significant change and ↑ indicates that PGE formation was significantly increased. LPS represents lipopolysaccharide and IL = 1 represents interleukin-1β *(3)*.

the membrane and allowed to incubate at 4°C overnight. The membranes are incubated separately with a polyclonal anti-goat COX-1 antibody and a polyclonal anti-COX-2 rabbit antibody, both diluted 1:100 at room temperature for 3 h with gentle shaking. The membranes are then washed with TBS-T + 0.5% milk three times for 5 min, 10 min with TBS-T *M* NaCl + 0.5% milk, and washed twice with TBS-T + 0.5% milk for 5 min to remove any unbound antibodies.

4. The membranes are then transferred into solutions containing 7.5 mL of secondary antibodies and incubated at room temperature for 1 h. For COX-1 detection, rabbit anti-goat IgG alkaline phosphatase conjugate, 1:3000 dilution is employed and for COX-2, goat anti-rabbit IgG alkaline phosphatase conjugate 1:3000 dilution is used as the secondary antibody. The membranes are then washed with TBS-T + 0.5% milk three times for 5 min, 10 min with TBS-T + 0.5% M NaCl + 0.5% milk and washed twice with TBS-T + 0.5% milk for 5 min to remove any unbound antibodies. The membranes are then exposed with a chemiluminescent detection system for 1 min, blotted, and then exposed for 1 min to Kodak X-O mat and X-ray film. The described technique has been employed to identify COX enzymes in various cell types. The COX antibodies invariably labeled the standard with no cross-reactivity between COX-1 and COX-2. COX proteins migrate to the 72 kDA range. Quantitation of enzyme level can be estimated visually or measured by densitometry.

3.4. Quantifying Prostanoids

1. The evaluation of COX activity and the determination of the metabolic pathway for a specific prostanoid is associated with the quantitation of the prostanoid

before and after exposure to a specific COX-1 or -2 inhibitor. Individual prostanoids are produced by the various cells in response to the same stimuli through the activity of different COX enzymes (**Table 1**; **ref. 3**). It may be intended to study a specific prostanoid associated with a well-defined response. A specific response, such as the production of mucin by colonic mucosal cells, occurs via a well-defined PGE_2 and PGE_2 receptor EP4 pathway *(18)*. The determination of whether this PGE_2 is produced by a COX-1 or -2 synthetic pathway would require stimulation of prostaglandin E_2 and mucin formation by a secretagogue *(19)* in the presence of a COX-1 or -2 inhibitor and measuring the formation of prostaglandin E_2. The cell line would require pretreatment with the inhibitor before exposing the cells to the stimulus.

2. In other systems the primary prostanoid produced may not be known and the entire spectrum may need to be evaluated by assaying for the production of PGE_2, PGD_2, $PGF_{2\alpha}$, and 6 keto $PGF_{1\alpha}$ (prostacyclin, I_2) in order to ascertain the primary prostanoid produced by a specific cell in response to a specific stimulus. We formerly would evaluate the prostanoid spectrum by thin-layer chromatography (TLC); however, recently the information for individual prostanoids is more readily obtained by extracting with organic solvents such as ethyl acetate-cyclohexane mixture, then purifying the preparation with Sep-Pak columns *(20)* and then using commercially available EIA kits that have excellent specificity and relatively good shelf life (*see* **Notes 2,3**).

3. Prostanoid levels should be expressed per mg cell protein or per number of cells. To determine the cell-protein concentration the media is removed and the cells washed with Kreb's Ringers bicarbonate buffer. The cells are scraped from the wells or when cells are growing on a specific substrate the cells are freed by incubation with a 1% collagenase solution for 20 min and then washed again with buffer and collected by centrifugation at 5000*g* for 10 min. Protein concentration is determined on cell specimens solubilized with 0.1 N NaOH for 1 h at 37°C and then sonicated for 10 s. Protein is quantitated utilizing the method of Bradford *(14)* using BSA as the standard.

4. PG assays are performed using a competitive enzyme assay that utilizes an acetylcholinesterase tracer. The prostanoid concentrations are determined by spectrophotometric analysis after the addition of Ellman's reagent and comparison to a standard curve. The enzyme-linked immunosorbent assay (ELISA) kit instructions are very detailed and easily followed (*see* **Note 5**).

5. The EIA employed to quantitate a prostanoid is based on the competition between the prostanoid and a prostanoid-acetylcholinesterase conjugate (tracer) for a limited amount of prostanoid monoclonal antibody (MAb). Because the concentration of prostanoid tracer is held constant while the concentration of unknown prostanoid varies, the amount of prostanoid tracer that is able to bind to the prostanoid MAb will be inversely proportional to the concentration of prostanoid in the sample. This antibody-prostanoid complex binds to a goat anti-mouse PAb that has been previously attached to the sample. The plate is washed to

remove any unbound reagents and then Ellman's reagent is added to the well. This reagent consists of acetylthiocholine with 5, 5′-dithio-bis(2-nitrobenzoic acid). Hydrolysis of acetylthiocholine by acetylcholinesterase produces thiocholine. The nonenzymatic reaction of thiocholine with 5,5′-dithio-bis (2 nitrobenzoic acid) produces a compound that has a strong absorbance at 412 nm.

6. The assay should be confirmed in the system by adding progressively increasing quantities of the specific prostanoid being measured to the culture media and ascertaining that the assay will quantitate the added prostanoid in a linear fashion. Cross-reactivity between antibodies is less than 1% and the intraassay and interassay coefficient of variation should be less than 5% (*see* **Note 3**).

7. Utilizing the aforementioned techniques, the opportunity exists to ascertain the metabolic pathways associated with the production of a specific prostanoid (*see* **Note 4**). Such information may lead to improved drugs to alter the production of a specific prostanoid involved in a particular response rather than the universal inhibition of prostanoid production now associated with the use of first generation nonsteroidal anti-inflammatory agents.

4. Notes

1. Cell populations that produce ng quantities of prostanoids should be studied at minute intervals for arachidonic acid release while cells that produce pg/mg cell protein of prostanoids should be evaluated at hourly intervals.

2. If the samples for prostanoids come from buffer solutions or cell-culture media, then it is not necessary to extract and purify the samples; however, if they are from plasma or tissues, then one needs to extract the samples with organic solvents and purify on Sep-Pack columns.

3. The necessity for extraction should be evaluated by measuring added known amounts of the specific prostanoid to the buffer solutions or culture media used to perform the experiments and measuring the concentration of the prostanoid in extracted and unextracted samples.

4. While the antibodies effectively discriminate and measure specific prostanoid species, within species, there may be cross-reactivity. For example, the assay may not discriminate well PGE_1 and E_2 levels.

5. Different pipet tips should be used to pipet the buffer solutions, standards, samples, tracers, and antibodies, and the pipet tips should not be exposed to the reagents already in the wells.

References

1. Eberhart, C. E. and DuBois, R. N. (1995) Eicosanoids and the gastrointestinal tract. *Gastroenterology* **109,** 285–301.
2. Fu, J. Y., Masferrer, J. L., Seibert, K., Raz, A., and Needleman, P. (1990) The induction and suppression of prostaglandin H2 synthase (cyclooxygenase) in human monocytes. *J. Biol. Chem.* **265,** 16,737–16,740.

3. Longo, W. E., Panesar, N., Mazuski, J., and Kaminski, D. L. (1998) Contribution of cyclooxygenase-1 and cyclooxygenase-2 to prostanoid formation by human enterocytes stimulated by calcium ionophore and inflammatory agents. *Pros. Other Lipid Med.* **56,** 325–339.
4. Dennis, E. A. (1994) Diversity of group types, regulation and function of phospholipase A2. *J. Biol. Chem.* **269,** 13,057–13,060.
5. Hara, S., Kudo, I., Chang, H. W., Matsuta, K., Miyamoto, T., and Inoue, K. (1989) Purification and characterization of extracellular phospholipase A2 from human synovial fluid in rheumatoid arthritis. *J. Biochem.* **105,** 395–399.
6. Huang, Z., Payette, P., Abdullah, K., Cromlish, W. A., and Kennedy, B. P. (1996) Functional identification of the active site nucleophile of the human 85-kDa cytosolic phospholipase A2. *Biochem.* **35,** 3712–3721.
7. Takayama, K., Kudo, I., Kim, D. K., Nagata, K., Nozawa, Y., and Inoue, K. (1991) Purification and characterization of human platelet phospholipase A_2 which preferentially hydrolyzes an arachidonyl residue. *FEBS* **282,** 326–330.
8. Capdevila, J. H., Falck, J. R., and Eastbrook, R. W. (1992) Cytochrome P450 and the arachidonic acid cascade. *FASEB J.* **6,** 731–736.
9. Peplov, P. V. (1996) Actions of cytokines in relation to arachidonic acid metabolism and eicosanoid production. *Pros. Leuk. Ess. Fatty Acids* **54,** 303–317.
10. Xie, W. L., Chipman, J. G., Robertson, D. L., Erikson, R. L., and Simmons, D. L. (1991) Expression of a mitogen responsive gene encoding prostaglandin synthase is regulated by mRNA splicing. *Proc. Natl. Acad. Sci. USA* **88,** 2692–2696.
11. Miyamoto, T. N., Ogino, N., Yamamoto, S., and Hayaishi, O. (1976) Purification of prostaglandin endoperoxide synthase from bovine vesicular gland microsomes. *J. Biol. Chem.* **251,** 2629–2636.
12. Williams, C. S. and DuBois, R. B. (1996) Prostaglandin endoperoxide synthase: why two isoforms? *Am. J. Physiol.* **270,** (Gastrointest. Liver Physiol., 33), G393–G400.
13. Seibert, K. and Masferrer, J. L. (1994) Role of inducible cyclooxygenase (COX-2) in inflammation. *Receptor* **4,** 17–23.
14. Bradford, M. M. (1976) A rapid and sensitive method for the quantitation of microgram qauntities of protein utilizing the principle of protein-dye binding. *Anal. Biochem.* **72,** 248–254.
15. Sharp, J. D., White, D. L., Chiou, G., Goodson, T., Gamboa, G. C., McClure, D., et al. (1991) Molecular cloning and expression of human Ca_2-sensitive cytosolic phospholipase A2. *J. Biol. Chem.* **266,** 14,850–14,853.
16. Grossman, E. M., Longo, W. E., Mazuski, J. E., Panesar, N., and Kaminski, D. L. (2000) Role of cytosolic phospholipase A_2 in cytokine stimulated prostaglandin release by human gallbladder cells. *J. Gastrointest. Surg.* in press.
17. Thommesen, L., Sjursen, W., Gasvik, K., Hanssen, W., Brekke, O. L., Skattebol, L., et al. (1998) Selective inhibitors of cytosolic or secretory phospholipase A_2 block TNF-induced activation of transcription factor nuclear factor-κB and expression of ICAM-1. *J. Immunol.* **161,** 3421–3430.

18. Belley, A. and Chadee, K. (1999) Prostaglandin E2 stimulates rat and human colonic mucin exocytosis via the EP4 receptor. *Gastroenterology* **117,** 1352–1362.
19. McCool, D. J., Marcon, M. A., Forstner, J. F., and Forstner, G. G. (1990) The T84 human colonic adenocarcinoma cell line produced mucin in culture and releases it in response to various secretagogues. *Biochem. J.* **267,** 491–500.
20. German, D., Barcia, J., Brems, J., Merenda, G., and Kaminski, D. L. (1989) Effects of bradykinin on feline gallbladder water transport and prostanoid formation. *Dig. Dis. Sci.* **34,** 1770–1776.

7

Mass Spectrometric Quantification of F_2-Isoprostanes as Indicators of Oxidant Stress

Jason D. Morrow and L. Jackson Roberts, II

1. Introduction

Free radicals derived primarily from molecular oxygen are believed to play an important role in a variety of disease processes including atherosclerosis, cancer, and neurodegenerative disorders (1). Much of the evidence for this association, however, is indirect largely because of limitations in techniques currently available to quantify free radicals or their products in biological systems (2,3). This is a particular problem when assessing oxidant injury in vivo in humans.

Measures of lipid peroxidation are frequently employed to implicate free radicals in pathophysiological processes. Techniques include measurement of malondialdehyde, conjugated dienes, short chain alkanes, and lipid hydroperoxides, among others (3). These methods suffer, however, from inherent problems related to sensitivity and specificity, particularly when applied to in vivo situations. In addition, artifactual generation of lipid-peroxidation products can occur ex vivo, and factors such as metabolism can markedly affect levels of various substances (3).

Previously, we described a series of prostaglandin (PG)F_2-like compounds, termed F_2-isoprostanes (F_2-IsoPs), that are produced in vivo in humans by a non-cyclooxygenase free radical-induced mechanism involving the peroxidation of arachidonic acid (4). Generation of these compounds initially involves formation of four positional peroxyl radical isomers that endocyclize to form PGG_2-like intermediates that are subsequently reduced to PGF_2-like compounds. Theoretically, four F_2-IsoP regioisomers are formed, each consist-

From: *Methods in Molecular Biology, vol. 186: Oxidative Stress Biomarkers and Antioxidant Protocols*
Edited by: D. Armstrong © Humana Press Inc., Totowa, NJ

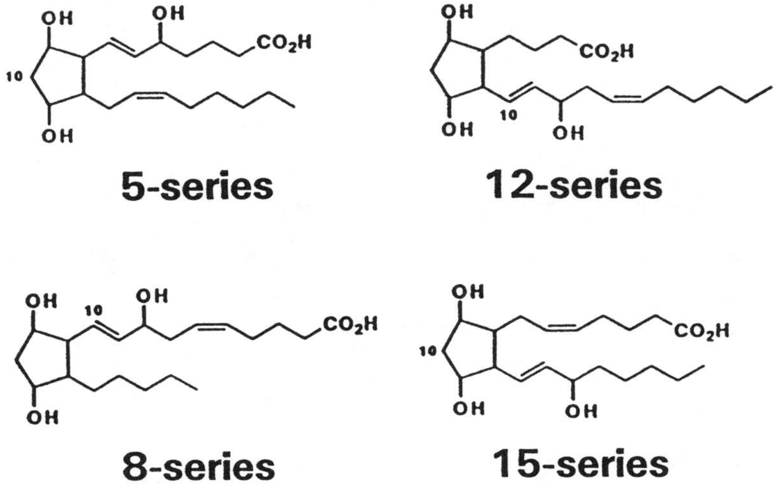

Fig. 1. Structure of the four F_2-IsoP regioisomers. Stereochemistry is not indicated.

ing of eight racemic diastereomers. Compounds are named as either 5-, 8-, 12-, or 15-series regioisomers according to which carbon the side chain hydroxyl group is attached (**Fig. 1**) *(5)*.

It has been found, over the past decade, that quantification of the F_2-IsoPs represents a reliable and useful approach to assess lipid peroxidation and oxidant stress in vivo *(6)*. For example, we have shown that the formation of F_2-IsoPs increases dramatically in animal models of oxidant stress and correlates with the degree of tissue damage. Further, F_2-IsoPs are present in easily detectable concentrations in normal human biological fluids, such as urine and plasma, allowing for a normal range to be defined and also for relatively small increases in these compounds to be measured accurately in settings of minimal oxidant stress. In addition, F_2-IsoPs can be detected in all types of biological fluids and tissues we have examined to date, providing an opportunity to assess the formation of these compounds at local sites of oxidant injury (*see* **Note 1**).

As noted, IsoPs are derived from arachidonic acid. The majority of arachidonic acid present in vivo exists esterified in phospholipids. We have previously reported that F_2-IsoPs are initially formed *in situ* from arachidonic acid esterified in phospholipids and are subsequently released preformed by the action of phospholipases *(7)*. This discovery raises an important issue regarding the assessment of IsoP formation in vivo in that depending on the studies that one undertakes, total isoprostane production may be more accurately assessed by measuring levels of both free and esterified IsoPs. Further, the fact that F_2-IsoPs are formed *in situ* in phospholipids can be utilized in an advantageous way to assess oxidant injury in specific animal or human organs by analyzing levels of these compounds esterified in phospholipids from tissue biopsy specimens that may be obtained for diagnostic purposes.

The purpose of this chapter is to discuss methods for the analysis of F_2-IsoPs from biological sources by gas chromatography (GC)/negative ion chemical ionization (NICI) mass spectrometry (MS). Procedures are outlined for the analysis of both free and esterified F_2-IsoPs. F_2-IsoPs from biological sources can only be quantified as free compounds using GC/MS. Therefore to measure levels of IsoPs esterified in phospholipids, the phospholipids are first extracted from a tissue or fluid sample and subsequently subjected to alkaline hydrolysis to liberate the F_2-IsoPs from the phospholipids. Following hydrolysis, the free F_2-IsoPs are measured using the same techniques as free compounds in biological fluids. Thus, the following methods first outline the extraction and hydrolysis of F_2-IsoP-containing phospholipids from tissue or fluid samples. Subsequently, techniques for the analysis of free compounds are discussed.

2. Materials

2.1. Equipment

1. Blade homogenizer-PTA 10S generator (Brinkmann Instruments, Westbury, NY).
2. Table-top centrifuge.
3. Rotavap unit (organic solvent evaporator) (Brinkmann Instruments).
4. Tank of nitrogen gas.
5. Analytical evaporation unit (such as Meyer N-Evap, Organomation Associates Inc., Berlin, MA).
6. Microcentrifuge (Fisher Scientific, Pittsburgh, PA).
7. Water bath at 37°C.
8. Capillary gas chromatography column (DB-1701, Fisons Inc., Folsom, CA).
9. Gas chromatograph/mass spectrometer (with capabilities for negative ion chemical ionization mass spectrometry) (Hewlett Packard 5890 instrument, Palo Alto, CA).
10. Sep-Pak cartridges, C-18 and silica (Waters Associates, Milford, MA).
11. Disposable plastic syringes (10 mL).
12. Glass scintillation vials (20 mL).
13. Reactivials (5 mL) (Pierce Scientific, Rockford, IL).
14. TLC plates: LK6D silica (Whatman, Maidstone, UK).
15. Microcentrifuge tubes.
16. Conical glass centrifuge tubes (40 mL).
17. Conical-bottom flasks (100 mL).

2.2. Reagents

1. Ultrapure water (triply distilled or its equivalent).
2. High-purity organic reagents: methanol, chloroform (containing ethanol as a preservative), ethyl acetate, acetonitrile, heptane, ethanol (Burdick and Jackson brand, VWR Scientific Inc., McGaw Park IL).
3. Butylated hydroxytoluene (BHT) (Aldrich Chemical Co., Milwaukee, WI).
4. Sodium chloride.
5. Triphenylphosphine (TPP) (Aldrich Chemical Co.).
6. Magnesium chloride.
7. Potassium hydroxide pellets.
8. HCl (American Chemical Society (ACS) certified or equivalent grade).
9. Pentafluorobenzylbromide (PFBB) (Aldrich Chemical Co.).
10. N,N′-Diisopropylethylamine (DIPE) (Aldrich Chemical Co.).
11. N,O-bis(trimethylsilyl)trifluoroacetamide (BSTFA) (Supelco Inc, Bellefonte, PA).
12. Dimethyformamide (DMF) (Aldrich Chemical Co.).
13. Undecane (Aldrich Chemical Co.).
14. Phosphomolybdic acid (Sigma Chemical Co., St. Louis, MO).
15. Na_2SO_4 (anhydrous).
16. $[^2H_4]$ 15-F_{2t}-IsoP (8-iso-PGF$_{2\alpha}$) internal standard (Cayman Chemical, Ann Arbor, MI).

3. Method of Assay

3.1. Handling and Storage of Biological Fluids and Tissues for Quantification of F$_2$-Isoprostanes

As noted earlier, we have detected measurable levels of F$_2$-IsoPs in all biological fluids that we have examined. An important consideration when quantifying F2-IsoPs as a measure of lipid peroxidation, however, is the fact that these compounds can be artifactually generated ex vivo in biological fluids, such as plasma, in which arachidonyl-containing lipids are present *(6,8)*. In this regard, we have previously reported that:

1. F$_2$-IsoPs can be generated ex vivo in biological fluids that are allowed to remain at room temperature for greater than 2 h or are stored at 0°C or –20°C.
2. Ex vivo generation does not occur, however, when fluids or tissues are stored at –70°C or colder for up to 1 yr.
3. Formation of IsoPs does not occur to a significant extent if biological fluids or tissues are processed immediately after procurement and if the free-radical scavenger BHT, and/or the reducing agent TPP, is added to the organic solvent during extraction of lipids.

Thus, samples obtained for analysis should either be processed immediately or stored at –70°C. In addition, fluids or tissues that are stored should be rapidly cooled to –70°C or lower; this can be accomplished by snap-freezing the samples in liquid nitrogen prior to storage at –70°C.

3.2. Extraction and Hydrolysis of F$_2$-IsoP-Containing Phospholipids

Esterified F$_2$-IsoPs can be analyzed in either biological fluids or tissues. The method for tissues differs somewhat from fluids and thus each is discussed separately (*see* **Note 2**).

3.2.1. Extraction of Tissue Lipids For Analysis of F$_2$-IsoPs

The method outlined herein is applicable to all animal and human tissue samples.

1. Weigh out 0.05–1 g of either fresh tissue or tissue frozen at –70°C.
2. Add tissue to 20 mL of ice-cold Folch solution (chloroform/methanol, 2 : 1 v/v) containing 0.005% BHT in a 40-mL glass centrifuge tube.
3. Homogenize tissue with blade homogenizer at full speed for 60 s.
4. Purge airspace in centrifuge tube with nitrogen and cap. Let solution stand at room temperature for 1 h to effect maximum extraction of lipids from ground tissue. Shake tube occasionally for several seconds during this period of time.
5. Add 4 mL aqueous NaCl (0.9%). Vortex or shake vigorously for 2 min.
6. Centrifuge on table-top instrument 800*g* for 10 min at room temperature.
7. After centrifugation, carefully pipette off top aqueous layer and discard. Remove the lower organic layer carefully from under the intermediate semisolid protein-aceous layer and transfer to a 100-mL conical bottom flask. Evaporate under vacuum on rotavap unit to dryness. When dry, immediately proceed to the hydrolysis step (**Subheading 3.3.**).

3.2.2. Extraction of F$_2$-IsoP-Containing Lipids From Biological Fluids

The technique used to extract F$_2$-IsoP-containing phospholipids from biological fluids differs somewhat from methods used for tissue because we have found that arachidonic acid in plasma is much more susceptible to autoxidation during lipid extraction than arachidonate in tissue samples (*see* **Note 2**). Thus, to suppress ex vivo oxidation of arachidonic acid in plasma, TPP, in addition to BHT, is added to the Folch solution.

1. Add 1 mL of a biological fluid to 20 mL ice-cold Folch solution containing 0.005% BHT and 5 mg TPP in a 40-mL glass conical centrifuge tube.
2. Shake vigorously for 2 min.
3. Add 10 mL 0.043% MgCl$_2$ and shake vigorously for 2 min.
4. Centrifuge 800*g* for 10 min at room temperature.

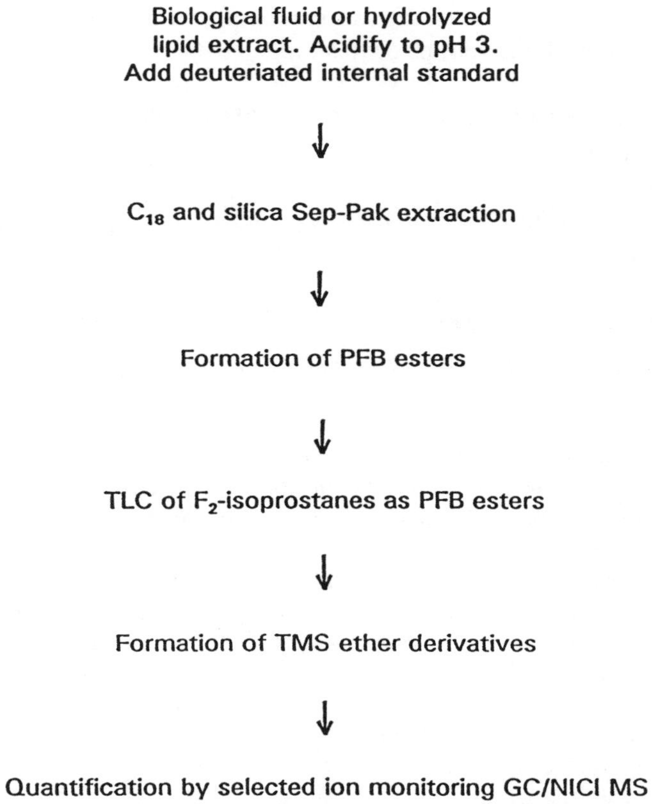

Biological fluid or hydrolyzed
lipid extract. Acidify to pH 3.
Add deuteriated internal standard

↓

C$_{18}$ and silica Sep-Pak extraction

↓

Formation of PFB esters

↓

TLC of F$_2$-isoprostanes as PFB esters

↓

Formation of TMS ether derivatives

↓

Quantification by selected ion monitoring GC/NICI MS

Fig. 2. Outline of the procedure for extraction, purification, derivatization, and MS quantification of F$_2$-IsoPs.

5. Separate lower organic layer from other layers and transfer to conical bottom flask (*see* **Subheading 3.2.1.**).
6. Dry organic layer at room temperature under nitrogen.
7. Proceed immediately to hydrolysis procedure (**Subheading 3.3.**).

3.3. Hydrolysis of Lipid Extracts

1. To lipid residue in conical flask, add 4 mL methanol containing 0.005% BHT and then add 4 mL aqueous KOH (15%). Purge flask with nitrogen and cap (*see* **Note 3**).
2. Incubate mixture at 37°C for 30 min.
3. After incubation, acidify the mixture to pH 3.0 with 1 *M* HCl. Then dilute the mixture to 80 mL with H$_2$O (pH 3.0) in preparation for extraction of free F$_2$-IsoPs. It is important to dilute the methanol in this solution to 5% or less

to ensure proper column extraction of F$_2$-IsoPs in the subsequent purification procedure (**Subheading 3.4.**).

3.4. Purification and Derivatization of F$_2$-IsoPs

The assay method for purification and derivatization of free F$_2$-IsoPs in biological fluids or in hydrolyzed lipid extracts is outlined in **Fig. 2**. For purposes of discussion, the method used to purify and derivatize F$_2$-IsoPs in plasma is detailed below but the technique is equally adaptable to other biological fluids or tissue extracts.

1. Acidify 1–3 mL of plasma to pH 3.0 with 1 *M* HCl.
2. Add 1–5 pmol of deuteriated [^{2}H$_4$]15-F$_{2t}$-IsoP (8-iso-PGF$_{2\alpha}$) internal standard and vortex.
3. Apply mixture to a C-18 Sep-Pak column connected to a 10-mL syringe preconditioned with 5 mL methanol and then 5 mL pH 3.0 water.
4. Wash column with 10 mL pH 3.0 water and then 10 mL heptane.
5. Elute F$_2$-IsoPs from column with 10 mL ethyl acetate/heptane (50:50, v/v) into a glass scintillation vial.
6. Add 5 g anhydrous Na$_2$SO$_4$ to the vial and swirl gently. This step removes residual water from the eluate.
7. Apply eluate to silica Sep-Pak preconditioned with 5 mL ethyl acetate.
8. Wash cartridge with 5 mL ethyl acetate.
9. Elute F$_2$-IsoPs from silica Sep-Pak with 5 mL ethyl acetate/methanol (50:50, v/v) into reactivial.
10. Evaporate eluate under nitrogen.
11. To convert F$_2$-IsoPs to PFB esters, add 40 µL 10% (v/v) PFBB in acetonitrile and 10 µL 10% (v/v) DIPE in acetonitrile to resideue for 30 min at room temperature.
12. Dry reagents under nitrogen and resuspend in 50 µL methanol.
13. Apply mixture to a lane on a silica TLC plate (LK6B) that has been prewashed with ethylacetate/ethanol (90:10, v/v). Chromatograph to 13 cm in a solvent system of chloroform/ethanol (93:7, v/v). For a TLC standard, apply approximately 2–5 µg of the methyl ester of PGF$_{2\alpha}$ to another TLC lane. Visualize the TLC standard by spraying the lane with a 10% solution of phosphomolybdic acid in ethanol and heating.
14. Scrape silica form the TLC plate in the region of the methyl ester of PGF$_{2\alpha}$ (R$_f$ = 0.18) and adjacent areas 1 cm above and below.
15. Place silica in microfuge tube and add 1 mL ethyl acetate. Vortex vigorously for 30 s.
16. Pour off the ethyl acetate into another microcentrifuge tube taking care not to disrupt the silica pellet.
17. Dry organic layer under nitrogen.
18. Add 20 µL BSTFA and 6 µL DMF to residue.
19. Vortex well and incubate sample at 37°C for 20 min.
20. Dry reagents under nitrogen.

21. Add 10 μL undecane (which has been stored over calcium hydride to prevent water accumulation). Sample is now ready for mass spectrometric analysis.

3.5. Quantification of F_2-IsoPs by GC/NICI MS

1. For the analysis of F_2-IsoPs, we routinely use a Hewlett Packard 5890 mass spectrometer interfaced with an IBM Pentium II computer system although other mass spectrometers can be utilized. The F_2-IsoPs are chromatographed on a 15 m DB1701 fused silica capillary column because we have found this GC columns gives excellent separation of individual regioisomers compared to other columns *(8)*. The column temperature is programmed from 190°C to 300°C at 20°C per min. Methane is used as the carrier gas for NICI at a flow rate of 1 mL/min.
2. The ion source temperature is 250°C, the electron energy is 70 eV, and the filament current is 0.25 mA. The ion monitored for endogenous F_2-IsoPs is the carboxylate anion m/z 569 (M-181, loss of $CH_2C_6F_5$). The corresponding carboxylate anion for the deuteriated internal standard is m/z 573.

3.6. Results: Application of the Assay to Biological Samples

1. **Fig. 3** shows the selected ion current chromatogram obtained from the analysis of F_2-IsoPs in the plasma of a rat 4 h after treatment with CCl_4 to induce an oxidant stress. The series of peaks in the upper m/z 569 chromatogram represent endogenous F_2-IsoPs. This pattern of peaks is virtually identical to that obtained from all other biological fluids and tissues examined thus far *(8)*. The peak in the lower m/z 573 chromatogram represent the $[^2H_4]15\text{-}F_{2t}\text{-IsoP}$ (8-iso-$PGF_{2\alpha}$) standard. Although all of the peaks noted in the upper m/z 569 chromatogram represent F_2-IsoPs, we routinely quantify the peak denoted by the asterisk (*). Using the ratio of the intensity of this peak to that of the internal standard, the concentration of F_2-IsoPs in the plasma sample in **Fig. 3** was calculated to be approx 2200 pmol/L, which is approx 20-fold above normal. Normal plasma levels of F_2-IsoPs in rats and humans are in the range of 60–170 pmol/L. Levels in other tissues and fluids have been reported elsewhere *(6,8)*. The F_2-IsoP assay in biological fluids is highly precise and accurate. The precision is ±6% and the accuracy is 96% (*see* **Notes 4, 5**).
2. In summary, this chapter outlines methods to assess lipid peroxidation associated with oxidant injury in vivo by quantifying concentrations of free and esterified F_2-Isops in biological fluids and tissues. Quantification of F_2-IsoPs appears to overcome many of the shortcomings associated with other methods to assess oxidant status, especially in vivo in humans. Thus measurement of F_2-IsoPs likely represents an important advance in our ability to assess the role of oxidant stress in human disease.

4. Notes

1. We have successfully used this assay to quantify F_2-IsoPs in a number of diverse biological samples (including urine, plasma, cerebrospinal fluid, and lung lavage fluid) and tissues (such as heart, muscle, brain, liver, and kidney).

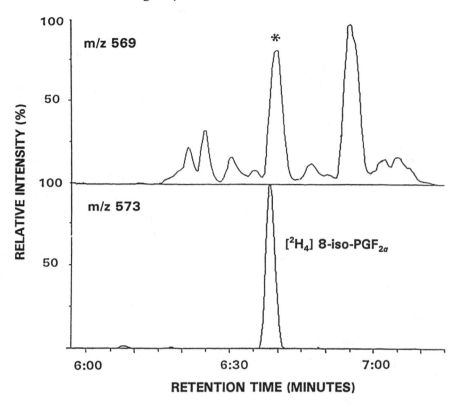

Fig. 3. Analysis of F_2-IsoPs in plasma from a rat 4 h after treatment with CCl_4 to induce lipid peroxidation. The peak in the m/z 573 chromatogram represents the $[^2H_4]15\text{-}F_{2t}\text{-IsoP}$ internal standard. The peaks in the m/z 569 chromatogram represent endogenous F_2-IsoPs. The peak denoted by the asterisk is the one routinely used to quantify the F_2-IsoPs.

2. Regardless of the biological sample being processed to quantify esterified F_2-IsoPs, it is extremely important that the utmost care be taken during the extraction procedure to prevent ex vivo generation of F_2-IsoPs. This is more likely to occur if reagents contain significant trace metal contamination. To avoid this, reagents should be of very high purity. Further, the quality of water used during lipid extraction is extremely important and we routinely used triply distilled water or its equivalent. In addition, agents including BHT and TPP are added to suppress autoxidation of lipids. Finally, all glassware and plasticware are hand washed by us and rinsed with ultrapure water prior to use.

3. Hydrolysis of lipids in tissue or biological fluids should be performed immediately after lipid extraction to avoid potential autoxidation of arachidonate contained in the phospholipids.

4. Quantification by GC/negative ion chemical ionization (NICI) MS is an extremely sensitive method to analyze F_2-IsoPs. Compounds are converted to pentafluorobenzyl (PFB) ester, trimethylsilyl (TMS) ether derivatives for this analysis. The lower limits of detection of F_2-IsoPs is in the range of 3–15 fmol using a deuteriated standard with a blank of less than 5 parts per thousand. Thus, it is not necessary to assay more than 1–3 mL of a fluid such as plasma, 0.2 mL urine, or 50–1000 mg tissue.

5. Quantification of F_2-IsoPs in biological samples using the methods outlined requires a mass spectrometer with NICI capabilities. The use of NICI MS enhances the sensitivity of the assay by orders of magnitude compared to the use of electron ionization MS and electrospray ionization MS, although the use of capillary column technology will likely make electrospray ionization MS as sensitive in the future.

Acknowledgements

Supported by NIH grants DK48831, GM42056, GM15431, DK26657, CA77839, and CA68435. J.D.M. is the recipient of a Burroughs Wellcome Fund Physician Scientist Award in Translational Research.

References

1. Halliwell, B. and Gutteridge, J. M. C. (1990) Role of free radicals and catalytic metal ions in human disease: an overview. *Methods Enzymol.* **186,** 1–85.
2. DeZwart, L. L., Meerman, J. H. N., Commandeur, J. N. M., and Vermeulen, P. E. (1999) Biomarkers of free radical damage applications in experimental animals and in humans. *Free Rad. Biol. Med.* **26,** 202–226.
3. Halliwell, B. and Grootveld, M. (1987) The measurement of free radical reactions in humans. *FEBS Lett.* **213,** 9–14.
4. Morrow, J. D., Hill, K. E., Burk, R. F., Nammour, T. M., Badr, K. F., and Roberts, L. J. (1990) A series of prostaglandin F_2-like compounds are produced in vivo in humans by a non-cyclooxygenase, free radical catalyzed mechanism. *Proc. Natl. Acad. Sci. USA* **87,** 9383–9387.
5. Taber, D. F., Morrow, J. D., and Roberts, L. J. (1997) A nomenclature system for the isoprostanes. *Prostaglandins* **53,** 63–67.
6. Morrow, J. D. and Roberts, L. J. (1997) The isoprostanes: unique bioactive products of lipid peroxidation. *Prog. Lipid Res.* **36,** 1–21.
7. Morrow, J. D., Awad, J. A., Boss, H. J., Blair, I. A., and Roberts, L. J. (1992) Non-cyclooxygenase-derived prostanoids (F_2-isoprostanes) are formed in situ on phospholipids. *Proc. Natl. Acad. Sci. USA* **89,** 10,721–10,725.
8. Morrow, J. D. and Roberts, L. J. (1999) Mass spectrometric quantification of F_2-isoprostanes in biological fluids and tissues as measure of oxidant stress. *Methods Enzymol.* **300,** 3–12.

8

Formation of Apolipoprotein AI-AII Heterodimers by Oxidation of High-Density Lipoprotein

Audric S. Moses and Gordon A. Francis

1. Introduction

Uptake of oxidized low-density lipoprotein (LDL) by scavenger receptors allows cholesterol to be accumulated in an unregulated fashion by cells in the artery wall, an event considered pivotal to the formation of atherosclerotic plaques (1). High-density lipoprotein particles (HDL) are believed to protect against atherosclerosis in part by removing excess cholesterol from the artery wall (2). HDL, however, is itself susceptible to oxidation by numerous in vitro methods to an extent equal to, or greater than, LDL (reviewed in **ref. 3**). In vitro oxidation of HDL by peroxidase-generated tyrosyl radical ("tyrosylated HDL"), one mechanism of phagocyte-mediated oxidation (4), results in a markedly enhanced ability of HDL to promote the mobilization of cholesterol from a variety of cultured cells (5–7). If occurring in vivo, this oxidative modification of HDL may actually enhance its ability to protect against atherosclerosis.

Oxidation of HDL results in the covalent crosslinking of its two major apolipoproteins, apoAI and apoAII. Isolation of the monomeric and crosslinked apolipoprotein subspecies of tyrosylated HDL allowed us to identify the component responsible for enhanced cholesterol mobilization by these particles to be apolipoprotein AI-AII heterodimers (8). Separation of the individual tyrosylated HDL protein subspecies could not be accomplished using size exclusion or reverse-phase high-performance liquid chromatography (HPLC), or fast-performance liquid chromatography (FPLC). This separation was eventually achieved using preparative electrophoresis combined with elution of

From: *Methods in Molecular Biology, vol. 186: Oxidative Stress Biomarkers and Antioxidant Protocols*
Edited by: D. Armstrong © Humana Press Inc., Totowa, NJ

isolated bands from the gel. We describe here the oxidation of HDL by tyrosyl radical and isolation of apoAI-AII heterodimers.

2. Materials
2.1. Equipment
2.1.1. Isolation of HDL$_3$

1. Refrigerated swinging bucket centrifuge for 50 mL polypropylene conical centrifuge tubes.
2. Refrigerated ultracentrifuge and fixed-angle rotor for 1" × 3.5" polycarbonate tubes.
3. Polypropylene beakers.
4. Volumetric pipets.
5. Dialysis tubing, 12,000–14,000 MW cutoff.
6. Biosafety hood.

2.1.2. Tyrosyl Radical-Oxidation of HDL$_3$, Preparative Electrophoresis, and Electroelution

1. Biorad 10 DG columns.
2. 37°C incubator.
3. Sodium-dodecyl sulfate polyacrylamide gel electrophoresis (SDS-PAGE) equipment, with a 1.5-mm preparative comb.
4. Fluorescence spectrophotometer.
5. Low-molecular weight centrifugal concentrators (10,000 and 5,000 MW cutoff).
6. Whole Gel Eluter (Biorad).

2.2. Reagents

1. Desiccated potassium bromide (KBr, obtained from BDH); phenylmethylsulfonyl fluoride (PMSF; Sigma); absolute ethanol (BDH); sodium chloride (NaCl, BDH); ethylenediaminetetraacetic acid (EDTA; BDH); Tris-HCl (Gibco BRL).
2. Potassium dihydrogen orthophosphate (KH$_2$PO$_4$, BDH); diethylenetriamine pentaacetic acid (DTPA; Sigma); horseradish peroxidase (HRP; Sigma); L-tyrosine (Sigma); hydrogen peroxide (H$_2$O$_2$; Fisher); Tris (Gibco BRL); CAPS (Fisher); Coomassie Brilliant Blue (Sigma), and other standard SDS-PAGE reagents.
3. Chelex 100 (Biorad); AG-11 (optional, Biorad); and heparin-sepharose (Pharmacia) resins for column chromatography.

3. Methods
3.1. HDL$_3$ Isolation from Plasma by Ultracentrifugation

1. Collect blood from donors fasted for 12 h.
2. Place 40 mL of blood into 50 mL sterile centrifuge tubes containing 1.0 mL 0.2 M EDTA, pH 7.4. Invert tubes to mix and place immediately on ice.
3. Spin blood at 3000 rpm 20 min, 4°C in swinging-bucket centrifuge.

4. In a cold room, combine plasma into a large plastic beaker, add 10 μL per 100 mL plasma of 0.1 M PMSF immediately (prepared fresh in absolute ethanol), while stirring for 30 s.

5. At room temperature, measure density of plasma using a volumetric pipet, then raise density to 1.125 g/mL using desiccated KBr (dissolve slowly to avoid foaming, adding KBr in small increments).

6. Load plasma into 1" × 3.5" polycarbonate centrifuge tubes and balance using a KBr solution of density 1.125.

7. Spin at 50,000 rpm for 24 h at 8°C in a Ti 50.2 (or equivalent) fixed-angle rotor.

8. In a biosafety hood, and on ice, remove caps from tubes and remove upper, bright orange layer (do not try to collect every last bit). This layer contains the very low-density lipoproteins (VLDL), LDL, and HDL_2 fractions.

9. Aspirate another 5 mm of fluid from the top of the tube and discard.

10. Using a pipet put to the bottom of the tube, and avoiding the protein pellet at the bottom of the tube, aspirate all but approx 1 cm (to avoid contamination with remaining LDL and associated fractions), and pool the bottom (HDL_3-containing) fractions from all tubes.

11. At room temperature, measure density and total volume, and adjust density to 1.21 g/mL with KBr.

12. Load into clean 1" × 3.5" centrifuge tubes and balance with a 1.21 KBr solution.

13. Spin at 50,000 RPM at 8°C for 24 h, or HDL and other plasma proteins are well separated (may take longer depending on the rotor properties).

14. Remove and pool all bright yellow upper fractions, measure volume and density, and adjust density again to 1.21 with KBr.

15. Spin again at 50,000 RPM at 8°C, for 24 h or until layers are well separated. This second ultracentrifugation at $d = 1.21$ may not be necessary if SDS-PAGE with silver staining of gel indicates HDL_3 is "clean" after the first separation at this density.

16. Pool top layers, and using 12,000–14,000 MW cutoff dialysis tubing, dialyze at 4°C against 50 mM NaCl, 5 mM Tris-HCl, 1 mM EDTA, pH 7.4, for 2–3 d, changing dialysis buffer at least 3 times.

17. To remove apolipoproteins E and B, pass HDL_3 over a 1 × 12 cm heparin-sepharose column *(9)*. The HDL_3 eluted from the column contains apoAI and apoAII as its major protein components, with small amounts of apoCII and apoCIII.

18. Filter HDL_3 through a 0.22-μm syringe filter. For storage, flush tube with overlay of argon gas to prevent oxidation, and store in a dark place at 4°C. Assess HDL_3 protein content using the method of Lowry et al. *(10)*.

3.2. Isolation of Lipid-Free Apolipoprotein AI and AII

1. Separation of HDL apolipoproteins can been achieved by ion-exchange FPLC using a Q-Sepharose Fast Flow column (Pharmacia), according to the method of Rye and Barter *(11)*. This method requires delipidation of the HDL protein prior to chromatography, and a desalting step at the end.

2. Isolation of free apolipoproteins can also be achieved by the method of Garner et al. *(12)*, which uses RP-HPLC to separate HDL proteins. The advantage of this method is that delipidation is not necessary, and the running buffers (water and acetonitrile) are removed by lyophilization, leaving no residual salt.

3.3. Tyrosyl Radical Oxidation of HDL₃ or of Free Apolipoproteins

1. Prepare a solution of 1 mg/ml HDL protein (if free apolipoproteins are used, combine apoAI and apoAII in a 1:1, monomer:monomer mole ratio, total protein 1 mg), 100 nM HRP, 100 μM H_2O_2, 100 μM L-tyrosine, in buffer A (66 mM KH_2PO_4, pH 8.0 passed over a 1 × 10 cm Chelex 100 [Biorad] column to remove free metal ions, followed by addition of 0.1 mM DTPA) *(5)*. Chelex treatment of buffers and addition of DTPA ensure the modification of apolipoproteins observed is not due to free metal ion-mediated oxidation. H_2O_2 is added last to the oxidation mixture.
2. Incubate mixture at 37°C in the dark for 16–20 h.
3. Stop reaction and separate modified HDL from low molecular-weight components (including free dityrosine formed during the reaction) by passing mixture over a Biorad 10 DG column. HDL₃ is typically recovered in the first mL after passage of the column void volume. Store modified HDL₃ under argon at 4°C in the dark.
4. Oxidation of HDL is confirmed by the presence of protein-associated dityrosine fluorescence (assessed by measuring fluorescence of 25 μL of sample in 975 μL buffer A at 328 nm excitation and 410 nm emission wavelengths), and by the presence of crosslinked protein bands at ~36, 45, 56, and 84 kDa (corresponding to apoAI-AII_monomer and apoAI-[AII]₂ heterodimers, apoAI dimers, and apoAI trimers, respectively) on 15% or 7–20% gradient polyacrylamide gel, or 15% mini-gel, followed by Coomassie blue or silver staining *(5)*.

3.4. Preparative 12% SDS-PAGE Gel

1. Concentrate HDL using low protein-binding centrifugal concentrators to approx 10–15 mg/mL.
2. Add 5X reducing SDS sample buffer (60 mM Tris-HCl, 25% glycerol, 2% SDS, 0.1% bromophenol blue, 0.5 mL β-mercaptoethanol, all in a total volume of 10 mL) to 5 mg HDL, and boil 5 min. Volume should be small enough to fit into preparative well of gel.
3. Prepare a 12% SDS-PAGE 1.5 mm gel, with a 3% acrylamide stacking gel, using a comb containing one large well and one small well (for standards). The gel itself does not need to contain SDS as long as the sample buffer and running buffer both do.
4. Load 10 μL low molecular-weight rainbow standards in small well, to allow monitoring of the progress of the HDL proteins running through the gel.
5. Run gel at a maximum current of 25 mA through the stacking gel, and 35 mA through the separating gel.

6. Stop the electrophoresis when the ~14 kDa standard has reached the bottom of the gel, or when maximum separation of the desired proteins has been achieved. If monomeric apoAII band is desired, stop electrophoresis when dye front is approx 1 cm above the bottom of the gel.

3.5. Elution of Proteins from the Gel by Whole Gel Eluter and Analysis by 7–20% Gradient SDS-PAGE

1. Prepare elution buffer (60 m*M* Tris, 40 m*M* CAPS, pH 9.4), and set up elution chamber according to manufacturers (Biorad) protocol.
2. Cut off stacking gel, and soak rest of gel in elution buffer for 15–60 min.
3. Place gel in eluter, with protein bands parallel to channels, and elute according to manufacturer directions, using a current of 200 mA for 30 min.
4. Following elution, efficiency of transfer can by assessed by Coomassie Blue staining of gel.
5. Run 80 µL aliquots of fractions from Whole Gel Eluter on 7–20% SDS-PAGE gels to determine which proteins are in each fraction.
6. Pool all fractions containing the same protein, e.g., the apoAI-AII heterodimer. Pooled fractions are then concentrated using 10,000 MW cutoff centrifugal concentrators (5,000 for apoAII monomer) and their purity determined by SDS-PAGE with silver staining *(8)*.
7. Removal of the majority of the residual free SDS can be achieved by repeated spins at the maximum speed recommended for the concentrator until a very small volume remains, followed by resuspension in SDS-free buffer. This is repeated at least 3 times.
8. A 1 × 10 cm Biorad AG-11 column may be used to remove most protein-bound SDS, if this is desired, however, up to 50% of the protein may be lost on the column.

3.6. Preparation of Reconstituted HDL Particles with Isolated Apolipoprotein Fractions

1. The whole lipid fraction of HDL is isolated by delipidation of HDL as previously described *(8)*.
2. To a glass tube, add HDL whole lipid fraction, and dry under N_2. Use a protein:phospholipid ratio of 1:80, with a final protein mass of 500 µg/tube to determine the amount of lipid extract to add.
3. Add 1 mL of reconstitution buffer (150 m*M* NaCl, 10 m*M* Tris-HCl, 0.1 m*M* DTPA, pH 8.0).
4. Add sodium deoxycholate (cholate:phospholipid ratio 1:1) and sonicate gently to dissolve all the lipid.
5. Add protein, and adjust total volume to 2 mL with reconstitution buffer. Overlay samples with argon, and stir vigorously for at least 2 h.
6. Spin in a 3 mL ultracentrifuge tube at 99,000 rpm at 4°C for 4 h. Remove upper lipid layer (not associated with protein) and discard.

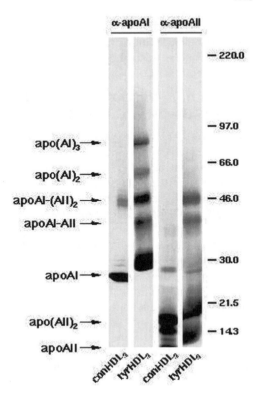

Fig. 1. Immunoblot analysis of apolipoprotein species of tyrosylated HDL. HDL (conHDL$_3$) and tyrosylated HDL (tyrHDL$_3$) proteins were separated on a 7–20% gradient SDS-PAGE gel (5 μg protein per lane) and transferred to nitrocellulose for immunoblotting with anti-apoAI (α-apoAI) or anti-apoAII (α-apoAII) antibodies (after **ref. 8**). Positions of monomeric and multimeric apoprotein species and molecular mass standards are indicated.

7. Adjust density to 1.21 g/mL with KBr, and bring volume up to 3 mL with a 1.21 KBr solution. Spin under same conditions for 20–24 h to pellet the free protein. Collect the upper 2/5 of the tube, and using 12,000–14,000 MW cutoff tubing, dialyze against reconstitution buffer at 4°C for 2–3 d, with a minimum of 4 buffer changes.

3.7. Results

1. *Western Blotting of Tyrosyl Radical-Oxidized HDL.* Representative Western blots of tyrosylated HDL using anti-apolipoprotein AI and anti-apolipoprotein AII antibodies are shown in **Fig. 1**. These demonstrate the formation of apoAI-AII and apoAI-(AII)$_2$ heterodimers as well as apoAI dimers and trimers in tyrosyl radical-oxidized HDL. Similar patterns of crosslinking are seen in HDL oxidized

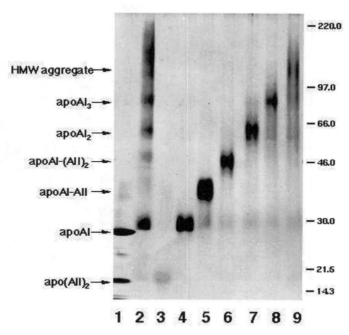

Fig. 2. Apolipoprotein species of tyrosylated HDL isolated by preparative electro-phoresis and whole gel elution. Isolated protein subspecies were pooled, concentrated, and run on nonreducing 7–20% SDS-PAGE along with HDL (lane 1) or tyrosylated HDL (lane 2) (5 µg protein/lane), and the gel stained with silver. Individual protein subspecies of tyrosylated HDL are indicated in lanes 3–9, as per labels on the left. Adapted with permission from **ref. 8**.

by other mechanisms (e.g., copper ion, aldehydes, cigarette smoke) (reviewed in **ref. 3**). An upward shift in molecular weight of the apoAI and apoAII bands is typical, and in the case of tyrosyl radical oxidation is presumed to be due to the formation of protein tyrosine-free tyrosine dimers.

2. *Isolated Tyrosylated HDL Apolipoprotein Subspecies.* Repeated preparative electrophoresis of tyrosylated HDL proteins, elution, and pooling of individual protein bands allows the isolation of pure apoprotein subspecies (**Fig. 2**) for use in further structural and cell-culture studies.

3. *Enhanced Cholesterol Mobilization by Reconstituted HDL Containing apoAI-AII Heterodimers of Tyrosylated HDL.* Reconstitution of the isolated apoprotein sub-species of tyrosylated HDL allowed us to demonstrate that the enhanced capacity of tyrosylated HDL to mobilize and prevent the accumulation of cellular choles-terol was due to the formation of apolipoprotein AI-AII heterodimers (**Fig. 3**) *(8)*. This enhanced cholesterol mobilization was seen with apoAII present as a monomer or dimer in the heterodimer complex. The enhanced cholesterol mobilization required the protein complex be reconstituted with lipids into

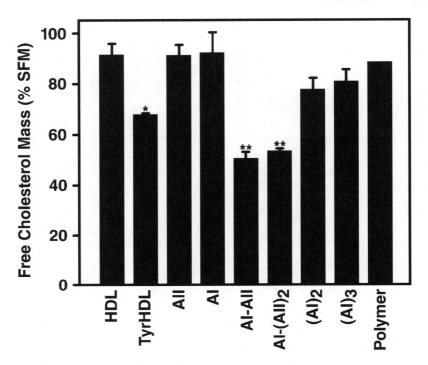

Fig. 3. Free cholesterol mass in fibroblasts incubated with LDL and reconstituted HDL containing individual tyrosylated HDL protein subspecies. Human fibroblasts were grown from 60% to full confluence in 10% lipoprotein-deficient serum to upregulate LDL receptors. Cells were then incubated for 24 h in serum-free medium (SFM) containing LDL (50 µg protein/mL) and 20 µg protein/mL HDL, tyrosylated HDL, or reconstituted HDL containing the indicated tyrosylated HDL protein species isolated as in Fig. 2 and reconstituted with the whole lipid fraction of HDL. Results are the mean ± SD of four determinations. $^*p < 0.01$; $^{**}p < .001$ as compared with cells incubated with HDL. Adapted with permission from **ref. 8**.

spherical HDL particles, suggesting that a particular conformation of the protein heterodimer present only when associated with this type of lipid particle is necessary for the enhanced effect.

4. Notes

1. The precautions indicated (use of EDTA, storage of isolated HDL under argon at 4°C in the dark), as well as the use of lab-ware free of any chemical (e.g., bleach) contamination, are extremely important to prevent nonspecific oxidation of HDL during isolation and storage.

2. As indicated, oxidation of HDL by other methods also results in the formation of apolipoprotein crosslinks including apoAI-AII heterodimers *(3)*. The enhanced cholesterol mobilization observed with tyrosylated HDL is owing to enhanced

translocation of cholesterol from a storage pool to an efflux-available pool in cultured cells, including fibroblasts, macrophages, and smooth-muscle cells *(5–7)*. HDL oxidized by other means also appear to enhance the mobilization of cholesterol from a pool available for storage in the cell (the "acyl-CoA:cholesterol acyltransferase-accessible pool") (W. Wang and G. Francis, unpublished observations). These results suggest HDL oxidized by other processes in vivo may also help prevent atherosclerosis by mobilizing cholesterol for removal from cells by other (nonoxidized) HDL particles *(3)*.

References

1. Heinecke, J. W. (1998) Oxidants and antioxidants in the pathogenesis of atherosclerosis: implications for the oxidized low density lipoprotein hypothesis. *Atherosclerosis* **141,** 1–15.
2. Fielding, C. J. and Fielding, P. E. (1995) Molecular physiology of reverse cholesterol transport. *J. Lipid Res.* **36,** 211–228.
3. Francis, G. A. (2000) High density lipoprotein oxidation: in vitro susceptibility and potential in vivo consequences. *Biochim. Biophys. Acta* **1483,** 217–235.
4. Heinecke, J. W., Li, W., Francis, G. A., and Goldstein, J. A. (1993) Tyrosyl radical generated by myeloperoxidase catalyzes the oxidative cross-linking of proteins. *J. Clin. Invest.* **91,** 2866–2872.
5. Francis, G. A., Mendez, A. J., Bierman, E. L., and Heinecke, J. W. (1993) Oxidative tyrosylation of high density lipoprotein by peroxidase enhances cholesterol removal from cultured fibroblasts and macrophage foam cells. *Proc. Natl. Acad. Sci. USA* **90,** 6631–6635.
6. Francis, G. A., Oram, J. F., Heinecke, J. W., and Bierman, E. L. (1996) Oxidative tyrosylation of HDL enhances the depletion of cellular cholesteryl esters by a mechanism independent of passive sterol desorption. *Biochemistry* **35,** 15,188–15,197.
7. Francis, G. A., Tsujita, M., and Terry, T. L. (1999) Apolipoprotein A-I efficiently binds and mediates cholesterol and phospholipid efflux from human but not rat aortic smooth muscle cells. *Biochemistry* **38,** 16,315–16,322.
8. Wang, W. Q., Merriam, D. L., Moses, A. S., and Francis, G. A. (1998) Enhanced cholesterol efflux by tyrosyl radical-oxidized high density lipoprotein is mediated by apolipoprotein AI-AII heterodimers. *J. Biol. Chem.* **273,** 17,391–17,398.
9. Weisgraber, K. H. and Mahley, R. W. (1980) Sufractionation of human high density lipoprotein by heparin-sepharose affinity chromatography. *J. Lipid Res.* **21,** 316–325.
10. Lowry, O. H., Rosebrough, N. J., Farr, A. L., and Randall, R. J. (1951) Protein measurement with the Folin phenol reagent. *J. Biol. Chem.* **193,** 265–275.
11. Rye, K. A. and Barter, P. J. (1994) The influence of apolipoproteins on the structure and function of spheroidal, reconstituted high density lipoproteins. *J. Biol. Chem.* **269,** 10,298–10,303.
12. Garner, B., Witting, P. K., Waldeck, R., Christison, J. K., Raftery, M., and Stocker, R. (1998) Oxidation of high density lipoproteins: formation of methionine sulfoxide in apolipoproteins AI and AII is an early event that accompanies lipid peroxidation and can be enhanced by alpha-tocopherol. *J. Biol. Chem.* **273,** 6080–6087.

9

Detection of Certain Peroxynitrite-Induced DNA Modifications

Hiroshi Ohshima, László Virág, Jose Souza, Vladimir Yermilov, Brigitte Pignatelli, Mitsuharu Masuda, and Csaba Szabó

1. Introduction

Nitric oxide and superoxide rapidly combine to form a toxic reaction product, peroxynitrite anion ($ONOO^-$) *(1,2)*. The oxidant reactivity of peroxynitrite is mediated by an intermediate with the biological activity of the hydroxyl radical. However, this product does not appear to be the hydroxyl radical *per se*, but peroxynitrous acid (ONOOH) or its activated isomer ($ONOOH^*$) *(2)*. Peroxynitrite readily reacts with proteins, lipids, and DNA under conditions of inflammation.

Peroxynitrite is a highly reactive species, which causes rapid oxidation of sulfhydryl groups and thioethers, as well as nitration and hydroxylation of aromatic compounds, such as tyrosine and tryptophan. The most widely studied and best characterized mechanism is tyrosine nitration *(3)*: measurement of 3-nitrotyrosine is often used as a marker of peroxynitrite in vivo and in vitro. Peroxynitrite also injures DNA via a number of mechanisms. Prominent DNA modifications induced by exposure to peroxynitrite include formation of 8-nitroguanine and 8-oxoguanine as well as the induction of DNA single-strand breakage *(4)*. DNA single-strand breakage induces a prominent secondary process, the activation of the nuclear enzyme poly(ADP ribose)synthetase [PARS, also known as poly(ADP-ribose) polymerase (PARP)]. PARS activation and subsequent cellular processes play an important role in the peroxynitrite-induced alterations in a variety of pathophysiological conditions *(5)*.

In the current chapter, we first describe the detection of 8-nitroguanine, which is probably one of the most prominent DNA modifications induced by

From: *Methods in Molecular Biology, vol. 186: Oxidative Stress Biomarkers and Antioxidant Protocols*
Edited by: D. Armstrong © Humana Press Inc., Totowa, NJ

peroxynitrite. Additionally, we describe methods to detect PARS activation, a common peroxynitrite-triggered downstream process.

8-Nitroguanine (nitro[8]Gua) is a unique base modification caused by reactive nitrogen species such as peroxynitrite *(6)* and the myeloperoxidase-H_2O_2-nitrite system *(7)*. Although nitro[8]Gua is not detectable by an electrochemical detector (ED), 8-aminoguanine (amino[8]Gua) has been reported to be electrochemically active *(8)*. Nitro[8]Gua is easily reduced by sodium hydrosulfite to amino[8]Gua. On the basis of these observations, we have developed a method to analyze nitro[8]Gua in DNA (or RNA) sensitively and selectively that involves acid-hydrolysis of DNA (or RNA), and chemical conversion of nitro[8]Gua to amino[8]Gua, which is then detected by high-performance liquid chromatography (HPLC)-ED *(9)*. RNA samples, but not DNA (*see* **Note 1**), can also be enzymatically hydrolyzed to nucleosides and analyzed similarly by HPLC-ED *(10)*.

Peroxynitrite is one of the most potent activator of the nuclear enzyme poly(ADP-ribose) synthetase (PARS) (also known as poly[ADP-ribose] polymerase [PARP], EC 2.4.2.30). PARS is a nick-sensor enzyme that becomes activated in response to DNA damage caused by various stimuli including ionizing radiation, alkylating agents, free radicals, reactive oxygen, and nitrogen intermediates (ROI-s and RNI-s) *(5,11)*. Upon activation, PARS cleaves NAD^+ into nicotinamide and ADP-ribose and polymerises the latter on nuclear acceptor proteins including histones and PARS itself. Poly ADP-ribosylation is involved in the maintenance of chromatine structure and genome integrity. However, excessive activation of PARS (by ROI-s and RNI-s) leads to necrotic cell death by depletion of cellular NAD^+ and ATP. The activity of the enzyme can be determined from peroxynitrite-treated cells (first protocol) and from tissue samples (e.g., vascular rings) (second protocol). Detection of poly(ADP-ribose) by immunofluorescence (third protocol), Western blotting *(12)* or by flow cytometry *(13)* can also be used to detect PARS activation.

A. Detection of 8-Nitroguanine in DNA and RNA by HPLC with an Electrochemical Detector

2. Materials

2.1. Equipment

1. Vortex/mixer.
2. Ultraviolet-visible (UV-VIS) absorbance spectrophotomer, Uvikon 922 spectrophotometer (Kontron Instrument, Saint Quentin Yvelines, France).
3. Centrifuge, Sigma SK10 (Bioblock Scientific, Illkirch, France).
4. Block heater.
5. Speed-Vac Concentrator (Savant, Farmingdale, NY).

6. HPLC pump, model SP8810 (Spectraphysics, San Jose, CA); ESA model 580 (Chelmsford, MA).
7. Electrochemical detector, ESA Coulochem II (Chelmsford, MA); Waters model M460 (Milford, MA).
8. Guard cell (part of the electrochemical detector ESA in **item 7**).
9. UV 100 detector (Thermoseparation Spectra series, San Jose, CA).
10. Ultrasphere ODS column (5 m two columns of 0.46 × 15 cm or one column of 0.46 × 25 cm (Beckman, Fullerton, CA).
11. HPLC Software (Knauer EuroChrom 2000 Integration Package (Chelmsford, MA).

2.2. Reagents

1. 8-Nitroguanine (nitro^{8}Gua) can be synthesized by the reaction of guanine with peroxynitrite and purified by HPLC using a preparative reversed-phase (RP) column *(6)*. Similarly, 8-nitroguanosine (nitro^{8}Guo) can be synthesized from guanosine with peroxynitrite. Both compounds are also commercially available from BIOLOG Life Science Institute (Flughafendamm 9a, P.O.B. 107125, D-28071 Bremen, Germany).
2. 8-Aminoguanosine (amino^{8}Guo) is obtained from Sigma Chemical Co. (St Louis, MO), and 8-aminoguanine (amino^{8}Gua) is prepared by acid hydrolysis of amino^{8}Guo *(9)*.
3. 8-Oxoguanine (oxo^{8}Gua) and 8-oxoguanosine (oxo^{8}Guo) are obtained from Aldrich (Milwaukee, WI) and Cayman Chemical Co. (Ann Arbor, MI), respectively.
4. Calf thymus DNA and calf-liver RNA (Sigma).
5. Peroxynitrite is synthesized in a quenched-flow reactor and excess hydrogen peroxide is destroyed by granular manganese dioxide *(14)*.
6. Sodium hydrosulfite (Sigma).
7. Nuclease P1 (*Penicillium citrinum*) and acid phosphatase type XA, sweet potato (Sigma).
8. Diethylenetriaminepentaacetic acid (Sigma).
9. Ethylenetriaminepentaacetic acid (Sigma).

3. Methods

3.1. Reaction of Calf Thymus DNA and Calf Liver RNA with Peroxynitrite and Other Reactive Nitrogen Species

1. Dilute peroxynitrite in a stock solution using cold 1 N NaOH and measure absorbance at 302 nm against 1 N NaOH. Calculate peroxynitrite concentration using $\varepsilon_{302\ nm} = 1670/M/cm$ *(14)*.
2. Add peroxynitrite prepared in 1 N NaOH at various concentrations (100 μL) to a reaction mixture containing an appropriate buffer (e.g., 0.1 *M* phosphate buffer, pH 7.4), calf thymus DNA or calf liver RNA (0.2 mg/mL), 100 μ*M* diethylenetriaminepentaacetic acid, a metal chelator, and an appropriate amount of HCl to neutralize the NaOH present in the peroxynitrite solution. Similar reactions can be performed with other reactive nitrogen species.

3. After the reaction, precipitate DNA with cold ethanol (two volumes), wash twice with 75% ethanol and once with ethanol, and dry briefly in a Speed Vac.
4. Dialyze the peroxynitrite-RNA reaction mixture against water. The dialyzed sample can be used directly or after precipitation with 2.5 volumes of ice-cold ethanol in the presence of 0.3 M sodium acetate, pH 5.2.
5. DNA and RNA isolated from cells and tissues using conventional methods can also be used for the following analyses.

3.2. Acid or Enzymatic Hydrolysis of DNA and RNA (see **Note 1**)

1. Acid hydrolysis: The DNA (or RNA) samples can be hydrolyzed in 0.1 N HCl (~1 mL/mg DNA or RNA) at 100°C for 30 min in order to depurinate purines (guanine, adenine, and their derivatives). Remove HCl in a Speed Vac and dissolve the residue in 100 μL of 0.1 M Tris-HCl buffer, pH 8.5.
2. Enzymatic hydrolysis: Dissolve the dialyzed or precipitated RNA sample (~100 μg) in 100 μL of 10 mM sodium acetate buffer, pH 4.5, containing 1 mM ethylendi-aminetetraacetate and incubate in the presence of 0.8 units nuclease P_1 and 1 unit acid phosphatase at 37°C for 30 min. Add 50 μL chloroform to the mixture, vortex, and centrifuge.
3. To 50 μL aliquots from acid (or enzymatic) hydrolysis, add a small amount of sodium hydrosulfite in order to reduce nitro[8]Gua (or nitro[8]Guo) to amino[8]Gua (amino[8]Guo).

3.3. HPLC Procedure

3.3.1. Analysis of Free Bases

1. HPLC system (the order of connection): pump (Spectraphisics model SP8810), injector, guard column, separation column, UV-detector (set at 254 nm), and ED (set at a potential +600 mV) (*see* **Note 2**).
2. Mobile phase: 12.5 mM citric acid-25 mM sodium acetate buffer containing 25 μM etheylenediaminetetraacetic acid (EDTA), pH 5.2, at a flow rate of 1 mL/min.
3. Standards: Oxo[8]Gua and amino[8]Gua are dissolved in water to prepare solutions in a range between 0.01 and 1 μM. Guanine standards (1–10 μM in water).
4. Inject 20 μL of each standard solution.
5. Using peak areas or peak heights, draw calibration curves for oxo[8]Gua and amino[8]Gua.
6. Using peak areas or peak heights detected by a UV-detector, draw calibration curves for guanine.
7. Centrifuge (1000g) for 5 min the samples prepared in **Subheading 3.2., step 1** before and after reduction with sodium hydrosulfite and inject 20 μL of the resultant supernatant.
8. **Fig. 1** shows representative chromatograms of acid-hydrolysates of calf-thymus DNA after incubation in vitro with 0.1 mM peroxynitrite (**Fig. 1B[a,b]**) or decomposed peroxynitrite (**Fig. 1C[a,b]**). Both samples show a peak correspond-

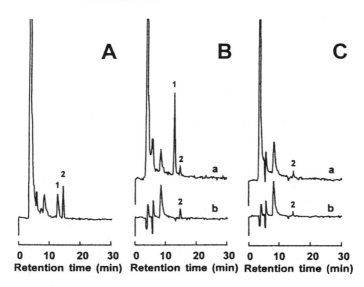

Fig. 1. Typical chromatograms obtained by HPLC-ED analyses of (**A**) standards, (**B**) acid-hydrolysates of DNA treated with peroxynitrite, and (**C**) those of DNA treated with decomposed peroxynitrite; (a,b) after and before reduction with sodium hydrosulfite, respectively. Peak 1 is nitro[8]Gua after reduction (i.e. amino[8]Gua) and peak 2 is oxo[8]Gua.

ing to oxo[8]Gua before reduction with sodium hydrosulfite (**Fig. 1B[b],C[b]**). After the reduction, bases from peroxynitrite-treated DNA show a new peak at a retention time corresponding to amino[8]Gua (**Fig. 1B[a]**), but this peak is not observed in samples from DNA treated with decomposed peroxynitrite (**Fig. 1C[a]**).

9. Concentrations of oxo[8]Gua are quantified by injecting the unreduced samples into HPLC-ED and using an oxo[8]Gua standard curve.
10. Concentrations of nitro[8]Gua are quantified by injecting the samples after reduction with sodium hydrosulfite using an amino[8]Gua standard curve.
11. Concentrations of unmodified guanine are measured by UV detection at 254 nm using a guanine standard curve.
12. The results are expressed as the amount of nitro[8]Gua or oxo[8]Gua (μmol or mmol) per mol guanine.

3.3.2. Analysis of Nucleosides from Enzymatic RNA Hydrolysis

1. HPLC system (the order of connection): pump (ESA model 580), guard cell (+350 mV), injector, separation column (Beckman Ultrasphere ODS columns, 5 μ, 0.46 × 25 cm), electrochemical detector (ESA Coulochem II; conditioning

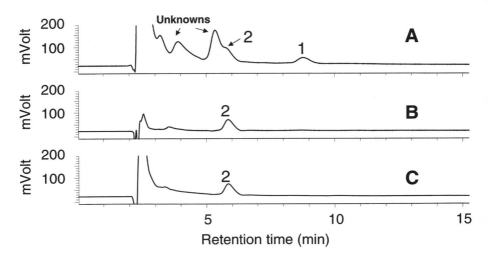

Fig. 2. Typical chromatograms obtained by HPLC-ED analysis of (**A**) nucleosides from enzymatic hydrolysis of peroxynitrite-treated RNA after reduction with sodium hydrosulfite; (**B**) those from the same RNA sample before reduction, and (**C**) those from RNA treated with decomposed peroxynitrite, but after reduction. Peak 1 is nitro[8]Guo after reduction (i.e., amino[8]Guo) and peak 2 is oxo[8]Guo.

cell, 0 mV; electrode 1, +150 mV; electrode 2, +300 mV), UV-detector (set at 254 nm) (*see* **Note 2**).

2. Mobile phase: 20 mM citric acid-22 mM sodium acetate buffer containing 12% methanol (HPLC-grade from Merck), pH 3.75, at a flow rate of 1 mL/min.
3. Standards: Oxo[8]Guo and amino[8]Guo are dissolved in water to prepare solutions in a range between 0.01 and 1 μM. Guanosine standards (10–100 μM in water).
4. Inject 20 μL of each standard solution.
5. Using peak areas or peak heights detected by electrode 2, draw calibration curves for oxo[8]Guo and amino[8]Guo.
6. Using peak areas or peak heights detected by a UV-detector, draw calibration curves for guanosine.
7. Centrifuge (1000g) for 5 min the samples prepared in **Subheading 3.2.**, **step 2** before and after reduction with sodium hydrosulfite and inject 20 μL of the resultant supernatant.
8. Representative chromatograms are shown in **Fig. 2**. After the reduction with sodium hydrosulfite, nucleosides from peroxynitrite-treated RNA show several peaks, one of which is at a retention time corresponding to amino[8]Guo (**Fig. 2A**). This peak is not observed in the same sample before reduction (**Fig. 2B**) or in the RNA treated with decomposed peroxynitrite even after reduction with sodium hydrosulfite (**Fig. 2C**).
9. Calculate concentrations of nitro[8]Guo, oxo[8]Guo and guanosine as described above (**Subheadings 3.3.1.**, **steps 7–9**) using standard curves for amino[8]Guo,

oxo[8]Guo and guanosine, respectively, and express the results as the amount of nitro[8]Guo or oxo[8]Guo (μmol or mmol) per mol guanosine.

4. Notes

1. Nitro[8]Gua formed in DNA is unstable. It is rapidly depurinated from DNA to generate free nitro[8]Gua (the half life of nitro[8]Gua in DNA incubated at 37°C in phosphate-buffered saline (PBS) has been estimated to be less than 4 h) *(9)*. In contrast, nitro[8]Gua formed in RNA is stable. No apparent depurination is observed when RNA containing nitro[8]Gua is incubated under similar conditions. It is therefore possible to hydrolyze RNA enzymatically to nucleosides, which can then be analyzed by HPLC-ED in a similar way to that described for free bases *(10)*.
2. There is a multitude of instruments available that fulfill the requirements for the analyses. We describe here two different EDs: amperometric and coulometric detectors. Either type can be used. However, in order to optimize the potentials applied, hydrodynamic voltammograms of each analyte should be generated for each ED. Inject a constant amount of the analyte while varying the potentials of an ED. Use the potential at which the maximum response from the detector is obtained.

B. Detection of Peroxynitrite-Induced Poly(ADP-Ribose) Synthetase Activation

1. In Vitro Technique *(15,16)*

2. Materials

2.1. Equipment

1. Liquid scintillation counter (e.g., Wallac 1409).
2. Incubator (set to 37°C).
3. Waterbath.
4. Microcentrifuge.

2.2. Reagents and Supplies

1. ^{3}H-NAD (NEN, Cat # NET-443).
2. NAD+ (Boehringer Mannheim).
3. Scintisafe scintillation fluid (Fisher).
4. Disposable cell scrapers (only if using adherent cells).

All other reagents including HEPES, sodium chloride, potassium chloride, magnesium chloride, digitonin and trichloroacetic acid (TCA) are from Sigma.

3. Methods

1. Prepare stock PARS buffer: 56 mM HEPES, pH 7.5, 28 mM KCl, 28 mM NaCl, 2 mM MgCl$_2$. (Store at 4°C. Stable for months.)

2. Treat cells with peroxynitrite in 12-well plates and incubate for 20–30 min in CO_2 incubator at 37°C.
3. While incubating the cells, make up 5% digitonin (in dimethyl sulfoxide [DMSO]) and 125 μM NAD^+ (in H_2O) and prepare assay buffer in a 50-mL conical tube by adding to 10 mL stock PARS buffer as follows: 20 μL digitonin (final concentration 0.01%); 10 μL NAD^+ (final concentration 0.125 μM); 50 μL ^3H-NAD^+.
4. You will need 0.5 mL assay buffer/sample. Place tube in 37°C water bath.
5. Aspirate off medium from all wells and immediately add 0.5 mL assay buffer.
6. Put plates in 37°C incubator for 10 min.
7. Scrape the cells with a disposable cell scraper and transfer each well contents into labeled Eppendorf tubes.
8. To each tube add 0.2 mL ice-cold 50% (w/v) TCA, cap tubes and refrigerate for 3 h.
9. Centrifuge samples in a microcentrifuge (10,000 rpm, 10 min).
10. Aspirate of supernatant and wash pellet twice with 1 mL ice-cold 5% TCA.
11. After aspirating of second TCA wash, add 0.5 mL/tube 2% SDS (in 0.1 N NaOH), cap tubes and solubilize pellet overnight in 37°C incubator.
12. Transfer the content of each tube to 7 mL scintillation vials previously filled with 6 mL scintillation fluid. Cap vials and vortex briefly.
13. Place samples into Beta scintillation counter and count 2 min in Tritium spectrum.
14. Calculate PARS activity using the following equation: PARS activity (nmol/min/μL) = total cpm × reaction time (min) $^{-1}$ × sample volume (μl) -1 x NAD specific activity (cpm/nmole)$^{-1}$.

4. Notes

1. When nonadherent cells are used instead of adherent ones, the assay is easier to perform. With nonadherent cells, it is advisable to transfer the peroxynitrite-treated cells into Eppendorf tubes. Quick spin (15 s, max speed) cells and aspirate medium. Resuspend cells in assay buffer and proceed to **step 5**.
2. The kinetics of PARS activation in response to peroxynitrite or other oxidative and nonoxidative stimuli needs to be determined for each cell type. In most cases 20–30 min incubation is sufficient.
3. It is advisable to add PARS inhibitors (e.g., 3 mM 3-aminobenzamide) to some samples of cells before peroxynitrite treatment and use these samples as negative controls.

B. Detection of Peroxynitrite-Induced Poly(ADP-Ribose) Synthetase Actuation (Continued)

1. Ex Vivo Technique (17)
2. Materials
2.1. Equipment

1. Tissue homogenizer.
2. Sonicator.

2.2. Reagents and Supplies

1. Phenylmethylsulfonyl fluoride (PMSF; Sigma).
2. 0.2-μm nitro-cellulose analytical test-filter funnels.

3. Method

1. Place freshly obtained tissue samples into 1–2 mL ice-cold stock PARS buffer (*see* composition in previous protocol) containing 0.1 mM PMSF. Stock = 100 mM in DMSO.
2. Homogenize tissues with a tissue tearer followed by sonication (3 × 15 s at maximal power) on ice.
3. Spin samples (3000g, 10 min, 4°C).
4. Use supernatants for protein and PARS activity assays.
5. Adjust the protein content of the samples to 0.5 mg/mL with stock PARS buffer.
6. Prepare PARS assay buffer by adding to 3 mL stock PARS buffer 5 μL 125 μM NAD$^+$ and 25 μL ^{3}H-NAD.
7. Combine 40 μL sample of tissue homogenate with 60 μL PARS assay buffer and allow the reaction to proceed for 1 min.
8. Stop the reaction by the addition of 900 μL ice-cold TCA.
9. Leave tubes on ice for 30 min.
10. Collect TCA insoluble precipitate by filtration through 0.2 μm nitro-cellulose analytical test filter funnels under vacuum and wash filters 5 times with 4 mL 5% ice-cold TCA.
11. Count the activity of membranes in 7-mL scintillation vials with ScintiSafe cocktail for 2 min beta spectrum in a scintillation counter.
12. Calculate PARS activity using the equation given in the previous protocol.

C. Immunohistochemical Detection of Poly(ADP-Ribose) (18–24)

2. Materials

2.1. Equipment

1. Fluorescent microscope.

2.1.2. Reagents and Supplies

1. Anti-poly(ADP-ribose) (monoclonal, Biomol # SA-216).
2. Fluoresceine isothiocyonate (FITC)-conjugated anti-mouse-IgG(goat) (Sigma).
3. Normal goat serum (Vector Laboratories).
4. Vectamount (Vector Laboratories).
5. Trichloroacetic acid (Sigma).
6. Triton-X 100 (Sigma).
7. Coverslips.
8. Coplin jars.

3. Method

1. Grow cells on coverslips in 12-well plates or in multichamber slides.
2. Treat cells with peroxynitrite.
3. After 15 min, aspirate off medium and wash cells with ice-cold-PBS.
4. Fix cells in ice-cold 10% TCA for 10 min.
5. Dehydrate in 70, 90%, and absolute ethanol for 3 min each at –20°C.
6. Rehydrate cells in PBS for 10 min at room temperature.
7. Incubate slides in blocking serum (2% normal goat serum in PBS-0.1% Triton-X-100) for 20 min.
8. Incubate slides with anti-poly(ADP-ribose) antibody (1:200) for 1 h at room temperature.
9. Wash slides six times for 5 min in PBS.
10. Incubate slides with FITC-conjugated goat anti-mouse immunoglobulin for 1 h at room temperature.
11. Wash slides six times for 5 min in PBS.
12. Mount in Vectamount and view under fluorescent microscope.

4. Notes

1. It is important to fix the cells in TCA as it inactivates poly(ADP-ribose) glycohydrolase (PARGH). PARGH, the enzyme that removes poly(ADP-ribose) from proteins is not inactivated during standard fixation procedures.
2. TCA fixation may interfere with the detection of antigens other than poly (ADP-ribose). This has to be considered when co-localization studies are performed *(20)*.
3. The anti-poly(ADP-ribose) monoclonal antibody (clone 10H) *(21)* can also be purchased from Serotec (Cat # MCA148). A rabbit polyclonal antibody (PAb) (commercially available from Biomol) can also be used for the immuno-cytochemical detection of poly(ADP-ribose).
4. Immunofluorescent detection of poly(ADP-ribose) in cells has been widely demonstrated, detection of poly(ADP-ribose) in frozen or paraffin-embedded tissue sections has also been described *(22–24)*.
5. This staining method has been successfully adapted to detect poly(ADP-ribosyl)ated proteins in a variety of tissues. However, immunolocalization of poly(ADP-ribose) in murine tissues is often difficult due to cross-reaction of secondary anti mouse antibodies with endogenous mouse immunoglobulins. In order to circumvent this difficulty, PARS activation can also be verified in frozen tissue sections by using biotinylated-NAD substrate *(25)*. This substrate also has additional applications such as immunocytochemistry and cellular ELISA.

References

1. Beckman, J. S., Beckman, T. W., Chen, J., Marshall, P. A., and Freeman, B. A. (1990) Apparent hydroxyl radical production by peroxynitrite: implications for

endothelial injury from nitric oxide and superoxide. *Proc. Natl. Acad. Sci. USA* **87,** 1620–1624.

2. Pryor, W. A. and Squadrito, G. L. (1995) The chemistry of peroxynitrite: a product from the reaction of nitric oxide with superoxide. *Am. J. Physiol.* **268,** L699–L722.

3. Ischiropoulos, H., Zhu, L., Chen, J., Tsai, M., Martin, J. C., Smith, C. D., and Beckman, J. S. (1992) Peroxynitrite-mediated tyrosine nitration catalyzed by superoxide dismutase. *Arch. Biochem. Biophys.* **298,** 431–437.

4. Szabo, C. and Ohshima, H. (1997) DNA damage induced by peroxynitrite: subsequent biological effects. *Nitric Oxide* **1,** 373–385.

5. Szabó, C. and Dawson, V. L. (1998) Role of poly(ADP-ribose) synthetase in inflammation and ischaemia-reperfusion. *Trends. Pharmacol. Sci.* **19,** 287–298.

6. Yermilov, V., Rubio, J., Becchi, M., Friesen, M. D., Pignatelli, B., and Ohshima, H. (1995) Formation of 8-nitroguanine by the reaction of guanine with peroxynitrite *in vitro. Carcinogenesis* **16,** 2045–2050.

7. Byun, J., Mueller, D. M., and Heinecke, J. W. (1999) 8-Nitro-2′-deoxyguanosine, a specific marker of oxidation by reactive nitrogen species, is generated by the myeloperoxidase-hydrogen peroxide-nitrite system of activated human phagocytes. *Biochemistry* **38,** 2590–2600.

8. Sodum, R. S., Nie, G., and Fiala, E. S. (1993) 8-Aminoguanine: a base modification produced in rat liver nucleic acids by the hepatocarcinogen 2-nitropropane. *Chem. Res. Toxicol.* **6,** 269–276.

9. Yermilov, V., Rubio, J., and Ohshima, H. (1995) Formation of 8-nitroguanine in DNA treated with peroxynitrite *in vitro* and its rapid removal from DNA by depurination. *FEBS Lett.* **376,** 207–210.

10. Masuda, M., Nishino, H., and Ohshima, H. (2002) Formation of 8-nitroguanosine in cellular RNA as a biomarker of exposure to reactive nitrogen species. *Chem. Biol. Interact,* in press.

11. de Murcia, G. and Menissier, de Murcia (1994). Poly(ADP-ribose) polymerase: a molecular nick-sensor. *Trends. Biochem. Sci.* **19,** 172–176.

12. Rosenthal, D. S., Ding, R., Simbulan-Rosenthal, C. M., Vaillancourt, J. P., Nicholson, D. W., and Smulson, M. (1997) Intact cell evidence for the early synthesis, and subsequent late apopain-mediated suppression, of poly(ADP-ribose) during apoptosis. *Exp. Cell Res.* **232,** 313–321.

13. Affar, E. B., Duriez, P. J., Shah, R. G., Winstall, E., Germain, M., Boucher, C., et al. (1999) Immunological determination and size characterization of poly(ADP-ribose) synthesized in vitro and in vivo. *Biochim. Biophys. Acta* **1428(2–3),** 137–146.

14. Reed, J. W., Ho, H. H., and Jolly, W. L. (1974) Chemical syntheses with a quenched flow reactor. Hydroxytrihydroborate and peroxynitrite. *J. Am. Chem. Soc.* **96,** 1248–1249.

15. Virág, L., Scott, G. S., Cuzzocrea, S., Marmer, D., Salzman, A. L., and Szabó, C. (1998) Peroxynitrite-induced thymocyte apoptosis: the role of caspases and poly (ADP-ribose) synthetase (PARS) activation. *Immunology* **94,** 345–355.

16. Virág, L., Scott, G. S., Antal-Szalmás, P., O'Connor, M., Ohshima, H., and Szabó, C. (1999) Requirement of intracellular calcium mobilization for peroxynitrite-induced poly(ADP-ribose) synthetase activation and cytotoxicity. *Mol. Pharmacol.* **56,** 824–833.

17. Pulido, E. J., Shames, B. D., Selzman, C. H., Barton, H. A., Banerjee, A., Bensard, D. D., and McIntyre, R. C. (1999) Inhibition of PARS attenuates endotoxin-induced dysfunction of pulmonary vasorelaxation. *Am. J. Physiol.* **277,** L769–L776.

18. Kupper, J. H., Van Gool, L., Muller, M., and Burkle, A. Detection of poly(ADP-ribose) polymerase and its reaction product poly(ADP-ribose) by immunocyto-chemistry. *Histochem. J.* **28,** 391–395.

19. Kupper, J. H., Muller, M., Jacobson, M. K., et al. (1995) Trans-dominant inhibition of poly(ADP-ribosyl)ation sensitizes cells against gamma-irradiation and N-methyl-N′-nitro-N-nitrosoguanidine but does not limit DNA replication of a polyomavirus replicon. *Mol. Cell Biol.* **15,** 3154–3163.

20. Lankenau, S., Burkle, A., and Lankenau, D. H. (1999) Detection of poly(ADP-ribose) synthesis in Drosophila testes upon γ-irradiation. *Chromosoma* **108,** 44–51.

21. Kawamitsu, H., Hoshino, H., Okada, H., Miwa, M., Momoi, H., and Sugimura, T. (1984) Monoclonal antibodies to poly(adenosine diphosphate ribose) recognize different structures. *Biochemistry* **23(16),** 3771–3777.

22. Love, S., Barber, R., and Wilcock, G. K. (1999) Increased poly(ADP-ribosyl)ation of nuclear proteins in Alzheimer's disease. *Brain* **122,** 247–253.

23. Scott, G. S., Jakeman, L. B., Stokes, B. T., and Szabó, C. (1999) Peroxynitrite production and activation of poly(ADP-ribose) synthetase in spinal cord injury. *Ann. Neurol.* **45,** 120–124.

24. Garcia, Soriano F., Virag, L., Jagtap, P., Szabo, E., Mabley, J. G., Liaudet, L., et al. (2001) *Nat. Med.* **7,** 108–113.

25. Bakondi, E., Bai, P., Szabo, E., Hunyadi, J., Gergely, P., Szabo, C., and Virag, L. (2002) Detection of Poly(ADP-ribose) Polymerase Activation in Oxidatively Stressed Cells and Tissues Using Biotinylated NAD Substrate. *J. Histochem. Cytochem.* **50,** 91–98.

10

Hydroxyl and 1-Hydroxyethyl Radical Detection by Spin Trapping and GC-MS

José A. Castro and Gerardo D. Castro

1. Introduction

The detection of free radicals is often carried out by electron spin resonance (ESR) techniques. The procedure is specific and very sensitive (provided the radicals have a sufficiently long half-life to be measured). In the case of extremely short-lived free radicals, which are difficult to directly detect by ESR, spin-trapping procedures have been developed based on reaction of highly reactive radicals with a "spin trap," generating an adduct free radical having a longer half-life and suitable for detection by regular ESR procedures *(1–4)*. A variety of spin traps have been described in the literature, each optimal for given experimental conditions or specific free radicals to be detected *(1–4)*. Hydroxyl (OH) radical detection by spin trapping and ESR procedures often has been done using 5,5-dimethylpyrroline N-oxide (DMPO) as the spin trap *(5–9)*. This spin trap very rapidly reacts with OH to produce a stable free radical adduct having a characteristic ESR spectrum *(5–9)*.

ESR equipment required for such OH detection is, however, very expensive and not widely available, which led to the development of several other simple and/or economic procedures *(1,10)*. Halliwell and Gutteridge *(1)* have made very detailed comments about the principles involved in these methods and suggestions to avoid interferences or ensure specificity. Studies by other laboratories revealed that gas chromatography-mass spectrometry (GC-MS) analysis of spin trap-free radical adducts may specifically and sensitively detect several free radicals *(11–14)*. However, to our knowledge no GC-MS procedures have been reported for hydroxyl or 1-hydroxyethyl (1HEt) free radical detection equivalent to those previously developed for ESR. Today,

From: *Methods in Molecular Biology, vol. 186: Oxidative Stress Biomarkers and Antioxidant Protocols*
Edited by: D. Armstrong © Humana Press Inc., Totowa, NJ

economic GC-MS equipment is more readily available than ESR instruments. Consequently, we have attempted the detection of OH or 1HEt radicals using GC-MS and the results obtained are described here.

2. Materials

2.1. Equipment

1. Vortex/mixer.
2. Heater block.
3. Centrifuge.
4. Freezing device (dry ice or –80°C freezer).
5. Conical (5-mL) glass reaction vials (Pierce, Rockford, IL).
6. Analytical GC-MS instrumentation.
    a. Hewlett Packard 5890 gas chromatograph (Palo Alto, CA).
    b. Hewlett Packard 5970B mass selective detector interfaced to the GC.
    c. HP 9000 series 300 workstation and the corresponding GC-MS software.
    d. Hewlett Packard Ultra 2, crosslinked 5% phenyl methyl silicone capillary column (12 m × 0.2 mm × 0.33 μm film thickness).

2.2. Reagents

Chemicals and solvents are analytical grade (Sigma Co., St. Louis, MO or Aldrich Chemical Co., Milwaukee, WI). Gases are from AGA Argentina.

1. α-phenyl-N-t-butylnitrone.
2. 5,5-dimethylpyrroline N-oxide.
3. Hydrogen peroxide, 30 wt. % solution in water.
4. Ferrous sulfate heptahydrate.
5. N,O-bis(trimethylsilyl)-trifluoracetamide (BSTFA).
6. Chelex 100 chelating resin.
7. Desferrioxamine mesylate.
8. Acetonitrile.
9. Toluene.
10. Absolute ethanol.
11. Nitrogen (99.9%).
12. Helium (UHP, 99.999%).

3. Methods

3.1. Sample Processing

3.1.1. Reference Spin Adducts Characterization

Spin adducts of hydroxyl (OH) and 1-hydroxyethyl (1HEt) radicals are generated in a Fenton-type chemical system in order to obtain mass spectra suitable to select appropriate masses for selected ion monitoring (SIM) detection. This is essential for having good sensitivity in the case of real samples. A "Fenton system" is performed essentially as follows: 5 mM FeSO$_4$ (prepared freshly in N$_2$-purged water), 5 mM H$_2$O$_2$, 50 mM EtOH and 30 mM 5,5-dimethylpyrroline N-oxide (DMPO) (or α-phenyl-N-t-butylnitrone, PBN) in 50 mM phosphate buffer, pH 7.2. Final volume is 1 mL. After addition of ferrous sulfate and vortexing, the reaction volume is immediately extracted with 500 μL toluene. Samples (whole volume) are then frozen at –80°C in order to allow the aqueous layer to solidify. The organic phase is separated and evaporated under nitrogen and then silylated (under N$_2$) with a mixture of N,O-bis(trimethylsilyl)-trifluoracetamide (BSTFA):acetonitrile (1:2), at 60°C over 15 min (*see* **Note 1**).

3.1.2. Processing of Biological Samples

Microsomes are prepared as previously described *(10)* and resuspended in 10 mL of 0.15 M KCl, 0.1 mM phosphate buffer, pH 7.4 (3:2 v/v), containing 0.5 mM desferrioxamine and recentrifuged at 100,000g in order to remove contaminating iron *(15)*. All buffers and solutions are iron depleted by passage through a column filled with Chelex 100 resin. Incubations involving microsomes (4.0–4.5 mg protein per mL), are added to NADPH generating system, 0.15 M MgCl$_2$, 20 mM DMPO (or 24 mM PBN) and 0.21 M EtOH in 50 mM phosphate buffer, pH 7.5. After 1 h at 37°C, the volume (3 mL) is extracted with 500 μL toluene, centrifuged, and the organic layer evaporated under nitrogen. The residue is silylated with BSTFA and analyzed by GC-MS (SIM detection) as described earlier.

3.2. GC-MS Procedure

3.2.1. GC and MS Operating Conditions

1. Chromatographic conditions are as follows: Ultra 2 column, programmed from 100°C (initial time 1 min) to 300°C at a ramp of 10°C/min. Injection port is at 250°C and transfer line to MS, 300°C. Carrier gas (helium) is at 80 kPa. Injection volume is 1 μL, in the splitless mode (0.2 min purge off time).
2. Mass spectrometer conditions are: source temperature at ca. 200°C. Spectra are taken at 70 eV, scanning quadrupole from 50–550 amu (0.86 scan/s). Solvent delay is performed to 7.2 min.

3.2.2. GC-MS Analysis

Data collection begins at 7.2 min after sample injection. Raw data is processed and stored by the on-line computer and resident GC-MS software. This allows post-analysis processing of data. Typically, electron multiplier voltage setting is at 1,400 volt for scan mode runs or 2,000 volt for SIM analysis. Total run time is 12 min and separation period between runs is 2 min. In experiments involving microsomal activation of ethanol, SIM mode of detection is employed to increase sensitivity. Selected masses are: for DMPO-OH, 275 ($M^{+\bullet}$) and 260 (M-15), for DMPO-1HEt, 288 (M-15), for PBN-1HEt, 250 (M - $\bullet CHCH_3OTMS$) and 194 (m/z 250 - C_4H_8).

3.3. Chromatographic Separation

3.3.1. GC-MS of Fenton-DMPO Mixtures

The capillary GC analysis with total ion current (TIC) detection of reaction products arising when hydroxyl radicals are generated in the presence of ethanol and DMPO in a Fenton system is depicted in **Fig. 1A** *(16)*. **Figure 1B,C** correspond to the spectra of DMPO-OH and DMPO-1HEt adducts, respectively. The adduct with 1HEt is formed in low yield, despite the different ethanol concentrations tested. In addition, its spectrum shows poor quality in terms of providing fragments for SIM; mass 288 has low abundance and m/z 186 should be avoided due to interference leading to artifacts. Two other peaks can be observed as corresponding to dimeric DMPO derived structures (**Fig. 2A,B** with proposed structures on spectra). The formation of these compounds as interference in ESR experiments involving DMPO has been postulated previously *(17)*.

3.3.2. GC-MS of PBN-1HEt Adduct Fenton System

Figure 3A shows capillary GC analysis with TIC detection of reaction products when free radicals derived from a Fenton system attack ethanol in the presence of the spin-trap PBN. Fragmentation pattern of the spectrum taken from peak at 9.1 min (**Fig. 3B**) is consistent with the presence of a 1-hydroxyethyl moiety attached to the spin trap structure (*see* **Note 2**). Fragments at m/z 250 and 194 are selected for further SIM analysis because both combined high-mass and high-abundance characteristics, necessary for detection with good sensitivity *(16,18)*.

3.3.3. GC-MS of Hydroxylated PBN Adducts Fenton System

Two peaks (I and II in **Fig. 3A**) related to the interaction between PBN and hydroxyl radicals are observed, but not the adduct corresponding to the addition of the hydroxyl radical to the nitrone. In both cases, molecular mass ($M^{+\bullet}$) and fragmentation patterns do not agree with what it would be expected from a hydroxyl-PBN adduct (PBN-OH) structure. This is not a surprise, in view of the low stability of the PBN-OH in aqueous media, decomposing to benzaldehyde and t-butylhydroaminoxyl radical, as reported previously by others *(19–21)*. Compounds I and II were not observed in those experiments, and the reason may rest on the nonradical nature of these molecules, therefore they are not evidenced in ESR. These compounds are formed also when ethanol

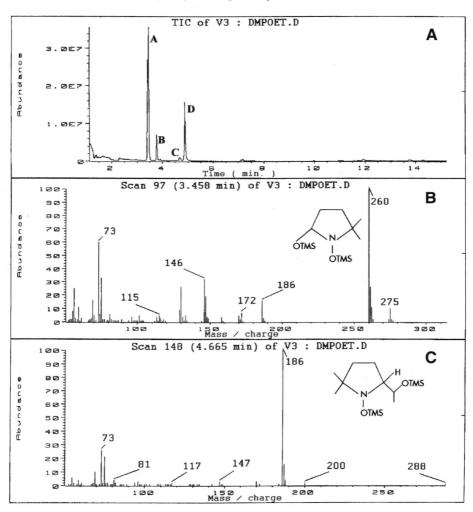

Fig. 1. (**A**) Total ion chromatogram obtained from a sample of the Fenton reaction system, in the presence of ethanol and DMPO. Peaks: A, hydroxyl adduct of DMPO; B, dimeric product of DMPO (monosilylated); C, 1-hydroxyethyl adduct of DMPO; D, dimeric product of DMPO (disilylated). (**B**) Mass spectrum taken from peak A in (**A**). $M^{+\bullet}$ is at m/z 275 and it is confirmed by m/z 260 (M - 15) and m/z 186 (M - $\bullet$OTMS). (**C**) Mass spectrum taken from peak C in (**A**). $M^{+\bullet}$ is deducted from m/z 288 (M-15). Other relevant fragment is at m/z 186 (loss of $\bullet$CHCH$_3$OTMS from $M^{+\bullet}$).

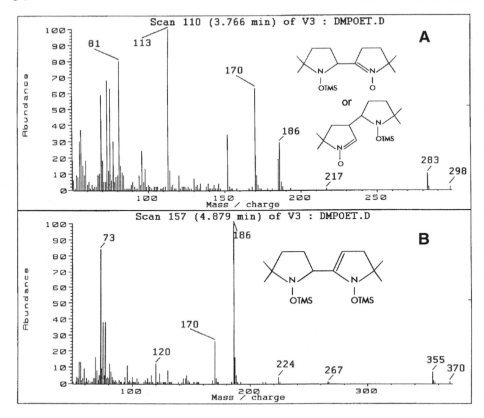

Fig. 2. **(A)** Mass spectrum taken from peak B in Fig. 1A. $M^{+\bullet}$ is found at m/z 298 and confirmed by M-15. Other relevant fragments are at m/z 186, 170 (loss of oxygen from 186), 113 and 81 (loss of TMSO• from m/z 170). **(B)** Mass spectrum taken from peak d in **Fig 1A**. $M^{+\bullet}$ is at m/z 370 and (M-15) at m/z 355 respectively. Half molecule is represented by the base peak at m/z 186.

is not present in the Fenton reaction (**Fig. 4A**). In addition, both can be observed when OH are generated from UV-mediated photolysis of H_2O_2, in the presence of PBN. Spectra taken from these peaks are depicted in **Fig. 4B,C**. From them, it can be assigned a molecular mass of 265 for both compounds and this value corresponds to an increment of 88 amu over the mass of PBN (*see* **Note 3**). Hydroxyl addition to the aromatic ring of the spin trap can give account for the proposed structures in the case of compounds I and II. Specific place of linkage of the hydroxyl group in the aromatic ring can not be deduced from mass spectra.

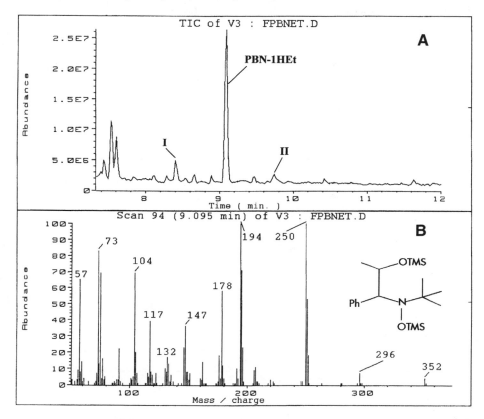

Fig. 3. **(A)** Gas chromatogram obtained from a sample of the reaction mixture containing PBN and ethanol in a Fenton system, after trimethylsilylation. Peaks: I and II, derived from PBN-hydroxyl adduct; PBN-1HEt, trimethylsilylated spin adduct of 1-hydroxyethyl radical. **(B)** Mass spectrum taken from peak PBN-1HEt in **Fig. 1A**. $M^{+\bullet}$ is not present but it can be deducted from m/z 250 (loss of •CHCH$_3$OTMS) and confirmed by other ions from a fragmentation pattern, typical of many PBN adducts: m/z 296 (loss of isobutylene from M-15), m/z 194 (base peak corresponding to the loss of isobutylene from m/z 250) and m/z 104 (loss of TMSOH from m/z 194). The presence of the 1-hydroxyethyl group on the molecule is confirmed by the fragment at m/z 117 (CH$_3$CH = O$^+$-TMS).

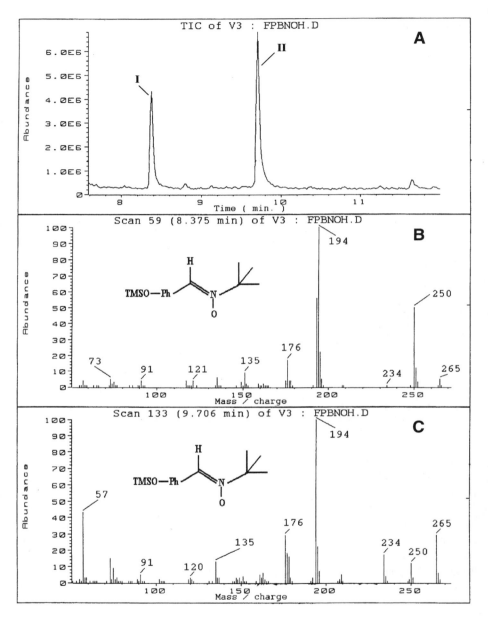

Fig. 4. **(A)** Gas chromatogram obtained from a sample of the reaction mixture containing PBN in a Fenton system in the absence of ethanol, after trimethylsilylation. Peaks: I and II, two isomers of hydroxylated PBN. **(B)** Mass spectrum taken from peak I in **(A)**. M$^{+\bullet}$ is assigned to m/z 265. Other relevant fragments are at m/z 250 (M-15), m/z 194 (loss of C$_4$H$_8$ from m/z 250) and m/z 176 (Me$_2$Si = O$^+$ -C$_6$H$_4$CN). **(C)** Mass spectrum taken from peak II in **(A)**. M$^{+\bullet}$ is assigned to m/z 265. Other fragments are at m/z 250 (M-15), m/z 234 (loss of oxygen from M-15), m/z 209 (loss of C$_4$H$_8$ from M$^{+\bullet}$), m/z 194 (loss of C$_4$H$_8$ from m/z 250) and m/z 176 (Me$_2$Si = O$^+$ -C$_6$H$_4$CN).

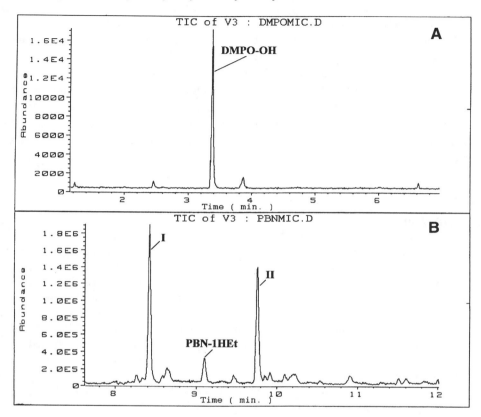

Fig. 5. **(A)** Selected ion current (SIM) profile obtained from the analysis of a sample of an incubation mixture containing microsomes, NADPH and DMPO. Masses selected for SIM are 260 and 275. **(B)** SIM profile obtained from the analysis of an incubation sample containing microsomes, NADPH, ethanol and PBN, after trimethylsilylation. Masses selected are 250 and 194.

3.3.4. Detection of Spin Adducts Formation in Rat Liver Subcellular Fractions

When rat liver microsomes metabolize ethanol, both radicals can be detected in the toluene extracts of incubation mixtures *(16)*. Hydroxyl radicals are efficiently trapped by DMPO but the 1-hydroxyethyl adduct of DMPO is observed in the same experiment only in a very low amount, probably due to the low efficiency of the SIM detection in this case (**Fig. 5A**). In general, experiments using DMPO as the trapping agent for hydroxyl free radicals generated from microsomes plus NADPH have poor reproducibility. This was previously reported by Rashba-Step et al. *(8)* using ESR. By contrast,

PBN-1HEt detection is more efficient *(16,18)*. Even so, only in the case of microsomal activation of ethanol in the presence of NADPH spin adduct of the 1-hydroxyethyl radical is detected (**Fig. 5B**). Peaks I and II are also observed in this microsomal system as reported earlier for the case of the Fenton reaction. They can be observed with good response and with better yield than that of the highly unstable PBN-OH adduct. Whether their formation could be used as an indicator of OH production in biological systems is at present under study. The reason is that these products might also result from mixed function oxygenase action on PBN, as already suggested by other workers in ^{14}C-PBN experiments *(22)*.

4. Notes

1. Reduction of the adduct (aminoxyl) radicals to the corresponding hydroxyl-amines is accomplished by BSTFA in the derivatization step. No difference in yield of products is observed when using hydrogen donor reagents, like 2-mercaptoethanol, previous to silylation.
2. Mass fragmentation pattern was confirmed by running Fenton reaction samples with d_6-ethanol or n-propanol as the alcohol and observing mass shifts in spectra.
3. Molecular mass at m/z 265 was confirmed by silylating a Fenton reaction sample with N-(tert-butyldimethylsilyl)-N-methyltrifluoracetamide (MTBSTFA) and looking for the mass shifts in spectra. $M^{+\bullet}$ was found at m/z 307 as predicted.

References

1. Halliwell, B. and Gutteridge, J. M. C. (1991) *Free Radicals in Biology and Medicine* (2nd ed.). Clarendon Press, Oxford, UK, pp. 47–58.
2. Aust, S. D., Chignell, C. F., Bray, T. M., Kalyanaraman, B., and Mason, R. P. (1993) Free radicals in toxicology. *Toxicol. Appl. Pharmacol.* **120,** 168–173.
3. Janzen, E. G. (1984) Spin trapping. *Methods Enzymol.* **105,** 188–198.
4. Knecht, K. T. and Mason, R. P. (1993) In vivo spin trapping of xenobiotic free radical metabolites. *Arch. Biochem. Biophys.* **303,** 185–194.
5. Buettner, G. R. (1993) The spin trapping of superoxide and hydroxyl free radicals with DMPO (5,5-dimethylpyrroline-N-oxide): more about iron. *Free Rad. Res. Commun.* **19,** S79–S83.
6. Burkitt, M. J. (1993) ESR spin trapping into the nature of the oxidizing species formed in the Fenton reaction: pitfalls associated with the use of 5,5-dimethyl-1-pyrroline-N-oxide in the detection of the hydroxyl radical. *Free Rad. Res. Commun.* **18,** 43–57.
7. Rashba-Step, J., Turro, N. J., and Cederbaum, A. L. (1993) Increased NADPH and NADH-dependent production of superoxide and hydroxyl radical by microsomes after chronic ethanol treatment. *Arch. Biochem. Biophys.* **300,** 401–408.
8. Rashba-Step, J., Turro, N. J., and Cederbaum, A. L. (1993) ESR studies in the production of reactive oxygen intermediates by rat liver microsomes in the presence of NADPH or NADH. *Arch. Biochem. Biophys.* **300,** 391–400.

9. Reinke, L. A., Rau, J. M., and McCay, P. B. (1994) Characteristics of an oxidant formed during iron (II) autoxidation. *Free Rad. Biol. Med.* **16**, 485–492.

10. Castro, G. D., Lopez, A. J., and Castro, J. A. (1988) Evidence for hydroxyl free radical formation during Paraquat but not for Nifurtimox liver microsomal biotransformation. A dimethyl sulfoxide scavenging study. *Arch. Toxicol.* **62**, 355–358.

11. Janzen, E. G., Krygsman, P. H., and Haire, D. L. (1988) The application of gas chromatographic/mass spectrometric techniques to spin trapping. Conversion of α-phenyl N-tert-butylnitrone (PBN) spin adducts to stable trimethylsilylated derivatives. *Biomed. Environ. Mass Spectrom.* **15**, 111–116.

12. Janzen, E. G., Towner, R. A., Krygsman, P. H., Haire, D. L., and Poyer, J. L. (1990) Structure identification of free radicals by ESR and GC/MS of PBN spin adducts from the in vitro and in vivo rat liver metabolism of halothane. *Free Rad. Res. Commun.* **9**, 343–351.

13. Janzen, E. G., Towner, R. A., Krygsman, P. H., Lai, E. N., Poyer, J. L., Brueggemann, G., and McCay, P. B. (1990) Mass spectroscopy and chromatography of the trichloromethyl radical adduct of phenyl tert-butyl nitrone. *Free Rad. Res. Commun.* **9**, 353–360.

14. Hong Sang, Janzen, E. G., Poyer, J. L., and McCay, P. B. (1997) The structure of free radical metabolites detected by EPR spin trapping and mass spectroscopy from halocarbons in rat liver microsomes. *Free Rad. Biol. Med.* **22**, 843–852.

15. Albano, E., Tomasi, A., Goria-Gatti, L., and Dianzani, M. U. (1988) Spin trapping of free radical species produced during the microsomal metabolism of ethanol. *Chem. Biol. Interact.* **65**, 223–234.

16. Castro, G. D., Delgado de Layño, A. M. A., and Castro, J. A. (1997) Hydroxyl and 1-hydroxyethyl free radicals detection using spin traps followed by derivatization and gas chromatography-mass spectrometry. *Redox Report* **3**, 343–347.

17. Finkelstein, E., Rosen, G. M, and Rauckman, E. J. (1980) Spin trapping of superoxide and hydroxyl radical: Practical aspects. *Arch. Biochem. Biophys.* **200**, 1–16.

18. Castro, G. D., Delgado de Layño, A. M. A., and Castro, J. A. (1998) Liver nuclear ethanol metabolizing system (NEMS) producing acetaldehyde and 1-hydroxyethyl free radicals. *Toxicology* **129**, 137–144.

19. Abe, K., Suezawa, H., Hirota, M., and Ishii, T. (1984) Mass spectrometric determination of spin adducts of hydroxyl aryl free radicals. *J. Chem. Soc. Perkin Trans. II.* 29–34.

20. Janzen, E. G., Kotake, Y., and Hinton, R. D. (1992) Stabilities of hydroxyl radical spin adducts of PBN-type spin traps. *Free Rad. Biol. Med.* **12**, 169–173.

21. Kotake, Y. and Janzen, E. G. (1991) Decay and fate of the hydroxyl radical adduct of α-phenyl-N-tert-butylnitrone in aqueous media. *J. Am. Chem. Soc.* **113**, 9503–9506.

22. Chen, G., Bray, T. M., Janzen, E. G., and McCay, P. B. (1991) The role of mixed function oxidase (MFO) in the metabolism of the spin trapping agent α-phenyl-N-tert-butyl-nitrone (PBN) in rats. *Free Rad. Res. Comms.* **14**, 9–16.

11

Analysis of Aliphatic Amino Acid Alcohols in Oxidized Proteins

Bénédicte Morin, Shanlin Fu, Hongjie Wang, Michael J. Davies, and Roger T. Dean

1. Introduction

Radical-mediated protein oxidation was first studied at the beginning of the 20th century by Henry Dakin (*[1]*; reviewed in **ref. 2**), but it is only recently that the use of the products of these reactions as specific markers of oxidative damage in in vivo situations has been established. Generic markers of protein oxidation such as protein carbonyls have been used longer as indices of protein oxidation in vivo *(3)* (reviewed recently in **ref. 4**), but it is unclear whether the measurement of such materials accurately reflects damage to proteins (as opposed to oxidation of, for example, lipids or sugars that are, or become, bound to proteins). Our purpose here is to discuss methods for analysis of alcohols of aliphatic amino acids in proteins, since there are few pathways known to generate these molecules other than oxygen radicals such as hydroxyl radicals/Fenton systems (*see* discussion in **ref. 5,6**).

HO• and a range of other radicals, in the presence of O_2, can oxidize the side-chains of aliphatic amino acids to hydroperoxides, alcohols, and carbonyl compounds, as a result of the formation of an initial carbon-centered radical (via hydrogen atom abstraction) and addition of oxygen to give a peroxyl radical. The chemistry of such peroxyl radicals is relatively well-understood and gives rise to oxygenated products via standard hydrogen abstraction, fragmentation, and dimerization reactions *(5,7,8)*. The hydroperoxides are known to be unstable; these decompose slowly in the absence of light, heat, reducing agents, or metal ions, but are lost rapidly in the presence of any of these agents *(9–12)*. These materials are therefore poor quantitative markers

From: *Methods in Molecular Biology, vol. 186: Oxidative Stress Biomarkers and Antioxidant Protocols*
Edited by: D. Armstrong © Humana Press Inc., Totowa, NJ

of protein oxidation, while the resultant alcohols are often stable, and suitable as markers. Exceptions are those alcohols that are natural enzymatic-reaction products (e.g., 4-hydroxyproline, 4- and 5-hydroxylysines), those that undergo further reaction (e.g. the internal cyclization of the two stereoisomers of 5-oxoleucine to give methylprolines *[13]*), and those that co-elute under the chosen conditions with other products (e.g., the two stereoisomers of 4-hydroxyvaline *[14]*).

Although these hydroperoxides and alcohols are known to be generated with a range of oxidants, these are mainly reactive radical species such as HO$^{\bullet}$, alkoxyl and peroxyl species; much less is known about the formation of such products with other oxidants. Valine and leucine alcohols are not generated significantly by HOCl or long wavelength UV light (>310 nm) *(15,16)*. They are also unlikely to be formed to any major extent during the reactions of peroxynitrite, NO$^{\bullet}$ or $^1O_2$ with proteins *(5)*, though this has not been investigated in detail. The use of these materials as markers of protein oxidation is mainly in relation to the damage induced by reactive radicals. The absence of such products, and the presence of other markers of protein oxidation (reviewed in **ref. 6**) may give valuable information as to the nature of the oxidant(s) which generate damage.

1.1. General Comments

Protein carbonyls as generic markers can be assayed by use of a number of different methods; these have been extensively reviewed elsewhere *(17–19)*, and will not be covered further here. Three major methods have been developed for the measurement of the known specific oxidation products: immunological methods on tissue or isolated protein samples, gas chromatography mass spectometry (GC-MS) and high-performance liquid chromatography (HPLC) with various different detectors. Issues of specificity are central to the use of immunological methods and these do not readily allow determination of multiple products present in a single sample, unlike either the GC or HPLC methods. Multiple product methods can permit measurement of both product and parent amino acid (and hence allow compensation for differences in the amounts of total protein present).

HPLC and GC-MS methods require the isolation, purification, and hydrolysis of the proteins under study before analysis of levels of particular oxidized side-chain lesions. Developments in MS should eventually permit such measurements to be done on intact proteins; at present protein hydrolysis is a necessary antecedent (e.g., **ref. 20,21**). As for MS, this is usually done enzymatically (e.g., using trypsin); this may convey advantages over the current acid hydrolysis techniques (gas phase, high temperature) used in many GC-MS and HPLC methods. This report concentrates on the HPLC approach, which we have developed and validated *(13–16,23,26)*. The preparation of purified materials for use as internal standards has been described in the same papers, and in **ref.** *22*.

2. Materials
2.1. Reagents and Equipment

1. High-purity (analytical-grade) reagents: sodium borohydride (NaBH$_4$), butylated hydroxytoluene, ethylenediaminetetraacetic acid (EDTA), sodium deoxycholate, trichloroacetic acid, 6 *M* HCl, 2-mercaptoacetic acid, phenol, incomplete *o*-phthaldialdehyde reagent (OPA, Sigma; some batches can be troublesome due to impurities), 9-fluorenyl-methyl chloroformate (FMOC; Sigma), Chelex-100 resin.
2. High-purity (HPLC grade) solvents: acetone, diethyl ether, acetonitrile, methanol, tetrahydrofuran.
3. Buffers: phosphate-buffered saline (PBS), 10 m*M* NaH$_2$PO$_4$ buffer, pH 4.3, 20 m*M* sodium acetate buffer, pH 5.2, 20 m*M* sodium acetate buffer, pH 4.05.
4. High-purity nitrogen or argon gas.
5. Nanopure water.
6. One-mL brown round-bottomed glass vials (Alltech), Pico-Tag reaction vessels, screw-topped 2-mL vials with 8-mm screw (Activon), 8-mm Teflon septa (Shimadzu), 200-μL flat-bottom inserts (Edwards Instruments, Australia).

2.2. Equipment

1. Analytical HPLC gradient system with UV and fluorescence detectors, autosampler, sample cooler, column oven, and pre-mixing system. For 3-hydroxyvaline and 5-hydroxyleucine analyses, a LC-NH$_2$ HPLC column (25 cm × 4.6 mm, 5-μm particle size, Supelco) and Pelliguard NH$_2$ HPLC guard column (2 cm). For 3-hydroxylysine analysis, a LC-18 HPLC column (15 cm × 4.6 mm, 3 μm particle size, Supelco) and Pelliguard C18 HPLC guard column (2 cm).
2. Speedy-vac freeze drier or equivalent.

3. Methods
3.1. Preparation of Samples

Extreme care must be taken in the storage, handling, and processing of the samples to avoid their artefactual oxidation. The extent of further oxidation induced by hydroperoxides present in the samples can be minimized by the use of a reductive step (usually involving treatment with sodium borohydride) early in the isolation and purification procedure *(13,14,22)*. The extent of artifactual oxidation can be assessed by taking purified proteins through similar handling and processing; the extent of such oxidation can be restricted *(15)*. Delipidation of the samples is also a prerequisite for successful protein hydrolysis. Incomplete recovery of protein during these (necessary) steps can be compensated for by measurement of both the levels of the oxidative lesion and the parent amino acid from which it is generated and expressing the concentration of the oxidized material with respect to the parent.

3.2. Recovery and Interassay Variability During Sample Preparation and Analysis

The recovery of many of the aliphatic alcohols, unlike their hydroperoxides, taken through the isolation and preparation procedures is usually >70% *(13,14,23)*. The known cases where the recovery is poor are 4-hydroxyVal and 4-hydroxyLeu, where internal cyclization to give the γ-lactone is believed to occur under the acidic conditions used in the hydrolysis step *(13)*. The accuracy of determinations of both the oxidized and parent amino acids obtained using this HPLC methodology has been assessed; the largest inter-assay variations for the wide range of oxidised amino acids tested (DOPA, di-tyrosine, *o-* and *m*-Tyr, 3-hydroxyvaline, 5-hydroxyleucine) are in the range 9–15%, whereas those of the parent amino acids (Tyr, Phe, Val, Leu) are 2–5% *(16)*.

The use of GC-MS, and particularly the use of isotope dilution MS, as a method for determining some protein-oxidation products, though not the aliphatic alcohols, is covered elsewhere *(24,25)*; the reader is referred to these articles for further details. We have discussed elsewhere *(6)* in more detail the general issues and approaches to comparing the extent of artefact with different methodologies and the importance of factors which confound the interpretation of the measured levels in terms of extent of protein oxidation: these include protein degradation, and re-incorporation of oxidized amino acids by protein synthesis.

3.3. Delipidation of Biological Samples for Protein Hydrolysis

All reagents are Chelex-treated *(27)*.

1. Fresh tissue is rinsed in PBS containing 100 μ*M*-butylated hydroxytoluene and 1 m*M* EDTA, at 4°C *(see* **Note 1**).
2. Tissue is homogenized in cold PBS (ca. 20% w/v), containing external standards *(see* **Note 2**).
3. The homogenate is then reduced by the addition of 1 mg/mL sodium borohydride *(see* **Note 3**) and then adjusted to approximately 1 mg/mL protein.
4. 700 μL of homogenate is placed in a 1-mL brown round-bottomed glass vial (Alltech), to which 50 μL of 0.3% (w/v) sodium deoxycholate and 100 μL of 50% (w/v) trichloracetic acid are added.
5. The suspension is thoroughly mixed, and then centrifuged at 9000*g* × 10 min at 4°C.
6. The precipitate is washed twice with 800 μL of cold acetone, thoroughly mixing and recentrifuging each time.

7. The precipitate is then washed with 800 µL of cold diethyl ether, resuspending thoroughly and recentrifuging.
8. The final delipidated precipitate is dried in a Speedy-vac freeze drier.

3.4. Protein Hydrolysis

The delipidated samples, plus appropriate controls, are subsequently hydro-lyzed to the free amino acids before analysis (*see* **Note 4**).

1. The freeze-dried delipidated protein samples (from **Subheading 3.3.**) are placed in Pico-Tag reaction vessels, containing 1 mL of 6 M HCl with 1% (v/v) phenol and 50 µL of 2-mercaptoacetic acid, with a maximum of 8 sample vials per reaction vessel (*see* **Note 5**).
2. Air is evacuated using a vacuum pump, and the vessel is filled with nitrogen or argon. The gassing and evacuation are repeated prior to hydrolysis in an oven at 100°C for 16 h.
3. The sample vials are then removed, and any residual acid is evaporated using the Speedy-Vac.
4. The product is normally re-dissolved in degassed water prior to further analysis.

3.5. Determination of 3-Hydroxyvaline and 5-Hydroxyleucine in Hydrolysates of Oxidised Proteins

1. It is usually necessary to concentrate the analyte from **Subheading 3.4.** prior to derivatization and final HPLC. This is done by redissolving the lyophilized protein hydrolysate in 25 µL of water, and injecting 20 µL onto an LC-NH$_2$ column (25 cm × 4.6 mm, 5 µm particle size, Supelco, with a Pelliguard NH$_2$ guard column (2 cm).
2. The column is pre-equilibrated with mobile phase (83% acetonitrile in 10 mM NaH$_2$PO$_4$ buffer, pH 4.3).
3. The column is developed at 1.5 mL/min and monitored at 210 nm.
4. 3-hydroxyvaline and 5-hydroxyleucine elute in a window between approx. 13.7 and 15.3 min (*see* **Fig. 1A**), and the pool is suitable for subsequent analysis by OPA derivatization and further HPLC. Separations can also be carried out on a semi-preparative scale; the higher flow rate and larger column size results in somewhat different elutions times (*see* **Fig. 1B**).
5. One mL of OPA reagent is reconstituted with 3 µL 2-mercaptoethanol in a vial, and placed in an autosampler pre-cooled to 5°C (*see* **Note 6**).
6. 40 µL aliquots of the HPLC pools or other amino acid samples in water (50 nM–100 µM in the case of valine alcohol), are placed in vial inserts in the auto-sampler. We use screw-topped 2-mL vial with 8-mm screw (Activon), 8-mm Teflon septa (Shimadzu), and 200-µL flat-bottom inserts (Edwards Instruments, Australia).
7. A pretreatment file is established that takes 20 µL of the OPA reagent and mixes it with the amino acid sample for precisely 2 min prior to injection of 15 µL on to the column.

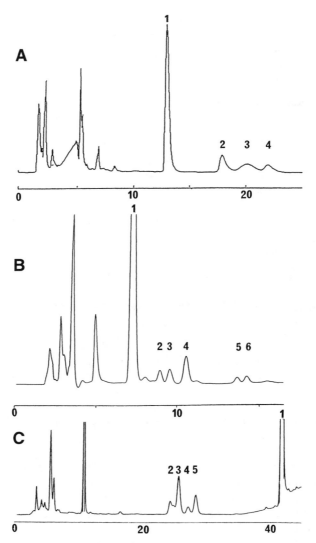

Fig. 1. HPLC chromatograms of γ-irradiated valine (**A**), leucine (**B**) and lysine (**C**).
(**A**) Chromatogram of γ-irradiated valine (1 mM) after reduction with $NaBH_4$. The
γ-irradiated solution was concentrated by 40-fold at room temperature under vacuum,
then treated with $NaBH_4$ (4 mg/mL). HPLC analysis of the reduced mixture (10 µL)
was performed using a LC-NH_2 column (25 cm × 4.6 mm, 5 µm particle size) with
82% CH_3CN in 10 mM sodium phosphate buffer, pH 4.3. Flow rate 1.5 cm^3/min, with
the eluent monitored at 210 nm. Peak 1, parent amino acid; peak 2, 3-hydroxyvaline;
peak 3, (3R) 4-hydroxyvaline; peak 4, (3S) 4-hydroxyvaline. (**B**) Chromatogram of
γ-irradiated leucine (1 mM) after reduction with $NaBH_4$. The γ-irradiated solution was

8. The OPA derivatized analytes are separated on an LC-18 column (15 cm × 4.6 mm, 3-μm particle size, Supelco), with a Pelliguard C18 guard column (2 cm).
9. The column is pre-equilibrated with solvent A (methanol/tetrahydrofuran/20 mM sodium acetate, pH 5.2, 20:2.5:77.5 by vol.).
10. The column is subsequently eluted at 1 mL/min with a gradient to solvent B (the same solvents at 80:2.5:17.5 by vol), as follows: 0 min, 0% B, to 25% B in 8 min; isocratic elution at 25% B for 5 min; then to 40% B in 10 min; then to 50% B in 2 min; isocratic elution at 50% B for 6 min, then to 100% B in 4 min; isocratic elution at 100% B for 3 min, then 0% B for 12 min prior to the next analysis.
11. The OPA derivatives are detected by fluorescence (λ_{ex} 340 nm, λ_{em} 440 nm), and a calibration curve for each analyte is established (*see* **Note 7**). As little as 0.1 pmol on the column can be detected.

3.6. Detection of 3-Hydroxylysine in Hydrolysates of Oxidized Proteins

While this method has been less extensively tested with biological samples, it has proved of utility.

1. After borohydride reduction and hydrolysis as described in **Subheadings 3.3.** and **3.4.**, samples are derivatized with FMOC (Sigma), which gives highly fluorescent and stable products.
2. These products are separated on a C18 column as for the OPA-products, but with a different system of mobile phases, again at 1 mL/min. Solvent A is acetonitrile: tetrahydrofuran: 20 mM sodium acetate, pH 4.05, 20:5:75 (v/v). Solvent B is the same solvents at 80:5:15 (v/v). The column is developed with 30 min of

---

Fig. 1. *(continued)* concentrated by 40-fold at room temperature under vacuum then treated with NaBH$_4$ (1 mg/mL). HPLC analysis of the reduced mixture (100 μL) was performed using a semi-preparative LC-NH$_2$ column (25 cm × 10 mm, 5 μm particle size) with 81% CH$_3$CN in 10 mM sodium phosphate buffer, pH 4.3. Flow rate 5 cm^3/min, with the eluent monitored at 210 nm. Peak 1, parent amino acid; peak 2,(4R) 4-methylproline; peak 3, (4S) 4-methylproline; peak 4, 4-hydroxyleucine; peaks 5 and 6, the two stereoisomers of 5-hydroxyleucine. (**C**) Chromatogram of γ-irradiated lysine (4 mM) after reduction with NaBH$_4$ (1 mg/mL). Ten μL of the γ-irradiated solution was incubated with 90 μL of borate buffer (0.1 M, pH 9.4) and 100 μL of a 15 mM solution of FMOC in acetone. The mixture was vortex mixed and extracted after 60 s with 3 × 1 cm^3 of pentane to remove the FMOC hydrolysis product. The aqueous solution was then diluted 10-fold for HPLC analysis. The FMOC derivatives were separated using a C$_{18}$ column with a solvent gradient and detected by fluorescence (*see* text). Peak 1, parent amino acid; peak 2, (5S) 5-hydroxylysine; peak 3, (5R) 5-hydroxylysine; peak 4, 3-hydroxylysine; peak 5, 4-hydroxylysine.

32% B in A, then a gradient to 70% B in 10 min, isocratic elution for 8 min, and re-equilibration at 32% B for the next 7 min.

3. FMOC derivatives are detected by fluorescence (λ_{ex} 265 nm, λ_{em} 310 nm) (*see* **Fig. 1C**).

3.7. Results

Sample analytical and semi-preparative HPLC chromatograms of the alcohols formed from valine, leucine, and lysine are shown in **Fig. 1**.

4. Notes

1. This mixture minimizes artifactual oxidation of the samples.
2. Valine, lysine, or leucine alcohols, when appropriate.
3. To convert all hydroperoxides to alcohols.
4. To control for the effects of the protein hydrolysis step itself, unoxidized amino acids, unoxidized proteins, and external standards should also be taken through this procedure.
5. With some biological samples the phenol is unnecessary for optimal recovery; the 2-mercaptoacetic acid will suffice.
6. Some batches of the incomplete OPA reagent can be troublesome due to impurities, and blank samples should be run.
7. The OPA derivatives of valine, leucine, and their respective alcohols have very similar fluorescent yields.

Acknowledgments

Work in the authors' laboratories is funded in part by the Australian Research Council, the National Health and Medical Research Council, the Wellcome Trust, the Clive and Vera Ramaciotti Foundations, Diabetes Australia Research Trust, and the Juvenile Diabetes Foundation International.

References

1. Dakin, H. D. (1906) The oxidation of amido-acids with the production of substances of biological importance. *J. Biol. Chem.* **1,** 171–176.
2. Dean, R. T. (1999) Henry Drysdale Dakin (1880–1952): Early studies on radical and 2-electron oxidation of amino acids, proteins and fatty acids. *Redox Report* **4,** 189–194.
3. Stadtman, E. R. (1988) Protein modification in aging. *J. Gerontol.* **43,** B112–B120.
4. Stadtman, E. R. and Berlett, B. S. (1988) Reactive oxygen-mediated protein oxidation in aging and disease. *Drug Metab. Rev.* **30,** 225–243.
5. Davies, M. J. and Dean, R. T. (1997) *Radical-Mediated Protein Oxidation: From Chemistry to Medicine.* Oxford University Press, Oxford, UK, pp. 1–443.
6. Davies, M. J., Fu, S., Wang, H., and Dean, R. T. (1999) Stable markers of oxidant damage to proteins and their application in the study of human disease. *Free Rad. Biol. Med.* **27,** 1151–1163.

7. Garrison, W. M. (1987) Reaction mechanisms in the radiolysis of peptides, polypeptides, and proteins. *Chem. Rev.* **87,** 381–398.
8. von Sonntag, C. (1987) *The Chemical Basis of Radiation Biology.* Taylor and Francis, London, pp. 1–515.
9. Simpson, J. A., Narita, S., Gieseg, S., Gebicki, S., Gebicki, J. M., and Dean, R. T. (1992) Long-lived reactive species on free-radical-damaged proteins. *Biochem. J.* **282,** 621–624.
10. Gebicki, S. and Gebicki, J. M. (1993) Formation of peroxides in amino acids and proteins exposed to oxygen free radicals. *Biochem. J.* **289,** 743–749.
11. Fu, S., Gebicki, S., Jessup, W., Gebicki, J. M., and Dean, R. T. (1995) Biological fate of amino acid, peptide and protein hydroperoxides. *Biochem. J.* **311,** 821–827.
12. Gebicki, J. M. (1997) Protein hydroperoxides as new reactive oxygen species. *Redox Report* **3,** 99–110.
13. Fu, S. L. and Dean, R. T. (1997) Structural characterization of the products of hydroxyl-radical damage to leucine and their detection on proteins. *Biochem. J.* **324,** 41–48.
14. Fu, S., Hick, L. A., Sheil, M. M., and Dean, R. T. (1995) Structural identification of valine hydroperoxides and hydroxides on radical-damaged amino acid, peptide, and protein molecules. *Free Rad. Biol. Med.* **19,** 281–292.
15. Fu, S., Davies, M. J., Stocker, R., and Dean, R. T. (1998) Evidence for roles of radicals in protein oxidation in advanced human atherosclerotic plaque. *Biochem. J.* **333,** 519–525.
16. Fu, S., Dean, R., Southan, M., and Truscott, R. (1998) The hydroxyl radical in lens nuclear cataractogenesis. *J. Biol. Chem.* **273,** 28,603–28,609.
17. Levine, R. L., Garland, D., Oliver, C. N., Amici, A., Climent, I., Lenz, A. G., et al. (1990) Determination of carbonyl content in oxidatively modified proteins. *Methods Enzymol.* **186,** 464–478.
18. Stadtman, E. R. (1993) Oxidation of free amino acids and amino acid residues in proteins by radiolysis and by metal-catalyzed reactions. *Annu. Rev. Biochem.* **62,** 797–821.
19. Levine, R. L., Williams, J. A., Stadtman, E. R., and Shacter, E. (1994) Carbonyl assays for determination of oxidatively modified proteins. *Methods Enzymol.* **233,** 346–357.
20. Chowdhury, S. K., Eshraghi, J., Wolfe, H., Forde, D., Hlavac, A. G., and Johnston, D. (1995) Mass spectrometric identification of amino acid transformations during oxidation of peptides and proteins: modifications of methionine and tyrosine. *Anal. Chem.* **67,** 390–398.
21. Finley, E. L., Dillon, J., Crouch, R. K., and Schey, K. L. (1998) Radiolysis-induced oxidation of bovine alpha-crystallin. *Photochem. Photobiol.* **68,** 9–15.
22. Dean, R. T., Fu, S., Gieseg, S., and Armstrong, S. G. (1996) Protein hydroperoxides, protein hydroxides, and protein-bound DOPA, in *Free Radicals: A Practical Approach* (Punchard N. and Kelly F. J., eds.), Oxford University Press, Oxford, pp. 171–183.

23. Morin, B., Bubb, W. A., Davies, M. J., Dean, R. T., and Fu, S. (1998) 3-Hydroxy-lysine, a potential marker for studying radical-induced protein oxidation. *Chem. Res. Toxicol.* **11,** 1265–1273.

24. Heinecke, J. W., Hsu, F. F., Crowley, J. R., Hazen, S. L., Leeuwenburgh, C., Mueller, D. M., et al. (1999) Detecting oxidative modification of biomolecules with isotope dilution mass spectrometry: sensitive and quantitative assays for oxidized amino acids in proteins and tissues. *Methods Enzymol.* **300,** 124–144.

25. Heinecke, J. W. (1999) Mass spectrometric quantification of amino acid oxidation products in proteins: insights into pathways that promote LDL oxidation in the human artery wall. *FASEB J.* **13,** 1113–1120.

26. Fu, S., Fu, M.-X., Baynes, J. W., Thorpe, S. R., and Dean, R. T. (1998) Presence of DOPA and amino acid hydroperoxides in proteins modified with advanced glycation end products: amino acid oxidation products as possible source of oxidative stress induced by age proteins. *Biochem. J.* **330,** 233–239.

27. Van Reyk, D. M., Brown, A. J., Jessup, W., and Dean, R. T. (1995) Batch-to-batch variation of Chelex-100 confounds metal-catalysed oxidation. Leaching of inhibitory compounds from a batch of Chelex-100 and their removal by a pre-washing procedure. *Free Rad. Res.* **23,** 533–535.

12

Rapid Determination of Glutamate Using HPLC Technology

Aqeela Afzal, Mohammed Afzal, Andrew Jones, and Donald Armstrong

1. Introduction

Glutamate is an excitatory transmitter released from neurons. It communicates information rapidly by activating receptors in other neurons *(1)*. Physiologic concentrations of glutamate are 1 μM extracellularly and several millimolar intracellularly. Higher extracellular levels lead to cytotoxicity in the neurons *(2)*. Glutamate exerts its excitatory action via ligand-gated ion channels to enhance Na^+/Ca^{++} conductance. There are two types of glutamate receptors: ionotropic and metabotropic. The ionotropic receptors are comprised of three families with intrinsic cation permeable channels (Na^+ and K^+): N-methyl-D-aspartate (NMDA), 2-amino-3-hydroxy-5-methyl-4-isoxazole propionate (AMPA), and kainate. The latter two families are known as the nonNMDA receptors *(3,4)*. The metabotropic receptors, on the other hand, are G protein-coupled subunits, which release secondary messengers in the cytoplasm or influence ion channels through the release of G-protein subunits within the membrane *(4)*. Vesicles in presynaptic terminals release glutamate via a calcium-dependent mechanism. Glutamate released from pre-synaptic structures is not enzymatically degraded. Instead, it is taken up by transporters, which directly regulate extracellular glutamate concentrations and limit excitotoxicity *(5,6)*.

Glutathione is an antioxidant and a redox modulator of NMDA receptor activity. The NMDA receptor fluctuates between fully reduced and fully oxidized states. Glutathione protects neurons against glutamate induced excitotoxicity and can function either as an excitatory transmitter or a neuro-

From: *Methods in Molecular Biology, vol. 186: Oxidative Stress Biomarkers and Antioxidant Protocols*
Edited by: D. Armstrong © Humana Press Inc., Totowa, NJ

modulator. As an excitatory neurotransmitter, glutathione depolarizes neurons and as a neuromodulator, glutathione binds to ionotropic glutamate receptors and forestalls the excitatory input to target neurons, thus fine-tuning the neurotransmission *(4)*.

Diseases attributed to glutamate cytotoxcitiy include Amyotrophic Lateral Sclerosis (ALS) *(5)*, epilepsy *(7)*, Huntington's disease (HD), Parkinson's disease (PD), and Alzheimer's disease (AD) *(3)*. Various techniques have been employed to measure glutamate toxicity levels in these diseases, including, immunohistochemistry *(2)*, autoradiography, positron emission tomography (PET), and nuclear magnetic resonance (NMR) *(8)*. However, these methods use expensive equipment, are labor-intensive, lack specificity and sensitivity, and do not provide quantitative information. We have developed a simple, rapid high-performance liquid chromatography (HPLC) technique to overcome the shortcomings of the above mentioned techniques.

2. Materials

2.1.1. Analytical HPLC System

1. Shimadzu LC-10 Atv pump system (max flow rate = 9.99 mL/min) (Shimadzu Inst. Co., Columbia, MD).
2. Shimadzu SCL-10A vp system controller.
3. Shimadzu FCV-10AL vp low-pressure gradient flow-control valve.
4. Shimadzu DGU-14A on-line degasser.
5. Shimadzu SPD-M10A vp UV/VIS photodiode array detector.
6. Shimadzu SIL-10A automatic sample injector.
7. Micron electronics SE440 BX online computer.

2.1.2. HPLC Column

1. Supelco LC-18 reverse-phase (RP) column: 25 cm × 4.6 mm, 5 μm particle size (Bellefonte, PA).
2. Supelco guard C-18 guard column: 4.6 × 20 mm, 5 μm particle size.
3. Supelco guard column cartridge holder.

2.2. Reagents

All solvents are HPLC-grade and filtered through 0.22-μm nylon membranes prior to use.

1. Methanol (Fisher Scientific, Fair Lawn, NJ).
2. Trifluoroacetic acid (TFA) (0.15%): in HPLC-grade water (Fisher).
3. Tetrahydrofuran (Fisher).
4. Hydrochloric acid (5 mM): in HPLC-grade water (Fisher).
5. L-Glutamic acid. (Sigma Chemical, St. Louis, MO).

3. Methods

3.1. Samples

Biological fluids are collected and frozen immediately at –80°C until time for the assay.

3.2. Preparation of Glutamate Standards

1. Prepare a 2 mg/mL solution of glutamate in 0.15% TFA (*see* **Note 1**).
2. Make serial dilutions of the stock solution in 0.15% TFA to obtain the following the concentrations: 1 mg/mL, 500 µg/mL, 250 µg/mL, and 125 µg/mL.
3. Inject 20 µL of the glutamate standard onto the C-18 HPLC column. Each chromatogram will have two peaks. The first peak is glutamate and the second peak is the solvent.
4. Use the chromatograms from the serial dilutions and make a standard curve of glutamate vs peak height for each dilution.

3.3. HPLC Sample Preparation

1. Sonicate 250 µL vitreous sample in 250 µL of 5 mM HCl.
2. Wet a C-18 Sep-Pack with methanol.
3. Load the sample on the Sep-Pak with a syringe (deproteinize samples prior to analysis).
4. Attach the Sep-Pak with a syringe and elute the with 500 µL of 5 mM HCl (*see* **Note 2**).
5. Elute the Sep-Pack again with 500 µL of methanol (*see* **Note 3**).
6. Centrifuge any samples that form a precipitate and decant the supernatant into an HPLC vial for injection.
7. Inject 20 µL onto the C-18 HPLC column at a flow rate of 1 mL/min; using TFA (0.15% in water): tetrahydrofuran (70:30 v/v) as the mobile phase.

3.4. Results

Standard curve and elution profile from aqueous or vitreous humor from the eye are shown in **Figs. 1** and **2**. Glutamate elutes at 3.1 min either from aqueous (**Fig. 1A**) or vitreous (**1B**) from the eye. The solvent peak is at 3.5 min.

A standard curve of glutamate from 15.625–1000 ng/mL is shown in **Fig. 2**. The curve is reproducible with a small standard deviation. The concentration of glutamate in **Fig. 1A,B** is 1299 ng/mL and 1917 ng/mL, respectively.

Aspartate, a related amino acid, elutes much later at 4.3 min.

4. Notes

1. Glutamate standards should be made fresh at time of the experiment.
2. Elute Sep-Pak slowly while filtering the samples and retain unwanted protein.
3. Use Sep-Pak once per sample and discard.

A

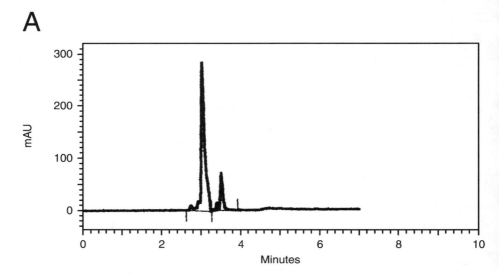

B

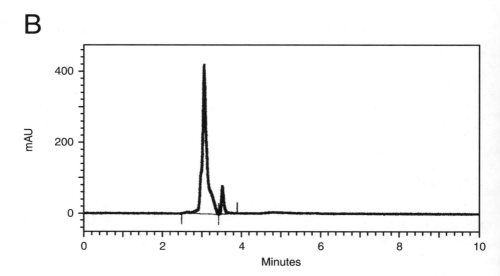

Fig. 1. Elution profiles of glutamate from **(A)** aqueous solution (a solvent peak appears at 3.5 min) and from **(B)** vitreous humor of eye (a solvent peak appears at 3.5 min).

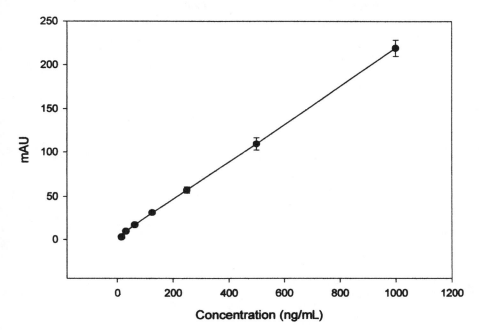

Fig. 2. A standard curve of glutamate.

References

1. Gegelashvili, G., Dehnes, Y., Danbolt, N. C., and Schousboe, A. (2000) The high affinity glutamate transporters GLT1, GLAST, and EAAT4 are regulated via different signaling mechanisms. *Neurol. Int.* **37,** 163–170.
2. Yoles, E., Friedmann, I., Barouch, R., Shani, Y., and Schwartz, M. (2001) Self-protective mechanism awakened by glutamate in retinal ganglion cells. *J. Neurotrauma* **18(3),** 339–349.
3. Meldrum, B. S. (2000) Glutamate as a neurotransmitter in the brain: review of physiology and pathology. *J. Nutr.* **130,** 1007S–1015S.
4. Oja, S. S., Janaky, R., Varga, V., and Saransari, P. (2000) Modulation of glutamate receptor functions by glutathione. *Neurochem. Int.* **37,** 299–306.
5. Tanaka, K. (2000) Functions of glutamate transporters in the brain. *Neurosci. Res.* **37,** 15–19.
6. Takahashi, M., Billups, B., Rossi, D., Sarantis, M., Hamann, M., and Attwell, D. (1997) The role of glutamate transporters in glutamate homeostasis in the brain. *J. Exp. Biol.* **200,** 401–409.
7. Chapman, A. G. (2000) Glutamate and epilepsy. *J. Nutr.* **130,** 1043S–1045S.
8. Cruz, F. and Cerdan, S. (1999) Quantitative ^{13}C NMR studies of metabolic compartmentation in the adult mammalian brain. *NMR Biomed.* **12,** 451–462.

13

A Rapid Method for the Quantification
of GSH and GSSG in Biological Samples

**Mohammed Afzal, Aqeela Afzal, Andrew Jones,
and Donald Armstrong**

1. Introduction

Measurement of O_2 consumption, conjugated dienes, and decomposition products of lipid hydroperoxide (LHP) such as malondialdehyde have been extensively used as markers of lipid peroxidation (*1*). Glutathione (GSH) and glutathione disulfide (GSSG) are biologically important intracellular thiols, and alterations in the GSH/GSSG ratio are often used to assess exposure of cells to oxidative stress. GSH plays an important role in scavenging free radicals and protects cells against several toxic oxygen-derived chemical species. A constant supply of reduced glutathione is necessary to repair the effects of spontaneous oxidation of sulfhydryl groups, which results in cell-membrane damage.

GSH is also an important co-enzyme for glutathione peroxidase activity and also provides protection for sulfhydryl groups of proteins from oxidation. During these reactions, GSH is oxidized to GSSG and recycled back by glutathione reductase. In this way LHP and hydrogen peroxide levels are regulated in the cells and the ratio of GSH to GSSG is indicative of oxidative stress (*2*). GSH is also involved in many other functions of the cell including regulation of protein and DNA synthesis and as an essential cofactor of many enzymes.

GSH and GSSG have been measured by a number of methods. Measurement of GSH and GSSG by their derivatization, with UV detection, is commonly used (*3–5*). A fluorometric method is also available to measure GSH and GSSG (*6*). Many workers (*7,8*) have used a high-performance liquid chromatography (HPLC) procedure for quantification of GSH and GSSG using phosphate

From: *Methods in Molecular Biology, vol. 186: Oxidative Stress Biomarkers and Antioxidant Protocols*
Edited by: D. Armstrong © Humana Press Inc., Totowa, NJ

buffer with more complex electrochemical detection. Measurement of GSH and GSSG using ion-exchange chromatography has also been described *(9)*. Knowledge of the thiols such as cystine, cysteine, homocysteine, and disulfides is equally important in defining the total oxidative protection afforded to tissue *(10)*.

Although several methods are available for measuring GSH and GSSG, all have disadvantages including the need to generate derivatives, the inability to conveniently measure both GSH and GSSG, and lack of sufficient sensitivity to allow detection in very small samples. Our method presents a simple and validated HPLC-UV detection procedure for determining GSH and GSSG in small plasma samples. This method is a fast, sensitive, and selective HPLC-method for the quantification of picomole quantities of GSH and GSSG in plasma, which does not require prior derivatization.

2. Materials
2.1. Equipment (Shimadzu Inst. Co., Columbia, MD)

1. SDP-M10A VP Photodiode Array Detector.
2. SIL-10A Auto-injector with 100 µL injection loop.
3. LC-10AT VP pump.
4. FCV-10ALVP, DGU-14-A degasser.
5. SCL-10A VP system controller.
6. Data station with class-VP, V5.03 chromatography software.

2.2. Reagents and Supplies

1. HPLC-grade acetonitrile (Fisher Scientific, Fair Lawn, NJ).
2. Trifluoroacetic acid, peptide synthesis grade (Sigma-Aldrich, St. Louis, MO).
3. Reverse-phase (RP) Sep-Pak cartridges for solid phase extraction (Waters Millipore, Milford, MA).
4. Suplecosil LC-18 column (25 cm × 4.6 mm ID 5 µm) (Belfonte, PA).
5. Target vials # C4011-1 (National Scientific Company, Duluth, GA).
6. Caps for target vials # C4011-1A (National Scientific Company Product).
7. 0.2-µm membrane PVDF filters for mobile-phase degassing.
8. Metaphosphoric acid (3%) (Fisher Scientific, Fair Lawn, NJ).

3. Methods
3.1. Standards Preparation

A 2 m*M* solution of GSH and GSSG is prepared in 0.1% trifluoroacetic acid (TFA). Serial dilutions are made of each standard for the standard curves. 10 µL of each dilution are injected onto a C-18 column using a mobile phase of acetonitrile: (0.1%) TFA (70:30) at a flow rate of 1.3 mL/min.

3.2. Sample Preparation

Plasma (0.2 mL) is diluted 1:2 with metaphosphoric acid (*see* **Note 1**) and incubated at room temperature for 15 min. Centrifuge at 10,000*g* for 5 min and pass the supernatant through a C-18 Sep-Pak cartridge (*see* **Note 2**). The cartridge is eluted with 0.1% TFA (*see* **Note 3**) and injected into a C-18 (ODS) column for HPLC analysis using an isocratic solvent system (acetonitrile: (0.1%) TFA (70:30) with a flow rate of 1.3 mL/min.

3.3. HPLC Separation

The method does not require prior derivatization of glutathiones. GSH and GSSG eluted from the column are detected at 210 nm. The retention time for the oxidized and reduced glutathione are recorded at 2.8 and 4.1 min, respectively. The detection limit of glutathione is 100 pmoles with acceptable accuracy and without need to generate derivatives. The recovery of GSH and GSSG added to the samples is 97–100%.

3.4. Specificity

Addition of standard GSH or GSSG (*see* **Note 4**) to samples increased the peak areas appropriately, without altering the peak shape, retention time, or peak area of the corresponding oxidized (reduced) thiol. Quantification was carried out by plotting a standard curve.

3.5. Results

Standard curves and elution profiles of GSH, GSSG in pure form and biological matrix are given in **Figs. 1** and **2**.

4. Notes

1. Metaphosphoric acid is used to inhibit oxidation of GSH to GSSG.
2. Cartridge is pre-washed with methanol.
3. TFA is very corrosive. It must be handled in a fume hood.
4. Standard samples of oxidized and reduced glutathione must be prepared fresh.

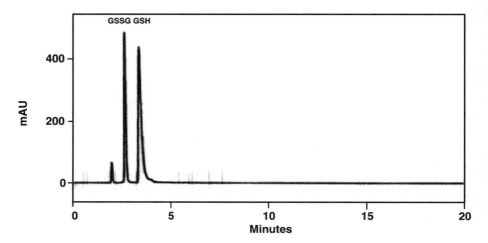

Fig. 1. Elution profile of GSSG standards at 2.8 min and GSH standards at 4.1 min.

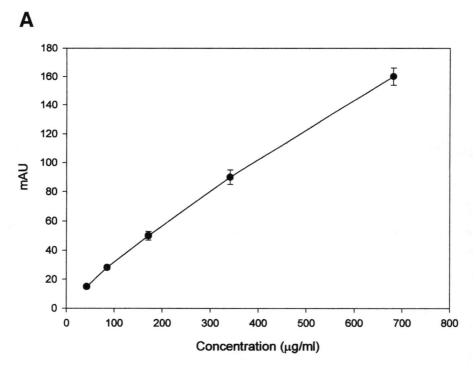

Fig. 2A. Standard curves of GSSG, a given concentration of either standard, GSSH omits less fluorescence than GSH.

B

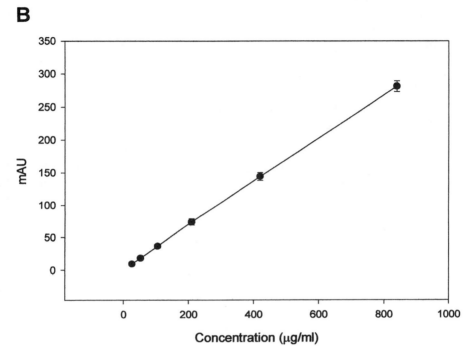

Fig. 2B. Standard curve for GSH.

References

1. Lima, M. H., Willmore, W. G. and Storey, K. B. (1995) Quantification of lipid peroxidation in tissue extracts based on Fe(III) xylenol orange complex formation. *Free Rad. Biol. Med.* **19(3),** 271–280.
2. Armstrong, D. and Browne, R. (1994) The analysis of free radicals, lipid peroxides, antioxidant enzymes and compounds related to oxidative stress as applied to the clinical chemistry laboratory, in *Free Radicals in Diagnostic Medicine, Adv. Exp. Med. Biol., vol. 366* (Armstrong D., ed.), Plenum Press, NY, pp. 43–58.
3. Ridnour, L. A., Winters, R. A., Ercal, N., and Spitz, D. R. (1999) Measurement of glutathione, glutathione disulfide in mammalian cell and tissue homogenates using high-performance liquid chromatography separation of N-(1-pyrenyl)maleimide derivatives. *Methods Enzymol.* **299,** 258–267.
4. Winters, R. A., Zukowski, J. Ercal, N., Mathew, R. H., and Spitz, D. R. (1995) Analysis of glutathione, glutathione disulfide, cysteine, homocysteine and other biological thiols by high-performance liquid chromatography following derivatization by n-(1-pyrenyl)maleimide. *Anal. Biochem.* **227(1),** 14–21.
5. Leroy, P., Nicolas, A., Thioudellet, C., Oster, T., Wellman, M., and Siest, G. (1993) Rapid liquid chromatographic assay of glutathione in cultured cells. *Biomed. Chromatogr.* **7(2),** 86–89.

6. Browne, R. and Armstrong, D. Reduced glutathione and glutathione disulfide, in *Free Radicals in Diagnostic Medicine, Methods in Molecular Biology, vol. 108* (Armstrong D., ed.), Humana Press, Totowa, NJ, pp. 347–352.
7. Liu, S., Ansari, N. H., Wang, C., Wang, L., and Srivastava, S. K. (1996) A rapid HPLC method for the quantification of GSH and GSSG in ocular lens. *Curr. Eye Res.* **15(7),** 726–732.
8. Lakritz, J., Plopper, C. J., and Buckpitt, A. R. (1997) Validated high performance liquid chromatography-electrochemical method for determination of glutathione and glutathione disulfide in small tissue samples. *Anal. Biochem.* **247(1),** 63–68.
9. Alpert, A. J. and Gilbert, H. F. (1985) Detection of oxidized and reduced glutathione with a recycling postcolumn reaction. *Anal. Biochem.* **144(2),** 553–5562.
10. Richie, J. and Lang, C. (1987) The determination of glutathione, cyst(e)ine and other thiols and disulfides in biological samples using high-performance liquid chromatography with dual electrochemical detection. *Anal. Biochem.* **163,** 9–15.

14

Protein Carbonyl Measurement by ELISA

I. Hendrikje Buss and Christine C. Winterbourn

1. Introduction

Protein carbonyls are formed by a variety of oxidative mechanisms and are sensitive indices of oxidative injury *(1)*. The conventional assay for protein carbonyls is a colorimetric procedure that measures binding of dinitrophenylhydrazine (DNP) *(2,3)*. Protein-bound DNP can also be measured with increased sensitivity by enzyme-linked immunosorbent assay (ELISA) *(4,5)*. Samples containing protein are reacted with DNP, then the protein is nonspecifically adsorbed to an ELISA plate. Unconjugated DNP and nonprotein constituents are easily washed away and give minimal interference. The adsorbed protein is probed with a commercial biotinylated anti-DNP antibody followed by streptavidin-linked horseradish peroxidase. Absorbances are related to a standard curve prepared of serum albumin containing increasing proportions of hypochlorous acid (HOCl)- oxidized protein that is calibrated colorimetrically. This method requires only microgram quantities of protein and avoids the high and sometimes variable blanks owing to unbound DNP that are limitations for the colorimetric method *(6,7)*. We have found it to be highly sensitive and discriminatory in analyzing plasma and lung aspirates from both critically ill adult patients and premature infants *(8–10)*, and others have used it for cell extracts *(11)*. Results correlate well with the colorimetric assay. Absolute values can differ between the two assays and each method is best suited for comparing samples using a standard procedure. The assay is available as a kit (Zenith Technology, Dunedin, New Zealand) or can be performed as described below.

From: *Methods in Molecular Biology, vol. 186: Oxidative Stress Biomarkers and Antioxidant Protocols*
Edited by: D. Armstrong © Humana Press Inc., Totowa, NJ

2. Materials

2.1. Equipment

1. Microplate reader (SpectraMax 190, Molecular Devices, Sunnydale, CA).
2. Microplate washer (AM60, Dynatech Ltd., Hong Kong).
3. Costar High Binding ELISA plates (Corning Inc., Corning, NY Cat. # 3590).
4. Micropipets.
5. Vortex/mixer.
6. 37°C incubator.

2.2. Reagants and Buffers

1. 2,4-Dinitrophenylhydrazine (DNP; Riedel-de-Haen, Seelze-Hannover, Germany). 10 mM DNP in 6 M guanidine hydrochloride, 0.5 M potassium phosphate buffer, pH 2.5 *(3)*.
2. Phosphate-buffered saline (PBS): 10 mM sodium phosphate buffer, pH 7.4, in 0.14 M sodium chloride.
3. Tween 20 (Sigma Chemical Co., St Louis, MO, P1379). Dilution buffer: PBS containing, 0.1% Tween 20 solution (PBST).
4. Biotin-conjugated rabbit IgG polyclonal antibody raised against a DNP conjugate of keyhole limpet hemocyanin (anti-DNP) (Molecular Probes Inc. Eugene, OR); 1:1000 in dilution buffer.
5. Streptavidin-biotinylated horseradish peroxidase (HRP) (Amersham International, Buckinghamshire, UK), 1:3000 in dilution buffer; or streptavidin-HRP (DAKO Corporation, Carpinteria, CA), 1:5000 in dilution buffer.
6. DAKO TMB (3,3′,5,5′-tetramethylbenzidine) one-step substrate system (DAKO Corporation, Cat. # 51600).
7. Protein-assay dye reagent concentrate (BioRad laboratories, Richmond, CA, Cat. # 500-0006).
8. Other reagents for standard preparation: bovine serum albumin (BSA, Sigma A9647); sodium borohydride (BDH, Poole, UK, #30114); hypochlorous acid (HOCL, household bleach).

3. Methods

3.1. Preparation and Calibration of Albumin Standards

3.1.1. Prepare Reduced Protein

Serum albumin as purchased already contains carbonyls. Therefore fully reduced bovine (BSA) or human serum albumin (HSA) needs to be prepared for use in the standard curve.

1. A 0.5 g/100 mL albumin solution in PBS is reacted with 0.1 g solid sodium borohydride for 30 min. Since this reaction produces hydrogen (notable through intense foaming of the protein), it should be carried out in a fume hood. Sodium borohydride breaks down over time, so the amount used may need to be increased after prolonged storage.

2. The solution is then neutralized with HCl by adding a 2 M solution slowly until pH 7.0 is reached. After overnight dialysis against PBS, the protein concentration is checked and adjusted to 4 mg/mL.

3.1.2. Prepare Oxidized Protein

Oxidized BSA containing additional carbonyls is prepared for use in the standard curve by reacting with HOCl.

1. The HOCl concentration in a stock solution of household bleach is determined spectrophotometrically by diluting an aliquot in 0.01 M NaOH and measuring the absorbance at 290 nm ($\varepsilon = 350$ $M^{-1}cm^{-1}$)
2. BSA (50 mg/mL in PBS) is reacted with 5 mM HOCl for 24 h at 37°C. The reaction is complete after this time and the BSA typically contains about 9 nmol carbonyls per mg protein (*see* **Note 1**).
3. The protein concentration is adjusted to 40 mg/mL for calibration by the colorimetric assay and 4 mg/mL for use in the ELISA.

3.1.3. Colorimetric Assay for Calibration of Standards

The carbonyl content of the oxidized and reduced BSA is determined using a modification of the standard colorimetric method *(2)*.

1. 10 mg protein in 250 µL PBS is reacted with 1 mL 10 mM DNP in 2 M HCl for 45 min, precipitated with 1 mL 28% TCA and washed three times with 2.5 mL ethanol/ethyl acetate (1 : 1). A blank for each sample consisting of protein reacted with 2 M HCl containing no DNP is carried through the procedure.
2. Pellets should be broken up mechanically and by sonicating during the washing steps, then dissolved at 37°C in 1 mL of 6 M guanidine hydrochloride, 20 mM potassium phosphate, pH 2.5, and the absorbance at 375 nm measured.
3. The protein concentration of the final extract is determined using the Bio-Rad protein assay dye reagent, with dilutions of 1 mg/mL BSA solution used to produce a standard curve. The results can then be adjusted for the protein loss (about 10%) that occurs with this method *(6)*. The absorbance of the blank is subtracted.
4. Carbonyl content is determined as nmol/mg protein using ε_{375} 22000 M^{-1} cm^{-1} *(2)* after subtracting the value for reduced albumin.
5. The fully reduced BSA consistently gives an A_{375} of about 0.13 per 10 mg, (equivalent of 0.6 nmoles carbonyls/mg), which is unaffected by further treatment with sodium borohydride. It is assumed to be nonspecific and not due to carbonyls.

3.2. Preparation of the Standard Curve

Standard curves are constructed by mixing varying proportions of HOCl-oxidized BSA with fully reduced BSA at a constant protein concentration. A range of 0–2 nmoles carbonyl/mg protein is appropriate for most purposes. To prepare a standard curve in this range using oxidized BSA with 9 nmoles carbonyl/mg, prepare 6 solutions containing 0–22% (v/v) of 4 mg/ml oxidized BSA and 100–78% (v/v) of 4 mg/ml reduced BSA. For oxidized BSA containing more or less carbonyls, or if a more extended range is required, adjust the proportions accordingly.

3.3. Sample Preparation and Protein Derivatization

Samples can be used fresh or stored at –80°C. First a total protein assay is performed using the BioRad assay on all standards and samples, which are then diluted in PBS to a protein concentration of 4 mg/mL.

3.3.1. Plasma Samples

Centrifuge plasma after thawing, then use 5 µL to dilute with PBS to 4 mg/mL. Pipet 15 µL of this dilution into a 1.5-mL test tube.

3.3.2. Lung Aspirate Samples

1. Because lung aspirates generally have low protein concentrations, they need to be concentrated by precipitation with trichloracetic acid (TCA). A sample volume containing 60 µg protein is mixed with 0.8 volumes of 28% TCA, left on ice for 10 min and centrifuged at 10,000g for 5 min.
2. The supernatant is discarded and 15 µL PBS added to the protein precipitate, which dissolves completely after adding the DNP-reagent for derivatization.

3.3.3. Blank and Standards

As a reagent blank use 15 µL PBS only; for standards, mix reduced and oxidized BSA as in **Subheading 3.2.**

3.3.4. Protein Derivatization

Protein derivatization is carried out with 45 µL of DNP solution (as in **Subheading 2.2., step 1**) added to 15 µL of sample, standard or blank to give a final protein concentration of 1 mg/mL. Vortex all tubes and leave 45 min at room temperature.

3.4. ELISA Procedure

3.4.1. Coating of ELISA Plate

Dilute 5 µL of each derivatized protein-solution into 1 mL PBS. Mix well, then add 200 µL aliquots (containing 1 µg protein) in triplicate to ELISA plate wells and incubate the plate overnight at 4°C.

3.4.2. Incubation and Color Development

Wash plate 5 times with PBS before each of the following steps:

1. To reduce unspecific binding, add 250 µL/well of PBST (PBS + 0.1% Tween 20) and leave 30 min at room temperature.
2. Add 200 µL/well biotinylated anti-DNP antibody (1:1000 in PBST), incubate for 1 h at 37°C.
3. Add 200 µL/well streptavidin-biotinylated horseradish peroxidase (1:3000 in PBST), incubate for 1 h at room temperature.
4. Add 200 µL/well of DAKO TMB one-step substrate system, allow color to develop for the same time in each well (approx 5 min) then stop the reaction with 100 µL 2.5 *M* sulfuric acid.
5. Read absorbances at 450 nm and subtract the background absorbance for the DNP-reagent blank (typically about 0.08–0.1). Determine the carbonyl concentration (nmol/mg protein) from the standard curve (*see* **Note 2**).

3.5. Results

3.5.1. Validation

As described elsewhere *(4)*, the conditions described for the assay have been chosen for greatest sensitivity in our hands, after testing different ELISA plates, protein concentrations and antibodies. ELISA-standard curves are linear up to at least 10 nmoles carbonyl/mg protein. With the development time described, 0.1 nmol/mg gives a net absorbance of about 0.08 above a blank of about 0.08 absorbance units as measured against air. Reduced BSA usually gives an absorbance about 0.02 units above the reagent blank. The assay shows good reproducibility when tested with plasma samples *(4)* Good correlations ($r^2 = 0.70$, $n = 26$) have been found between the ELISA and colorimetric methods for plasma samples by us *(4)* and others *(13)*.

3.5.2. Clinical Samples

Plasma from normal adults contains protein carbonyls of ≤0.1 nmoles/mg protein by the ELISA method. Higher levels (up to 4 nmoles/mg) have been measured in plasma from critically ill patients *(8)* and lung aspirates from adults *(9)* and premature infants *(10)*. The TCA-precipitation step before DNP-derivatization that is generally necessary to concentrate the protein from aspirates causes a small (20%) increase in response. The assay has been used for tissue samples *(11)*, although these may require removal of RNA before analysis *(12)*.

4. Notes

1. The ELISA was calibrated with BSA treated with hypochlorous acid to generate stable DNP-reactive carbonyls. Carbonyls can also be generated with iron and ascorbate, but the response is not identical. Thus, carbonyl values depend to some extent on the oxidized protein standard used.

2. Absolute values are also affected by whether the blank is taken as the reagents without protein, or fully reduced BSA. The absorbances of these solutions are normally within 0.02 absorbance units, but this can make a difference for samples with low carbonyl content. The main strength of the ELISA assay, therefore, is for comparing samples in a standard system.

Acknowledgments

This work was supported by the Health Research Council of New Zealand.

References

1. Stadtman, E. R.(1990) Metal ion catalysed oxidation of proteins: biochemical mechanism and biological consequences. *Free Rad. Biol. Med.* **9,** 315–325.
2. Levine, R. L., Garland, D., Oliver, C. N., Amici, A., Climent, I., Lenz, A.-G., et al. (1990) Determination of carbonyl content of oxidatively modified proteins. *Methods Enzymol.* **186,** 464–478.
3. Levine, R. L., Williams, J. A., Stadtman, E. R., and Shacter, E. (1994) Carbonyl assays for determination of oxidatively modified proteins. *Methods Enzymol.* **233,** 346–357.
4. Buss, H., Chan, T. P., Sluis, K. B., Domigan, N. M., and Winterbourn, C. C. (1997) Protein carbonyl measurement by a sensitive ELISA method. *Free Rad. Biol. Med.* **23,** 361–366.
5. Winterbourn, C. C. and Buss, H. (1999) Protein carbonyl measurement by enzyme-linked immunosorbent assay. *Methods Enzymol.* **300,** 106–111.
6. Reznick, A. Z. and Packer, L. (1994) Oxidative damage to proteins spectrophotometric method for carbonyl assay. *Methods Enzymol.* **233,** 357–363.
7. Lyras, L., Evans, P. J., Shaw, P. J., Ince, P. G., and Halliwell, B. (1996) Oxidative damage and motor neurone disease difficulties in the measurement of protein carbonyls in human brain tissue. *Free Rad. Res.* **24,** 397–406.
8. Winterbourn, C. C., Buss, I. H., Chan, T. P., Plank, L. D., Clark, M. A., and Windsor, J. A. (2000) Protein carbonyl measurements show evidence of early oxidative stress in critically ill patients. *Crit. Care Med.* **28,** 143–149.
9. Pantke, U., Volk, T., Schmutzler, M., Kox, W. J., Sitte, N., and Grune, T. (1999) Oxidized proteins as a marker of oxidative stress during coronary heart surgery. *Free Rad. Biol. Med.* **27,** 1080–1086.
10. Buss, I. H., Darlow, B. A., and Winterbourn, C. C. (2000) Elevated protein carbonyls, lipid peroxidation products and myeloperoxidase in tracheal aspirates from premature infants. *Pediatr. Res.* **47,** 640–645.
11. Sitte, N., Merker, K., and Grune, T. (1998) Proteasome-dependent degradation of oxidized proteins in MRC-5 fibroblasts. *FEBS Lett.* **440,** 399–402.
12. Cao, G. and Cutler, R. G. (1995) Protein oxidation and aging. I. Difficulties in measuring reactive protein carbonyls in tissues using 2,4-dinitrophenylhydrazine. *Arch. Biochem. Biophys.* **320,** 106–114.
13. Marangon, K., Devaraj, S., and Jialal, I. (1999) Measurement of protein carbonyls in plasma of smokers and in oxidized LDL by an ELISA. *Clin. Chem.* **45,** 577–578.

15

Nᵉ-(carboxymethyl)lysine (CML) as a Biomarker of Oxidative Stress in Long-Lived Tissue Proteins

J. Nikki Shaw, John W. Baynes, and Suzanne R. Thorpe

1. Introduction

Glycoxidation products are a subset of advanced glycation endproducts (AGEs) that are formed by the nonenzymatic glycation and subsequent irreversible oxidation of proteins *(1,2)*. N^{ϵ}-(carboxymethyl)lysine (CML) is a well-characterized glycoxidation product that accumulates in tissues with age, and its rate of accumulation is accelerated in diabetes *(1–4)*. CML is now known to be formed from the oxidation of both carbohydrates and lipids, making it a biomarker of general oxidative stress *(5–7)*. Oxidative stress and resultant protein modification have been implicated in the pathogenesis of the chronic complications of diabetes, including nephropathy and atherosclerosis *(1,4,5,8)*. The accumulation of CML in long-lived tissue such as skin collagen reflects oxidative stress over an extended period of the life-span, and has been shown to be greater in patients with diabetic complications than those without complications *(8,9)*.

In this chapter, we describe in detail the procedure for the analysis of CML in skin collagen, but the technique can readily be adapted to other tissues *(3,10,11)*. Our gas-chromatography-mass spectrometry (GC-MS) method allows for sensitive and specific measurement of CML as the N-trifluoroacetyl methyl ester (TFAME) derivative (*see* **Fig. 1**). An alternative method for measurement by HPLC is described in **ref. 12**.

2. Materials
2.1. Equipment

1. Glass 13 × 100 mm screw-cap test tubes with teflon-lined caps.

From: *Methods in Molecular Biology, vol. 186: Oxidative Stress Biomarkers and Antioxidant Protocols*
Edited by: D. Armstrong © Humana Press Inc., Totowa, NJ

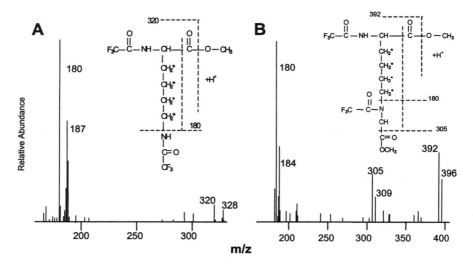

Fig. 1. Full-scan spectra of TFAME derivatives of lysine (**A**) and CML (**B**). The injected standard contained 1 μmol each of lysine and d_8-lysine, and 2 nmol each of CML and d_4-CML. Insets show fragmentation pattern. Asterisk (*) indicates heavy-labeled atoms in internal standards.

2. Water-bath sonicator (Branson Equipment, Shelton, CT).
3. Refrigerated centrifuge with rotor holding 13 × 100 mm glass screw-cap tubes.
4. Heating blocks with inserts for 13 × 100 mm glass screw-cap tubes (Barnstead/Thermolyne Corp., Dubuque, IA).
5. Speed-Vac centrifugal evaporator system capable of drying multiple 13 × 100 mm glass screw-cap tubes *in vacuo* (Savant Instruments, Farmingdale, NY).
6. N-Evap nitrogen-drying system for evaporation of organic solvent from multiple samples (Organomation, South Berlin, MA).
7. Microcentrifuge with rotor holding 1.5-mL conical plastic centrifuge tubes (Heraeus Instruments, Germany).
8. Conical glass inserts for autosampler vials, 300 μL volume (National Scientific, Lawrenceville, GA).
9. Analytical GC-MSD system (Hewlett-Packard, Palo Alto, CA).
   a. HP model 6890 gas chromatograph equipped with an autosampler and model 5872A mass selective detector.
   b. HP-Chrom data acquisition software.
   c. HP 5MS, 30 m, 0.25 mm i.d., 0.25 micron GC column.

2.2. Reagents

1. DL–Alanine (*see* **Note 1**).
2. DL–Valine.
3. DL– Leucine.

Table 1
Typical Amounts of Internal and External
Standards Used to Construct Standard Curves

CML1 (nmol)	d_4-CML (nmol)	Lysine (nmol)	d_8-Lysine (nmol)
0.0	0.6	0	60
0.2	0.6	200	60
0.4	0.6	400	60
0.6	0.6	600	60
0.8	0.6	900	60

4. DL–Serine.
5. Diethylenetetraamine-pentaacetic acid (DTPA).
6. Sodium borohydride ($NaBH_4$).
7. Trifluoroacetic anhydride (TFAA).
8. Deuterated lysine: DL–lysine–3,3,4,4,5,5,6,6-d_8•2HCl.

2.3. Standards and Calibrators

1. Standards and deuterated standards. Lysine standards are weighed out, and concentrations confirmed by amino acid analysis. Deuterated lysine (d_8-lysine) was used as an internal standard for lysine analysis. Unlabeled and heavy-labeled CML were synthesized by reaction of glyoxylic acid and N^α-acetyllysine and N^α-formyl-d_4-lysine, respectively, in the presence of $NaBH_3CN$ as described previously (13,14). Standard curves for each analysis are prepared from mixtures of increasing amounts of unlabeled lysine and CML and constant amounts of d_8-lysine and d_4-CML added as internal standards to each unlabeled standard (see **Table 1**). The same amounts of heavy-labeled standards are added to each collagen sample.
2. Standard reference materials. We use quality control samples consisting of two pooled human samples representing young (18 yr) and old (85 yr) individuals. These skin collagen samples were processed as described in **Subheading 3.1.**, then pulverized to a fine powder in order to ensure homogeneity. Duplicates of each pool are analyzed along with each batch of samples for quality control.

3. Methods
3.1. Processing of Skin Samples

1. Any visible hair, vascular or adipose tissue is scraped off the sample with a single-edge razor blade. The sample is then extracted in 50 volumes (v/w) of 1 M NaCl for 24 h at 4°C, replacing the solution at 8 h.

2. The extraction step is repeated with 0.05 *M* acetic acid for 24 h, followed by chloroform-methanol (2:1) for 24 h.
3. The residual insoluble collagen samples are rinsed in deionized water, lyophilized, and stored at –70°C until analyzed (*see* **Note 2**).

3.2. Sample Reduction and Hydrolysis

1. Weighed collagen samples (1–1.5 mg dry weight) are placed in 13 × 100 mm screw-cap tubes. Rehydrate samples in 500 μL deionized water for a 0.5 h at 4°C prior to sample reduction. The quality control pools are treated in the same manner.
2. To each sample, add 500 μL of borate buffer (0.2 *M* boric acid containing 2 m*M* DTPA, pH adjusted to 9.1 with NaOH). Reduce by addition of 50 μL freshly made 2 *M* NaBH$_4$ in 0.1 *N* NaOH for 16 h at 4°C. Alternatively, reduction may be carried out for 4 h at room temperature.
3. Following reduction, carefully remove the borate solution with a Pasteur pipet and wash samples twice with deionized water, removing the wash each time with a Pasteur pipet. In the case of pulverized samples, it is necessary to centrifuge the tubes to form a pellet (5 min at 2500*g* in a refrigerated centrifuge) in order to remove the reduction solution and water washes.
4. Prepare unlabeled standards for standard curve, using amounts shown in **Table 1**. To each tube add 1 mg of a mixture of amino acids containing equal proportions of alanine, valine, leucine, and serine as carrier. Add heavy-labeled internal standards to each of the samples and unlabeled standards.
5. Add 2 mL 6N HCl to each tube. Blow N$_2$ into each tube for 15 s, then cap quickly.
6. Hydrolyze samples in a heating block for 24 h at 110°C.
7. Dry hydrolysates *in vacuo* (Speed-Vac).

3.3. Preparation of TFAME derivatives for GC-MS Analysis

1. Add 1 mL freshly made 1 *M* HCl in methanol to each tube, cap, and heat at 65°C in a heating block for 30 min (methanolic HCl: to 28 mL methanol, add 2 mL acetyl chloride dropwise. Caution: this is a very exothermic reaction).
2. Dry samples completely under nitrogen (N-Evap).
3. Add 1 mL TFAA to each tube, working under a fume hood. Cap tightly, let stand at room temperature for 1 h, then dry under nitrogen (N-Evap).
4. Add ~100 μL of methylene chloride to each sample, and sonicate for a few seconds (any salts in the tube will not dissolve). Transfer liquid to 1.5-mL plastic, conical microcentrifuge tubes.
5. Centrifuge at 9,000*g* for 8 min. Transfer supernatant to GC autosampler vials equipped with 300 μL inserts and cap tightly.

3.4. GC/MS Analysis

1. GC parameters: Carrier gas: He, at 0.8 mL/min. Injector temp: 275°C. Injection volume: 1 μL.

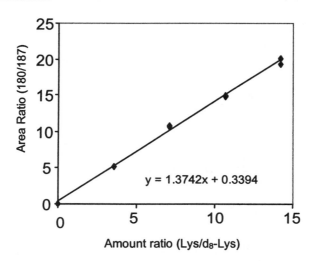

Fig. 2. Typical standard curve for calculation of lysine in collagen samples. After computing the area ratio of unlabeled lysine to d_8-lysine in the collagen sample, the amount of unlabeled lysine is computed using the equation for the line relating area ratio and amount ratio. The amount ratio is calculated from **Table 1**.

2. Temperature program: Initial temp = 130°C, hold for 3 min; Ramp 1 = 4°C/min for 12.5 min to 180°; Ramp 2 = 5°C/min for 12 min to 240°; Ramp 3 = 15°C/min to 290° (final temp); hold final temp 5 min.
3. MSD parameters (*see* **Note 3**). Operate in selected ion monitoring (SIM) mode. Ions: Lysine = 180 m/z; d_8-Lysine = 187; CML = 392; d_4-CML = 396.
4. Calculations. Standard curves are calculated for each of the unlabeled standards to give the relative response of the unlabeled vs deuterated forms. For example, the 187 ion for the deuterated lysine is not formed in a 1:1 ratio with the 180 ion for the unlabeled lysine (*see* **Fig. 1**). The standard curve also compensates for variations in sample dilution or derivatization efficiency (*6,13*). The absolute amounts of CML and lysine in the samples are computed by normalization to their respective internal standards, using the relative response factors determined from the standard curves (*see* **Fig. 2**). The concentration of CML is normalized to the lysine content of the sample to indicate the extent of lysine modification in the protein.

3.5. Results

3.5.1. Chromatography

Mass spectra of the lysine and CML standards are shown in **Fig. 1**. The dominant ions for lysine and d_8-lysine are 180 m/z and 187 m/z, respectively (*see* **Note 3**). The dominant ions for CML and d_4-CML are 392 m/z and 396 m/z, respectively. Extracted SIM chromatograms for CML and lysine in skin collagen

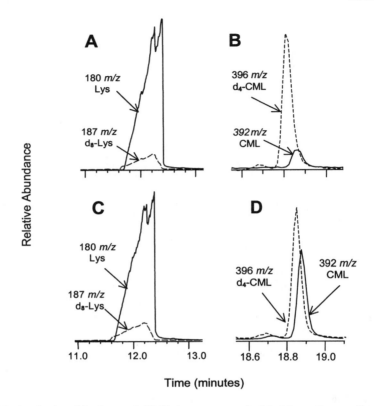

Fig. 3. Analysis of lysine and CML in young and old skin collagen. Samples are from 18-yr-old **(A,B)** and 85-yr-old **(C,D)** individuals, and were analyzed by SIM for lysine and CML content.

are shown in **Fig. 3**. The graphs emphasize that although the lysine content is comparable (compare A and C), the CML content is significantly lower in the 18 compared to the 85 year old skin collagen (compare B and D). The interassay coefficient of variation (CV) for CML is ~5% in our laboratory *(15)*.

3.5.2. Concentrations of CML in Human Tissues

Table 2 summarizes the range of CML concentrations found in human tissue. Skin collagen and lens crystallins are very long-lived proteins. CML increases linearly with age, from negligible concentration in infancy to the levels shown at 80 yr. In contrast, CML concentrations in shorter-lived proteins such as serum albumin and LDL do not show this age-dependent increase *(13)*. However, CML concentrations are elevated even in these shorter-lived proteins in individuals with uremia *(17)*.

Table 2
Typical CML Concentrations Found in Human Tissues

Tissue	CML Concentration (mmol CML/mol Lys)	References
Skin collagen		
Neonate	Not detected	*(3,15)*
18 yr	0.3–0.5	*(15,16)*
80 yr	1.5–1.7	*(3,15,16)*
Lens Crystallins		
Neonate	Not detected	*(3)*
80 yr	7.0	*(3)*
Serum	0.1	*(17)*
LDL	0.02–0.03	*(7,8)*
Erythrocyte membrane	0.2	*(6)*

4. Notes

1. All reagents were obtained from Sigma Chemical (St. Louis, MO), except for the deuterated lysine (Isotec, Inc., Miamisburg, OH).
2. Skin samples should be stored at −70°C until ready to scrape and extract. If frozen, allow to thaw prior to scraping. Once lyophilized, collagen can be stored indefinitely at −70°C.
3. The Hewlett-Packard 5872A mass selective detector allows for changes in detector sensitivity during the course of a run (i.e., a change in the ion-source electron multiplier voltage). Reducing the sensitivity during the period that the dominant lysine and d_8-lysine ions (180 and 187) elute prevents overloading the detector. Subsequently increasing the electron multiplier voltage while the less abundant CML ions (392 and 396) elute enhances sensitivity for CML. Alternatively, monitoring the less abundant lysine ions, 320 and 328, can also prevent overloading the detector while still having sensitivity settings high enough to measure CML.

References

1. Dyer, D. G., Blackledge, J. A., Katz, B. M., Hull, C. J., Adkisson, H. D., Thorpe, S. R., et al. (1991) The Maillard reaction in vivo. *Zeitschrift fur Ernahrungswissenschaft* **30,** 29–45.
2. Reddy, S., Bichler, J., Wells-Knecht, K. J., Thorpe, S. R., and Baynes, J. W. (1995) N^ε-(carboxymethyl)lysine is a dominant advanced glycation end product (AGE) antigen in tissue proteins. *Biochemistry* **34,** 10,872–10,878.

3. Dunn, J. A., Dyer, D. G., Knecht, K. J., Thorpe, S. R., McCance, D. R., Bailie, K., et al. (1990) Accumulation of Maillard reaction products in tissue proteins, in *The Maillard Reaction, Advances in Life Sciences.* Birkhauser Verlag, Basel, pp. 425–430.

4. Thorpe, S. R. and Baynes, J. W. (1996) Role of the Maillard reaction in diabetes mellitus and diseases of aging. *Drugs Aging* **9,** 69–77.

5. Fu, M.-X., Requena, J. R., Jenkins, A. J., Lyons, T. J., Baynes, J. W., and Thorpe, S. R. (1996) The advanced glycation end product, N^{ε}-(carboxymethyl)lysine, is a product of both lipid peroxidation and glycoxidation reactions. *J. Biol. Chem.* **17,** 9982–9986.

6. Requena, J. R., Ahmed, M. U., Fountain, C. W., Degenhardt, T. P., Reddy, S., Perez, C., et al. (1997) Carboxymethyl ethanolamine, a biomarker of phospholipid modification during the Maillard reaction in vivo. *J. Biol. Chem.* **28,** 17,473–17,479.

7. Requena, J. R., Fu, M.-X., Ahmed, M. U., Jenkins, A. J., Lyons, T. J., and Thorpe, S. R. (1996) Lipoxidation products as biomarkers of oxidative damage to proteins during peroxidation reactions. *Nephrol. Dialysis Transpl.* **11,** 48–53.

8. Lyons, T. J., Li, W., Wells-Knecht, M. C., and Jokl, R. (1994) Toxicity of mildly modified low-density lipoproteins to cultured retinal capillary endothelial cells and pericytes. *Diabetes* **43,** 1090–1095.

9. McCance, D. R., Dyer, D. G., Dunn, J. A., Bailie, K. E., and Thorpe, S. R. (1993) Maillard reaction products and their relation to complications in insulin-dependent diabetes mellitus. *J. Clin. Invest.* **91,** 2470–2478.

10. Requena, J. R., Fu, M. X., Ahmed, M. U., Jenkins, A. J., Lyons, T. J., Baynes, J. W., and Thorpe, S. R. (1997) Quantification of malondialdehyde and 4-hydroxynonenal adducts to lysine residues in native and oxidized human low-density lipoprotein. *Biochem. J.* **322,** 317–325.

11. Dunn, J. A., Patrick, J. S., Thrope, S. R., and Baynes, J. W. (1989) Oxidation of glycated proteins: age-dependent accumulation of N^{ε}-(carboxymethyl)lysine in lens protein. *Biochemistry* **28,** 9464–9468.

12. Monnier, V. M., Bautista, O., Kenny, D., Sell, D. R., Fogarty, J., Dahms, W., et al. (1999) Skin collagen glycation, glycoxidation, and crosslinking are lower in subjects with long-term intensive versus conventional therapy of type I diabetes. *Diabetes* **48,** 870–880.

13. Knecht, K. J., Dunn, J. A., McFarland, K. F., McCance, D. R., Lyons, T. J., Thorpe, S. R., and Baynes J. W. (1991) Effect of diabetes and aging on (carboxymethyl)lysine levels in human urine. *Diabetes* **40,** 190–196.

14. Ahmed, M. U., Thorpe, S. R., and Baynes, J. W. (1986) Identification of N^{ε}-(carboxymethyl)-lysine as a degradation product of fructoselysine in glycated protein. *J. Biol. Chem.* **261,** 4889–4894.

15. Dunn, J. A., McCance, D. R., Thorpe, S. R., Lyons, T. J., and Baynes, J. W. (1991) Age-dependent accumulation of N^{ε}-(carboxymethyl)lysine and N^{ε}-(carboxymethyl) hydroxylysine in human skin collagen. *Biochemistry* **30,** 1205–1210.

16. Dyer, D. G., Dunn, J. A, Thorpe, S. R., Bailie, K. E., Lyons, T. J., McCance, D. R., and Baynes, J. W. (1993) Accumulation of Maillard reaction products in skin collagen in diabetes and aging. *J. Clin. Invest.* **91,** 2463–2469.
17. Degenhardt, T. P., Grass, L., Reddy, S., Thorpe, S. R., Diamandis, E. P., and Baynes, J. W. (1997) The serum concentration of the advanced glycation end-product, N^ε-(carboxymethyl)-lysine is increased in uremia. *Kid. Intl.* **52,** 1064–1067.

16

Measurement of S-Glutathionated Hemoglobin in Human Erythrocytes by Isoelectric Focusing Electrophoresis

Haw-Wen Chen and Chong-Kuei Lii

1. Introduction

Protein thiol groups are susceptible to oxidation by reactive oxygen species *(1)*. The occurrence of a specific thiol modification named protein S-glutathionation has been proposed to protect proteins from oxidative damage owing to its reversible characteristics *(2–4)*. In addition, S-glutathionation modulates protein functions including enzyme activity *(4,5)*, cytoskeletal protein strength *(6)*, hemoglobin-gelling activity *(7)*, and chaperon activity *(8)*. Because an increase in cellular glutathione (GSSG) level is correlated with the formation of S-glutathionated protein, a mixed-disulfide exchange reaction has been proposed to be responsible for this protein modification *(2,5,9,10)*. However, glutathione (GSH) modified proteins are also formed under a circumstance without significant change in GSSG level *(11,12)*, and a radical-initiated process has therefore also been proposed *(3)*. S-Gluthionated hemoglobin formation owing to oxidative stress has been noted in the red blood cells from different species *(13–15)*. There are a total of six cysteine residues in the human hemoglobin molecule, but two of them (β93 residues) are highly reactive *(16)* and can form mixed disulfide with GSH *(14)*. However, the location of reactive cysteines in hemoglobin varies in different animal species. For example, the cysteines in chicken α-hemoglobin are more active than those in β hemoglobin *(15)*.

Owing to the presence of one net negative charge on GSH (an NH_3^+ and a COO^- group from glutamate residue and a COO^- group from the glycine residue of GSH), the addition of GSH to a protein usually changes the protein's

From: *Methods in Molecular Biology, vol. 186: Oxidative Stress Biomarkers and Antioxidant Protocols*
Edited by: D. Armstrong © Humana Press Inc., Totowa, NJ

139

net charge at physiological conditions by -1. Therefore, the isoelectric point of a protein is changed once it reacts with GSH. The reaction of additional GSH molecules with protein would generate further changes in charge *(17)*. Isoelectric focusing (IEF) can separate an isoelectric point-changed protein from its native form. Recently, we have applied this method to detect S-glutathionated hemoglobin formation in human red blood cells under oxidative stress *(13)*. Moreover, when IEF analysis is combined with immunoblotting, it provides superior specificity and simplicity in the detection of protein S-glutathionation compared to sodium borohydride reduction and radioisotope methods *(18)*.

2. Materials
2.1. Equipment

1. Water bath.
2. Centrifuge.
3. Vortex.
4. Orbital shaker.
5. IEF instrumentation.
   a. Amersham Pharmacia Biotech power supply (MultiDrive™ XL) (Piscataway, NJ).
   b. Amersham Pharmacia Biotech Electrophoresis system (Multiphor II).
   c. Thermostatic circulator.
   d. Electrode wicks ($300 \times 6 \times 1$ mm), Serva Biochemical Inc. (Westbury, NY).
   e. Applicator strip ($100 \times 6 \times 1$ mm) (Serva Biochemical).
6. Gel-casting tool.
   a. Thick glass plate ($125 \times 260 \times 3$ mm), Amersham Pharmacia Biotech.
   b. Thin glass plate ($125 \times 260 \times 1$ mm), Amersham Pharmacia Biotech.
   c. Spacer (0.75 mm)
   d. Gel support film (265×125 mm, Serva NetFixR® for polyacrylamide gel).

2.2. Reagents

1. Acrylamide, N,N'-methylenebisacrylamide, TEMED, and Coomassie blue G250 (BioRad Laboratories, Hercules, CA).
2. Diamide, GSSG, iodoacetic acid (IAA), iodoacetamine, tert-butyl hydroperoxide (TBH), and glucose (Sigma Co., St. Louis, MO).
3. Cathode, pH 10.0, and anode, pH 3.0, electrode fluids (Serva Biochemical).
4. Preblended pH 5.0–8.0 ampholyte (Amersham Pharmacia Biotech).

3. Methods
3.1. Blood Collection and Erythrocyte Preparation

1. Draw blood samples into Vacutainer tubes containing ethylenediaminetetraacetic acid (EDTA) as anticoagulant.
2. Prepare red blood cells by centrifugation at 500g for 10 min.

3. Decant plasma and then wash red blood cells three times with phosphate-buffered saline (PBS, containing 8.9 mM Na$_2$HPO$_4$, 1.1 mM NaH$_2$PO$_4$, and 140 mM NaCl, pH 7.4).
4. Resuspend the washed red blood cells at a hematocrit of 10% with PBS containing 10 mM glucose or prepared in 10% hemolysate with ddH$_2$O.

3.2. Erythrocyte Treatment

1. Treat erythrocytes in a hematocrit of 10% with 2 mM diamide (a thiol oxidizing agent) and 50 mM IAA (a thiol alkylating agent carrying a negative charge) at 37°C for 30 min. In cells treated with TBH, erythrocytes are incubated at 37°C with 1.5 mM TBH and samples are taken before TBH treatment or 5, 15, and 30 min after TBH treatment. For GSSG treatment, 10% erythrocyte hemolysate is incubated with 20 or 50 mM GSSG at 37°C for 30 min.
2. To stop agent-induced hemoglobin modification, erythrocytes are centrifuged and washed twice with cold PBS. Resuspend erythrocytes in PBS at a hematocrit of 10%.
3. Incubate the samples with 50 mM iodoacetamide (carrying no charge) at 37°C for 30 min to block protein sulfhydryl groups and to prevent the generation of artifacts during sample preparation (*see* **Note 1**). Samples for complete reduction with dithiothreitol contain no iodoacetamide. The difference in the bands focused on IEF gel before and after dithiothreitol reduction represents the extent of mixed disulfide formation between GSH and hemoglobin.

3.3. Casting IEF Gels

1. IEF gels are made in the laboratory. Lay the thick glass plate (125 × 260 × 3 mm) on a lab bench and fix spacers on both long ends of the glass plate. Put drops of water on the glass plate and then put gel supporting film. Roll the gel-supporting film with a glass rod to remove excess water.
2. Prepare IEF gel acrylamide solution (4% T–2.6% C) by mixing 2.53 mL 30% acrylamide/0.8% Bis solution, 10.76 mL H$_2$O, 0.95 mL preblended pH 5.0–8.0 ampholyte (final ampholyte is 2%), and 19 μL TEMED. De-gas for 15 min. Add 1.0 mL 0.8% ammonium persulfate and gently mix. This amount of solution (19 mL) is sufficient for making a 125 × 210 mm acrylamide gel.
3. Pour the IEF gel acrylamide solution on the center of the gel-supporting film. Put the thin glass plate (125 × 260 × 1 mm) between the spacers to allow the acrylamide to polymerize between the gel-supporting film and the glass plate. Cover the glass plate carefully to prevent trapping air bubbles in the gel.
4. Let the gels stand at room temperature for 90 min. Separate the thick glass plate from the gel-supporting film and remove the thin glass plate from the gel. The polymerized gel on the gel-supporting film is now ready for IEF use (*see* **Note 2**). If not used immediately, wrap the supporting film, gel, and thin glass plate together in a plastic film to prevent drying of the gel. The gel can be stored in this manner for up to 1 mo at 4°C.

3.4. IEF Analysis

1. Set the thermostatic circulator at 6°C for 1 h before the experiment.
2. Pour a few milliliters of water on the center of the cooling plate in the Multiphor II electrophoresis system, then apply the IEF gel to the cooling plate. A uniform thin layer of water is important to ensure efficient heat transfer from the gel during electrofocusing.
3. Wet the electrode wicks with electrode fluids and apply them at a distance of 0.5 cm from the edge of the gel at the anode and cathode. Put the sample strip on the site 1 cm from the electrode wick at the anode.
4. Set the power supply at a voltage limit of 1500 V and a current and power limit of 2.75 mA/cm and 1.25 W/cm gel width, respectively. Prefocus the gel for 10 min.
5. Load 2 µL of each sample onto the sample strip and focus the gel for 60 min using the same power as in prefocusing. Due to the red color of hemoglobin, the migration of hemoglobin to the cathode is visible during electrofocusing (*see* **Note 3**).
6. At the end of electrofocusing, turn off the power supply and remove the electrode wicks and sample strip. Carefully remove the gel and immerse the gel in 10% trichloroacetic acid fixing solution for 5 min, then wash the gel with ddH$_2$O for 10 min. Place the gel in destaining solution (containing 25% methanol and 12.5% acetic acid in H$_2$O) for 5 min. Place the gel in staining solution for 15 min at room temperature. The staining solution contains 400 mL methanol, 200 mL HCl, 400 mL H$_2$O, and 400 mg Coomassie blue R250 in a total volume of 1 L. Transfer the gel to destaining solution until the background is clear.
7. Leave the gel to dry at room temperature for 24–48 h.

3.5. Results

3.5.1. S-Glutathionation of Hemoglobin in the Presence of Thiol-Modifying Agents

Two thiol-modifying agents, GSSG and iodoacetic acid, are studied in this experiment (**Fig. 1**). As mentioned earlier, the addition of one molecule of GSH to protein changes the protein net charge by –1. This increase in negative charge leads to the focusing of S-glutathionated protein near the anode (bottom side of the gel). Before the treatment, hemoglobin shows one major (band a) and one minor (band b) band on the IEF gel. In the presence of GSSG, two extra bands (as arrows point) appear and are accompanied by the reduction in bands a and b. It is known that the extra band indicated by the upper arrow is the modified form of band b and the other extra band is the modified form of band a *(13)*. The change in either band a or band b is dose-dependent. Approximately 50% of hemoglobin in band a and 90% of hemoglobin in band b are modified in the presence of 50 m*M* GSSG. With dithiothreitol treatment, the modified bands disappear and reverse to their native states, which implies that these modifications are related to protein thiols. Similar changes in hemoglobin

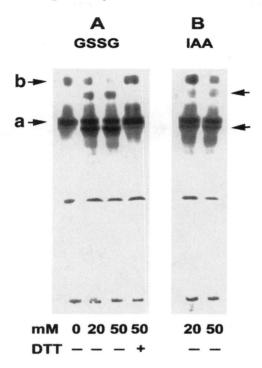

Fig. 1. The detection of hemoglobin S-glutathionation by isoelectric focusing in erythrocytes treated with glutathione disulfide (GSSG) and iodoacetic acid (IAA). **(A)** 10% erythrocyte hemolysate is incubated in the presence of 0, 20, and 50 mM GSSG. **(B)** Erythrocytes at a hematocrit of 10% are exposed to 20 and 50 mM IAA. The incubation time with GSSG or IAA is 30 min at 37°C. Dithiothreitol (DTT, 50 mM) treatment is used to confirm that the modification is due to protein-GSH mixed disulfide formation. Arrows indicate the S-glutathionated hemoglobin. Bands a and b are hemoglobin. The bottom of the gel is the anodic side and the top of the gel is the cathodic side.

are noted on alkylation with iodoacetic acid. Bands a and b migrate identically to the positions as noted in the GSSG-treated samples.

3.5.2. Oxidant-Induced Formation of S-Glutathionated Hemoglobin

tert-Butyl hydroperoxide, an organic hydroperoxide, and diamide, a thiol oxidant, are used to induce oxidative stress (**Fig. 2**). tert-Butyl hydroperoxide significantly induces hemoglobin S-glutathionation as noted by the increase in band intensity at the same positions as noted with GSSG treatment. The production of S-glutathionated hemoglobin is time-dependent on tert-butyl hydroperoxide. In addition to S-glutathionated hemoglobin formation, tert-butyl

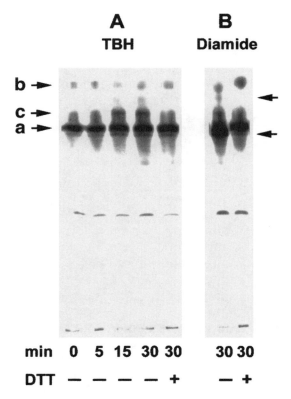

Fig. 2. Isoelectric focusing detects S-glutathionated hemoglobin formation in the oxidant-induced erythrocytes. Erythrocytes at a hematocrit of 10% are incubated with either 1.5 mM tert-butyl hydroperoxide (**A**) at 37°C for different time intervals or with 2.0 mM diamide (**B**) at 37°C for 30 min. Dithiothreitol (DTT, 50 mM) is used as the reducing agent in the designated samples. Arrows indicate the S-glutathionated hemoglobin. Bands a and b are hemoglobin. Band c is methemoglobin.

hydroperoxide also oxidizes hemoglobin to methemoglobin. Methemoglobin is also detected by IEF analysis. However, the increase in positive charge leads to the migration of methemoglobin toward the cathode, focusing at the position as indicated in band c *(13)*. A similar pattern to that observed in GSSG-treated erythrocyte hemolysate is also obtained in the presence of diamide, which causes cellular GSSG production and protein thiol oxidation.

4. Notes

1. To prevent the generation of artificial modification of protein sulfhydryl groups during sample preparation, a thiol-blocking agent is required. To reach this purpose, iodoacetamide is used. Incontrast to iodoacetic acid, iodoacetamide

carries no charge and thus does not change the protein net charge. Iodoacetamide-modified proteins can focus the same position as native proteins.

2. Owing to the presence of air in solution may interfere gel polymerization, especially the acrylamide level is low (4% T), a degas step of IEF polyacrylamide solution before making the gel is necessary.

3. The IEF gel can be cut into any size dependent on the number of samples. To do this, make sure the cutting edge of IEF gel is smooth and the width of the whole gel is uniform. Any defect may cause distortion of protein bands on gel. Sometimes, these defects and gel overdry (when gel is stored too long) may cause gel burning during electrophoresis.

References

1. Stadtman, E. R. (1992) Protein oxidation and aging. *Science* **257,** 1220–1224.
2. Thomas, J. A., Poland, B., and Honzatko, R. (1995) Perspectives in biochemistry and biophysics: protein sulfhydryls and their role in the antioxidant function of protein S-thiolation. *Arch. Biochem. Biophys.* **319,** 1–9.
3. Thomas, J. A., Park, E.-M., Chai, Y.-C., Brooks, R., Rokutan, K., and Johnston, R. B. Jr. (1990) S-thiolation of protein sulfhydryls, in *Biological Reactive Intermediate IV: Molecular and Cellular Effects, and Human Impact* (Synder, R. and Sipes, G., eds.), Plenum Publishing, New York, pp. 95–103.
4. Barrett, W. C., DeGnore, J. P., Konig, S., Fales, H. M., Keng, Y. F., Zhang, Z. Y., and Chock, P. B. (1999) Regulation of PTPIB via glutathionation of the active site cysteine 215. *Biochemistry* **38,** 6699–6705.
5. Zeigler, D. M. (1985) Role of reversible oxidation-reduction of enzymes thiol-disulfides in metabolic regulation. *Ann. Rev. Biochem.* **54,** 305–329.
6. Stournaras, C., Drewes, G., Blackholm, H., Merkler, I., and Faulstich, H. (1990) Glutathionyl(cysteine-374) actin forms filaments of low mechanical stability. *Biochim. Biophys. Acta* **1037,** 86–91.
7. de Llano, J. J. M., Jones, W., Schneider, K., Chait, B. T., Rodgers, G., Benjamin, L. J., et al. (1993) Biochemical and functional properties of recombinant human sickle hemoglobin expressed in yeast. *J. Biol. Chem.* **268,** 27,004–28,011.
8. Cherian, M., Smith, J. B., Jiang, X. Y., and Abraham, E. C. (1997) Influence of protein-glutathione mixed disulfide on the chaperon-like function of alpha-crystallin. *J. Biol. Chem.* **272,** 29,099–29,103.
9. Brigelius, R., Lenzen, R., and Sies, H. (1982) Increase in hepatic mixed disulfide and glutathione disulfide levels elicited by paraquat. *Biochem. Pharmacol.* **31,** 1637–1641.
10. Bellomo, G., Mirabelli, F., DiMonte, D., Richelmi, P., Thor, H., Orrenius, C., and Orrenius, S. (1987) Formation and reduction of glutathione mixed disulfides during oxidative stress. *Biochem. Pharmacol.* **36,** 1313–1320.
11. Rokutan, K., Thomas, J. A., and Johnston, R. B. Jr. (1991) Phagocytosis and stimulation of the respiratory burst by phorbol ester initiated S-thiolation of specific proteins in macrophages. *J. Immunol.* **147,** 260–264.

12. Chai, Y.-C., Hendrich, S., and Thomas, J. A. (1994) Protein S-thiolation in hepatocytes stimulated by t-butyl hydroperoxide, menadione, and neutrophils. *Arch. Biochem. Biophys.* **310,** 264–272.
13. Lii, C.-K. and Hung, C.-N. (1997) Protein thiol modifications of human red blood cells treated with t-butyl hydroperoxide. *Biochim. Biophys. Acta* **1336,** 147–156.
14. Di Simplicio, P., Lupis, E., and Rossi, R. (1996) Different mechanism of formation of glutathione-protein mixed disulfides of diamide and tert-butyl hydroperoxide in rat blood. *Biochim. Biophys. Acta* **1289,** 252–260.
15. Dafre, A. L. and Reischl, E. (1998) Oxidative stress causes intracellular reversible S-thiolation of chicken hemoglobin under diamide and xanthine oxidase treatment. *Arch. Biochem. Biophys.* **358,** 291–296.
16. Garel, M.-C., Domenget, C., Caburi-Martin, J., Prehu, C., Galacteros, F., and Beuzard, Y. (1986) Covalent binding of glutathione to hemoglobin. I. Inhibition of hemoglobin S polymerization. *J. Biol. Chem.* **31,** 14,704–14,709.
17. Chai, Y.-C., Jung, C.-H., Lii, C.-K., Ashraf, S. S., Hendrich, S., Wolf, B., et al. (1991) Identification of an abundant S-thiolated rat liver protein as carbonic anhydrase III: characterization of S-thiolation and dethiolation reactions. *Arch. Biochem. Biophys.* **284,** 270–278.
18. Lii, C.-K. and Hendrich, S. (1993) Selenium deficiency suppressed the S-glutathionation of carbonic anhydrase III in rat hepatocytes under oxidative stress. *J. Nutr.* **123,** 1480–1486.

17

Oxidation of Cellular DNA Measured with the Comet Assay

Andrew R. Collins and Mária Dušinská

1. Introduction

The comet assay, or alkaline single-cell gel electrophoresis, originated as a simple and sensitive method for detecting DNA strand breaks in cells such as peripheral blood lymphocytes, cultured cells, and disaggregated tissue, e.g., liver *(1,2)*. Cells embedded in a thin layer of agarose on a microscope slide are lysed with detergent and high salt, and the resulting nucleoids electrophoresed at high pH. DNA is drawn out of the nucleoid to form a "comet tail" (*see* **Note 1**). Comets are visualized by fluorescence microscopy, after staining with a suitable DNA-binding dye. The relative amount of DNA in the tail compared with the head reflects the number of DNA breaks present.

DNA breaks are not specific to oxidative damage, but arise from exposure of cells to various damaging agents, or as intermediates in cellular repair of the damage. The need for a specific assay for oxidative damage was met *(3,4)* by including in the assay bacterial repair endonucleases that recognize and remove oxidized bases, introducing additional breaks in the DNA and increasing the tail intensity (*see* **Note 2**). Endonuclease III and formamidopyrimidine glycosylase (FPG) detect, respectively, oxidized pyrimidines and altered purines including 8-oxoguanine. These enzymes possess two activities; glycosylase (removing damaged base) and AP endonuclease (breaking the DNA at the baseless sugar or AP site) (*see* **Note 3**).

This version of the comet assay (*see* **Fig. 1**) has been widely used to measure oxidative damage in human cellular DNA in relation to disease, nutrition, and occupational and environmental mutagen exposure. Further modifications have been made for the investigation in vitro of antioxidant activity and DNA repair.

From: *Methods in Molecular Biology, vol. 186: Oxidative Stress Biomarkers and Antioxidant Protocols*
Edited by: D. Armstrong © Humana Press Inc., Totowa, NJ

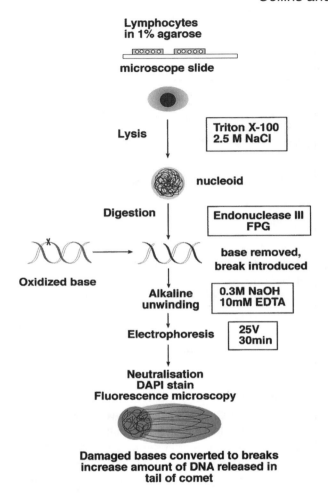

Fig. 1. Schematic representation of the comet assay incorporating lesion-specific endonucleases.

These methods and approaches will be described, and some examples of the findings will be given.

2. Materials

2.1. Equipment

1. Centrifuge.
2. Electrophoresis power unit, rated up to at least 500 mA, with voltage set at 25 V.
3. Horizontal electrophoresis tank, with platform approximately 20 cm × 20 cm.

4. Fluorescence microscope with 40× objective and filters appropriate for stain (*see* **Subheading 3.9.**).

2.2. Reagents

1. Standard analytical reagents; Na$_2$EDTA, tris(hydroxymethyl)aminomethane (Tris), KCl, KOH, NaCl, NaOH, N-(2-hydroxyethyl)piperazine-N′-(2-ethanesulphonic acid) (HEPES), DL-dithiothreitol (DTT), glycerol.
2. Bovine serum albumin, fraction V (BSA), t-octylphenoxypolyethoxyethanol (Triton X-100), and adenosine 5′-triphosphate (disodium salt), grade 1 (Sigma, St Louis, MO).
3. Agarose, electrophoresis grade (Gibco BRL 5510UA).
4. Agarose, low melting point (LMP) (Gibco BRL 5517US).
5. Fluorescent DNA-binding dye, 4′6-diamidine-2-phenylindol dihydrochloride (DAPI) (Boehringer Mannheim, Germany).

2.3. Solutions

Prepare solutions from appropriate stocks, such as 0.5 *M* Na$_2$EDTA, 1 *M* Tris, 1 *M* KCl, and so on. Keep solutions at 4°C.

1. Dissolve standard (electrophoresis grade) agarose at 1% in water by heating in an unsealed vessel in a microwave oven. This is used for precoating slides.
2. Dissolve LMP agarose (1%) as above but in phosphate-buffered saline (PBS) rather than water. Cool to 37°C before use. This is used for embedding cells.
3. Lysis solution: 2.5 *M* NaCl, 0.1 *M* EDTA, 10 m*M* Tris. Prepare 1 L. Set pH to 10.0 with either solid NaOH, or preferably concentrated (10 *M*) NaOH solution. (Add 35 mL of NaOH first to ensure that EDTA dissolves, and then add dropwise to pH 10.0.) Just before use, add 1 mL Triton X-100 per 100 mL and mix thoroughly.
4. Buffer for endonuclease III and FPG: 40 m*M* HEPES, 0.1 *M* KCl, 0.5 m*M* EDTA, 0.2 mg/mL BSA, pH 8.0, with KOH (make as 10X stock, adjust to pH 8.0, and freeze at –20°C).
5. Electrophoresis solution: 0.3 *M* NaOH, 1 m*M* EDTA; can be re-used several times.
6. Neutralizing solution: 0.4 *M* Tris, pH to 7.5 with concentrated HCl.
7. Stain for fluorescence microscopy: DAPI at 1 µg/mL in water (store aliquots at –20°C).
8. In vitro repair assay extract buffer: 45 m*M* HEPES, 0.4 *M* KCl, 1 m*M* EDTA, 0.1 m*M* DTT, 10% glycerol, adjusted with KOH to pH 7.8 (*see* **Note 4**).
9. In vitro repair reaction buffer: 45 m*M* HEPES, 0.25 m*M* EDTA, 2% glycerol, 0.3 mg/mL BSA, adjusted with KOH to pH 7.8, plus 2.5 m*M* ATP.

2.4. Supplies

1. Microscope slides: plain glass with frosted ends.
2. Enzymes: endonuclease III and FPG are prepared from bacteria containing overproducing plasmids with the appropriate cloned gene. The crude extract

is free of significant nonspecific nucleases. Enzymes are best obtained from a laboratory already using them, although commercial sources do exist.

3. Image-analysis software (optional): e.g., from Kinetic Imaging Ltd. (Sefton St, Liverpool L3 4BQ, UK); Perceptive Instruments (Haverhill, Suffolk CB9 7BN, UK); or Laboratory Imaging (Nad Upadem 901/63, 14900 Prague 4-Haje, Czech Republic).

3. Methods

3.1. The Basic Comet Assay

3.1.1. Preparation of Cells

1. Lymphocytes: Take ~30 µL blood from finger prick. Add to 1 mL PBS in a 1.5-mL Eppendorf tube. Mix and leave on ice for 30 min. Then place 100 µL of density gradient solution (Histopaque 1077, from Sigma, or Lympho-prep, from Nycomed) below the blood/PBS using a pipettor. Spin at 200*g*, 3 min, 4°C. Retrieve lymphocytes in 100 µL from just above boundary between PBS and Histopaque/Lymphoprep, using pipettor. Add to 1 mL PBS. Spin again. Remove as much supernatant as possible using pipettor.

   Alternatively, blood can be collected in larger quantity by venepuncture, into a 'Vacutainer' (or similar) tube with anticoagulant; isolate lymphocytes by standard density sedimentation procedure, wash with PBS, centrifuge, and either use at once or suspend in freezing medium (90% fetal calf serum [FCS], 10% dimethyl sulfoxide [DMSO]) and freeze slowly to –80°C for storage. On thawing, wash with PBS and centrifuge.

2. Cultured (monolayer) cells: Wash cells growing in dish with PBS, and add trypsin/ethylenediaminetetraacetic acid (EDTA); incubate until cells begin to detach, remove trypsin, add 1 mL of appropriate medium, remove cells by pipetting. Alternatively, use a silicon rubber scraper to remove cells. Transfer cells to Eppendorf tube and centrifuge at 200*g*, 3 min, 4°C. Remove supernatant, disperse pellet in 1 mL PBS. Centrifuge again, and remove supernatant using pipettor.

3. Cell number: The 30 µL sample of blood suggested in **step 1** of **Subheading 3.1.1.** should give enough lymphocytes for a reasonable density of comets. If cultured cells, or lymphocytes from frozen storage are to be used, the density should be such that each gel contains approx 2×10^4 cells.

3.1.2. Alkaline Single-Cell Gel Electrophoresis

1. Precoat ordinary, alcohol-cleaned clear glass microscope slides by dipping in a (vertical) staining jar of melted 1% standard agarose in H_2O; drain off excess agarose, wipe the back clean, and dry in a warm oven. Slides can then be stored at room temperature.

2. Embedding cells in agarose. Tap tube to disperse pelleted cells (*see* **Subheading 3.1.1.**) in the small volume of medium remaining. Quickly add 140 µL of 1% LMP agarose in PBS at 37°C and mix by tapping tube and quickly aspirating

agarose up and down with pipettor. Place as two drops (about 80 μL each) on a precoated slide and cover each with an 18 × 18 mm coverslip. Work quickly as the agarose sets at room temperature. Leave slides at 4°C for 5 min.

3. Lysis. Remove cover slips from gels and place slides in 100 mL of (cold) lysis solution with Triton X-100 in a horizontal glass staining jar. Leave at 4°C for at least 1 h (overnight is fine).

4. Enzymic digestion. Prepare 300 mL of enzyme reaction buffer, and use 1 mL for enzyme dilutions. The working concentration of the enzyme will depend on the source and should be notified by the supplier. The remainder of the buffer (at 4°C) is used for three 5-min washes of the slides, in a staining jar.

   Take slides from last wash, and drain excess buffer onto a tissue. Place 50 μL of enzyme solution (or buffer alone, as control) on each gel, and cover with a 22 × 22 mm cover slip. Incubate slides in a moist box at 37°C for 45 min (endonuclease III) or 30 min (FPG).

5. Alkaline treatment. Gently place slides (minus cover slips) on platform in electrophoresis tank containing pre-cooled electrophoresis solution, forming complete rows (gaps filled with blank slides). Gels must be (just) covered. Leave 40 min, preferably in cold room; the ambient temperature should preferably be 4°C, and certainly not >12°C.

6. Electrophoresis. 30 min at 25 V (constant voltage setting). If there is too much electrolyte covering the slides, the current may be so high that it exceeds the maximum. A high current also causes heating. 300 mA is recommended. If necessary, remove some solution.

7. Neutralization. Three 5-min washes with neutralising solution in staining jar at 4°C.

8. Storage. Slides may be analyzed immediately, or kept overnight in a dark, moist chamber before viewing. If the gels are dried by placing slides in a warm oven (either before or after staining), they can be kept indefinitely at room temperature.

9. Staining. Place 20 μL of DAPI solution onto each gel and cover with a 22 × 22 mm cover slip. Alternative stains: Propidium iodide (2.5 μg/mL), Hoechst 33258 (0.5 μg/mL), or ethidium bromide (20 μg/mL).

10. Quantitation: visual analysis. With practice, visual scoring is fast and accurate. Five classes of comet, from class 0 (undamaged, no discernible tail) to class 4 (almost all DNA in tail, insignificant head) can be distinguished (*see* **Fig. 2**). Score 100 comets selected at random from each slide (avoid the edges of the gel, where atypical comets are found). Each comet is given a value according to the class it is put into, so that an overall score can be derived for each gel, ranging from 0–400 arbitrary units *(5)*.

11. Quantitation: computer-image analysis. Several companies supply software that, linked to a closed-circuit digital camera mounted on the microscope, automatically analyzes individual comet images. The programs are based on separation of the comet head from the tail, and they measure various parameters including tail length; % of total fluorescence in head and tail; and "tail moment,"

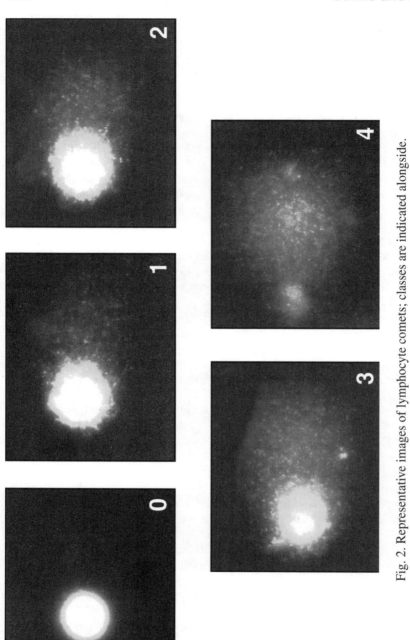

Fig. 2. Representative images of lymphocyte comets; classes are indicated alongside.

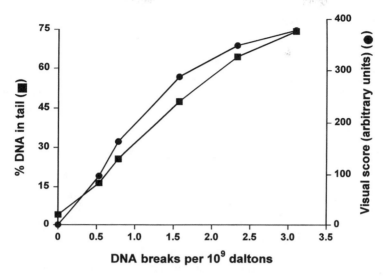

Fig. 3. X-ray calibration of the comet assay: standard curves derived from cells irradiated with different doses of X-rays (known to induce damage at the rate of 0.31 breaks per 10^9 daltons per Gray). Computer image analysis (% DNA in tail) and visual scoring (arbitrary units) give comparable results.

essentially representing the product of tail length and relative tail intensity. % DNA in tail is linearly related to DNA break frequency up to about 80% in tail, and this defines the useful range of the assay. Tail length tends to increase rapidly with dose at low levels of damage, but soon reaches its maximum. It may be a useful parameter at near-background levels of damage. Tail moment combines the information of tail length and tail intensity, but suffers from lack of linearity.

When slides are analyzed in parallel by visual scoring and by computer image analysis, the match is excellent *(6)* (*see* **Fig. 3**).

12. Calculation. Control gels, without enzyme treatment, provide an estimate of the background of DNA strand breaks (SB). Enzyme-treated gels reveal strand breaks and oxidized bases (SB + OX). Assuming a linear dose response, whether working in % DNA in tail or in arbitrary units, subtraction of (SB) from (SB + OX) gives a measure of the extent of base oxidation.

13. Calibration. Ionizing radiation produces strand breaks in DNA with known efficiency. For calibration, irradiate cells embedded in agarose with a range of doses, and immediately carry out comet assay analysis. The standard curve (**Fig. 3**) will relate break frequency expressed as Gray-equivalents, or as breaks per unit length of DNA, to the comet assay parameter (% tail DNA, or arbitrary units) *(7)*.

3.2. Measuring Antioxidant Protection Ex Vivo

To test for the effectiveness of antioxidant supplements or antioxidant-rich foods in humans, isolate lymphocytes from blood samples taken at intervals during the supplementation period. Suspend 5×10^4 cells in 1 mL of PBS with 100 μM H_2O_2, and incubate for 5 min on ice. Centrifuge at 4°C, suspend lymphocytes in agarose and proceed as in **Subheading 3.1.2.** Run control without H_2O_2. Antioxidant status is assessed from the level of damage induced by H_2O_2.

3.3. Modification to Measure Antioxidant Activity In Vitro

1. Principles of the method. This modification was devised in order to test putative antioxidants (such as fruit or vegetable juice) for ability to protect against DNA oxidation in vitro, in a model system as close as possible to the cell nucleus. Nucleoids are treated, in the gel, with H_2O_2 in the presence of putative antioxidant and oxidative damage measured as strand breaks.
2. Procedure. Prepare fruit/vegetable extract by homogenizing in PBS (adjust pH to 7.0 if necessary). Make serial dilutions with PBS. Embed cells (e.g., lymphocytes or cultured cells) in agarose as usual and immerse slides in solution of 1% Triton X-100 in PBS at 4°C for 15 min. Wash slides with cold PBS, drain excess buffer, lay slides flat, and add to each gel 25 µL of extract (two gels for each dilution) followed by 25 µL of H_2O_2 (in this order; if H_2O_2 is added first, substantial damage is done to DNA before the antioxidant is present). Incubate for 5 min on ice. Wash with cold PBS, and immerse slides in normal lysis solution (*see* **Subheading 2.3., item 3**) for 1 h. Continue with alkaline incubation and electrophoresis, etc. (from **Subheading 3.1.2., step 5**). Estimate DNA breaks produced by H_2O_2 alone and in the presence of extract; for comparison, run similar experiment with serial dilutions of vitamin C in PBS in place of extract.

This approach can also be used, of course, to assess the antioxidant efficacy of isolated, purified or synthesized compounds (if they are water-soluble or water-miscible).

3.4. Comet Assay Modified to Measure DNA Repair Capacity

1. Principles of the method. The assay as described in **Subheading 3.1.**) uses specific enzymes to reveal and measure an unknown amount of DNA damage. This modification, conversely, uses DNA with a defined amount of specific damage to estimate the activity of repair enzymes in a cell extract.
2. Preparing cell extract. Isolate lymphocytes from 10 mL of blood. Wash three times with 3× diluted extract buffer; *see* **Subheading 2.3., item 8**), centrifuging at 700*g*, 5 min, 4°C. Aspirate off as much as possible of the supernatant. Resuspend the pellet by vigorously tapping the tube; add 100 µL of (undiluted) extract buffer for each 10^7 cells. Divide cell suspension into 50 µL aliquots, freeze in liquid nitrogen and store at −80°C.

Just before the assay, thaw an aliquot, add 12 μL of a 1% solution of Triton X-100 (in extract buffer), mix, and centrifuge at 14,000g for 5 min at 4°C to remove nuclei and debris. Take 2.5 μL of the supernatant for determination of protein concentration using the bicinchonic acid kit (Sigma, St. Louis, MO), and mix remaining supernatant with 4 volumes of reaction buffer (*see* **Subheading 2.3., item 9**) with ATP. Keep on ice.

3. Preparing DNA substrate. Any convenient cultured cell line may be used to provide a damaged substrate. The treatment will depend on the repair pathway being studied. Ultraviolet (UV) light induces cyclobutane pyrimidine dimers (repaired by nucleotide excision repair); alkylating or oxidizing agents introduce base damage (repaired by base excision repair). The appropriate level of damage will be found by trial and error. It may be necessary to incubate cells to allow rejoining of strand breaks, leaving only damaged bases. Embed the treated cells in agarose on slides (with parallel untreated cells), place in lysis solution for 1 h at 4°C, wash 3 times with reaction buffer (no ATP), and allow excess to drain off onto tissue.

4. Assaying repair. Add 50 μL of cell extract with ATP to each gel, cover with cover-slip, and incubate for 10, 20, or 30 min at 37°C in a humid chamber. Continue with alkaline treatment and electrophoresis. Analyse gels for DNA breaks; the different incubation times allow calculation of the rate of incision at damage sites, adjusted for different protein concentrations of extracts.

3.5. Results

3.5.1. Quality Control

1. Slides should be scored 'blind.' Everyone involved in scoring should regularly undertake analysis of standard slides (representing a range of levels of DNA breakage) to ensure that values obtained by different workers are comparable. If visual scoring is used, it should ideally be calibrated against computer-based image analysis: score each comet visually before obtaining the % DNA in tail (or other image analysis parameter); calculate the mean % tail DNA for each class. The plot of % tail DNA against class number (0–4) should be approximately linear and similar for all scorers.

2. In human studies, especially if a cohort is studied over an extended period, it is important to include standard lymphocyte samples in every batch of analyses. For this purpose, at the start of the trial prepare a sufficient quantity of lymphocytes (*see* **Subheading 3.1.1., step 1**; pool isolated cells from several people if necessary), suspend in freezing medium at 3×10^6/mL, freeze 100 μL aliquots slowly to –80°C and store at this temperature until needed.

3.5.2. Intra- and Interindividual Variation

Lymphocytes were taken from a single individual on 19 occasions over a period of approx 1 yr. Strand breaks and endonuclease III-sensitive sites were measured; the coefficients of variation (CV) were 30 and 59%, respectively. Part of this variability is the result of experimental variation; parallel determinations of SBs carried out on a single lymphocyte sample gave a CV of 11%. The residue of variability presumably reflects seasonal effects, variations in diet, infection/inflammation, and general level of physiological stress.

Comparing individuals (a group of nine healthy men aged 40–50, sampled and assayed on the same day with slides in the same electrophoresis tank), the CVs were higher; 56% for SBs, and 88% for endonuclease III sites. Groups of men and women, again healthy and age-matched, showed the same CVs.

3.5.3. Comet Assay Results from Human Trials

1. The comet assay, with endonuclease III/FPG, has been used to demonstrate decreases in endogenous oxidative DNA damage in human lymphocytes, following supplementation for several days or weeks with vitamin C/β-carotene/vitamin E *(8)*, with vegetables rich in carotenoids *(9)*, and with soya milk *(10)*. Shorter-term trials (such as single large doses of antioxidants) typically show protective effects at the level of H_2O_2-induced damage, but not endogenous damage *(11)*. Endogenous base oxidation in lymphocyte DNA correlates negatively with plasma carotenoid concentrations *(12)*. DNA breaks and oxidized bases are elevated in patients suffering from insulin-dependent diabetes mellitus *(13)* or ankylosing spondylitis *(14)*.
2. Removal of DNA breaks and oxidized bases during incubation of H_2O_2-treated lymphocytes has been followed with the comet assay *(5)*. This is not a simple measure of DNA repair, since damage also accrues from exposure to the high oxygen tension of the atmosphere; carotenoid or vitamin C supplementation alleviates the damage and "repair" appears faster *(15,16)*.

3.5.4. Relative Advantages and Disadvantages

1. Advantages: The comet assay is sensitive, simple, economical, and requires little special equipment. It can be made selective if appropriate lesion-specific enzymes are included. Damage is examined at the level of single cells, so heterogeneity in a cell population is detected. It requires extremely small amounts of material, and is thus conveniently used in clinical and population studies, animal and cell culture experiments.
2. Disadvantages: The comet assay may be too sensitive, reaching saturation at a relatively low level of damage (*see* **Note 2**). Although it is quantitative, it requires careful calibration to express results in terms (for example, oxidized bases per million unoxidized bases) that allow comparison with results from other assays such as high-performance liquid chromatography (HPLC) and gas chromatography-mass spectrometry (GC-MS). (There is currently a lively debate over the relative merits of these approaches to the measurement of background levels of DNA oxidation in humans *[17]*.)

4. Notes

1. How do the comets arise? One explanation (discussed in **ref.** *18*) is in terms of DNA supercoiling. After lysis, when membranes, cytoplasm and most nuclear proteins have been removed, the DNA remains as a "nucleoid," consisting of loops of supercoiled DNA attached to the nuclear matrix. While the DNA is supercoiled, it remains tightly aggregated and does not move appreciably under an electrophoretic field. However, a single SB in a loop is enough to relax supercoiling, and this loop then extends towards the anode. The more loops are relaxed, the more intense is the tail of the comet. Consistent with this account:
   a. Tail intensity rather than length increases with dose of damage;
   b. breaks are detected with equal sensitivity under neutral or alkaline conditions;
   c. after alkaline electrophoresis and neutralization, tail DNA is single-stranded while head DNA is double-stranded *(18)*.

2. The assay becomes insensitive above a certain level of DNA breakage (when maximal movement into the tail has occurred). If enzyme-sensitive sites are estimated above an already high level of breaks, serious underestimation is likely. In this case, alkaline unwinding followed by neutral electrophoresis will render the assay less sensitive *(19)*.

3. AP (apurinic/apyrimidinic) sites may be present as lesions in DNA (not simply as intermediates in the in vitro enzymic digestion). They are alkali-labile, and so may be converted into DNA breaks under the alkaline conditions of the comet assay. It is not clear whether this is generally the case. Exonuclease III or endonuclease V (both of which have AP endonuclease acivity) can be used to check for the presence of residual AP sites.

4. DTT and other -SH reagents are tolerable only at low concentration in enzyme reaction buffers, since they have been shown to induce DNA damage.

5. The previous mention of proprietary brands does not constitute endorsement of particular products.

Acknowledgments

The support of the Scottish Executive Rural Affairs Department, the Slovak Ministry of Health, and the European Commission (IC15 CT96 1012) is gratefully acknowledged.

References

1. Singh, N. P., McCoy, M. T., Tice, R. R., and Schneider, E. L. (1988) A simple technique for quantitation of low levels of DNA damage in individual cells. *Exp. Cell Res.* **175,** 184–191.

2. McKelvey-Martin, V. J., Green, M. H. L., Schmezer, P., Pool-Zobel, B. L., de Méo, M. P., and Collins, A. (1993) The single cell gel electrophoresis assay (comet assay): a European review. *Mutation Res.* **288,** 47–63.

3. Collins, A. R., Duthie, S. J., and Dobson, V. L. (1993) Direct enzymic detection of endogenous oxidative base damage in human lymphocyte DNA. *Carcinogenesis* **14,** 1733–1735.
4. Dušinská, M. and Collins, A. (1996) Detection of oxidised purines and UV-induced photoproducts in DNA of single cells, by inclusion of lesion-specific enzymes in the comet assay. *Altern. Lab. Animals* **24,** 405–411.
5. Collins, A. R., Ma, A., and Duthie, S. J. (1995) The kinetics of repair of oxidative DNA damage (strand breaks and oxidised pyrimidines) in human cells. *Mutation Res.* **336,** 69–77.
6. Collins, A., Dušinská, M., Franklin, M., Somorovská, M., Petrovská, H., Duthie, S., et al. (1997) Comet assay in human biomonitoring studies: reliability, validation, and applications. *Environ. Mol. Mutagenesis* **30,** 139–146.
7. Collins, A. R., Dušinská, M., Gedik, C., and Štětina, R. (1996) Oxidative damage to DNA: do we have a reliable biomarker? *Environ. Health Perspect.* **104(Suppl. 3),** 465–469.
8. Duthie, S. J., Ma, A., Ross, M. A., and Collins, A. R. (1996) Antioxidant supplementation decreases oxidative DNA damage in human lymphocytes. *Cancer Res.* **56,** 1291–1295.
9. Pool-Zobel, B. L., Bub, A., Muller, H., Wollowski, I., and Rechkemmer, G. (1997) Consumption of vegetables reduces genetic damage in humans: first results of a human intervention trial with carotenoid-rich foods. *Carcinogenesis* **18,** 1847–1850.
10. Mitchell, J. H. and Collins, A. R. (1999) Effects of a soy milk supplement on plasma cholesterol levels and oxidative DNA damage in men: a pilot study. *Eur. J. Nutr.* **38,** 143–148.
11. Panayiotidis, M. and Collins, A. R. (1997) *Ex vivo* assessment of lymphocyte antioxidant status using the comet assay. *Free Rad. Res.* **27,** 533–537.
12. Collins, A. R., Olmedilla, B., Southon, S., Granado, F., and Duthie, S. J. (1998) Serum carotenoids and oxidative DNA damage in human lymphocytes. *Carcinogenesis* **19,** 2159–2162.
13. Collins, A. R., Rašlová, K., Somorovská, M., Petrovská, H., Ondrušová, A., Vohnout B., et al. (1998) DNA damage in diabetes: correlation with a clinical marker. *Free Rad. Biol. Med.* **25,** 373–377.
14. Dušinská, M., Lietava, J., Olmedilla, B., Rašlová, K., Southon, S., and Collins, A. R. (1999) Indicators of oxidative stress, antioxidants and human health, in *Antioxidants in Human Health* (Basu, T. K., Temple, N. J., and Garg, M. L., eds.), CABI Publishing, Oxford, pp. 411–422.
15. Collins, A. R., Duthie, S. J., Fillion, L., Gedik, C. M., Vaughan, N., and Wood, S. G. (1997) Oxidative DNA damage in human cells: the influence of antioxidants and DNA repair. *Biochem. Soc. Transact.* **25,** 326–331.
16. Torbergsen, A. C. and Collins, A. R. (2000) Recovery of human lymphocytes from oxidative DNA damage: the apparent enhancement of DNA repair by carotenoids is probably simply an antioxidant effect. *Eur. J. Nutr.* **39,** 80–85.

17. Collins, A., Cadet, J., Epe, B., and Gedik, C. (1997) Problems in the measurement of 8-oxoguanine in human DNA. Report of a workshop, DNA Oxidation, held in Aberdeen, UK, 19–21 January, 1997. *Carcinogenesis* **18,** 1833–1836.
18. Collins, A. R., Dobson, V. L., Dušinská, M., Kennedy, G., and Štětina, R. (1997) The comet assay: what can it really tell us? *Mutation Res.* **375,** 183–193.
19. Angelis, K. J., Dušinská, M., and Collins, A. R. (1999) Single cell gel electrophoresis; detection of DNA damage at different levels of sensitivity. *Electrophoresis* **20,** 1923–1933.

18

Measurement of DNA Double-Strand Breaks with Giant DNA and High Molecular-Weight DNA Fragments by Pulsed-Field Gel Electrophoresis

Yoshihiro Higuchi

1. Introduction

Reactive oxygen species (ROS) such as hydroxyl radicals ($^\bullet$OH), superoxide anions (O_2^-) and hydrogen peroxide (H_2O_2) have been shown to damage to chromosomal DNA and other cellular components, resulting in DNA degradation, protein denaturation, and lipid peroxidation *(1,2)*. We know a little about the in vivo action mechanism of ROS produced by anticancer drugs and by X-ray irradiation on chromatin DNA in the nuclei of intact cells. DNA damage induced by ROS in vivo or in the cultured cell system is classified into single- and double-strand breaks and nucleotide base oxidative modifications *(2–4)*. The application of gel electrophoresis to the measurement of DNA double-strand breaks has been described by some workers for DNA irradiated in vitro *(5–7)*. Double-strand breaks are generally thought to have a greater biological consequence than single-strand DNA breaks because they can lead directly to chromosomal aberrations, and more frequently to the loss of genetic information *(6,8)*. Ionizing radiation such as X-ray and γ-ray are, in general, thought to produce $^\bullet$OH radicals from water molecules in or around the target sites in the DNA, and these in turn attack DNA and break it down *(1,3)*. In addition, the involvement of such radicals in the induction of apoptosis has been suggested in several cell lines *(9–12)*.

Some research groups have used pulsed-field gel electrophoresis (PFGE) to analyze the precise molecular nature of DNA fragments produced by reactive oxygen species in cultured animal cells *(4,15–18)*. In this article, we introduce

From: *Methods in Molecular Biology, vol. 186: Oxidative Stress Biomarkers and Antioxidant Protocols*
Edited by: D. Armstrong © Humana Press Inc., Totowa, NJ

the methods for PFGE and agarose gel electrophoresis (AGE) and present the results obtained of DNA fragmentation produced by ROS.

2. Materials

2.1. Equipment for Electrophoresis

2.1.1. PFGE Apparatus (see **Note 1**)

1. Pharmacia LKB 2015 Pulsaphor electrophoresis unit (New version; Gene Navigator) consisting of lid, point, or hexagonal electrode; gel-support tray; electrophoresis tank; and some accessories such as a comb with 16 wells and a gel blockformer.
2. Pulsaphor Plus Control Unit (New version; Gene navigator control unit).
3. Power supply: Pharmacia EPS 1001 (power, 1–100W; current, 1–400 mA; Volt, 5-1000V).
4. Cooling circulator apparatus: Pharmacia Multi Temp III.
5. UV transilluminator with 302 or 312 nm of emission wavelength.
6. Incubator ranging from room temp to 100°C.
7. Gel knife (should be autoclaved).

2.1.2. AGE Apparatus (see **Note 2**)

1. Mini Mupid agarose electrophoresis apparatus, type II, submarine style from Advance Co., Tokyo, Japan.

2.2. Reagents

1. Agarose: Sigma type II agarose: medium EFO, gel strength >1000 g/cm^2 (1%) (Sigma Chemicals, St. Louis, MO) (*see* **Note 3**). Low melt temperature agarose. Agarose L, >450 g/cm^2 (1.5%), melt temp. <65°C (1.5%), gel temp. <30°C (1.5%) (Wako Chemicals, Osaka, Japan).
2. DNA size standard: Chromosomal DNA of *Saccharomyces cerevisiae* 225 kb–2.2 Mb and *Saccharomyces pomb* 3.5–5.7 Mb; λ ladder 0.05–1 *M* band; λ DNA digested with *Hind* III. These size markes were obtained from Bio-Rad Laboratories, Hercules, CA; Pulse Marker (λ DNA *Hind* III fragments plus uncut λ DNA and λ DNA concatemers) 0.1–200 kb from Sigma; pUC-19 prepared by digesting with restriction endonuclease *BSP* 1286I, 225–1161 bp.
3. Enzymes: Proteinase K (EC.3.4.21.14) from *Tritirachium album* (chromato-graphically purified and DNase-free grade, from Merck, Darmstadt, Germany); RNase A (DNase free) from Roche Diagnostics, Mannheim, Germany. If the RNase contains DNase, heat at 95°C for 5 min before use.
4. Buffers: PBS, Dulbecco's phosphate-buffered saline; TBE: 89 m*M* Tris-boric acid, 2 m*M* ethylenediaminetetraacetic acid (EDTA), pH 8.0; ESP: a mixture of 0.5 *M* EDTA, pH 9.0, 1% (w/v) N-lauroyl sarcosine sodium (Sigma) and 1 mg/mL of proteinase K; TE: 10 m*M* Tris-HCl, 1 m*M* EDTA, pH 7.4. PBS, TBE, TE, and 0.5 *M* EDTA buffers need to be autoclaved.

3. Methods
3.1. Cell Sample Processing

1. Preculture T-24 human bladder carcinoma cells to a density of 2×10^6 cells in 5 mL of Dulbecco's modified Eagles medium (DMEM) supplemented with 5% fetal calf serum (FCS) (DF-5) in culture dish (21 cm^2) at 37°C in humidified air containing 5% CO_2, and use the cells to examine DNA damage by H_2O_2 or X-ray irradiation.
2. Hydrogen peroxide treatment: Treat T-24 cells with H_2O_2 in the culture dish for at the indicated doses in 5 mL of DF-5 at 37°C in humidified air containing 5% CO_2.
3. X-ray irradiation: Irradiate T-24 cells grown in the culture dish with X-ray using an X-ray irradiation apparatus (Toshiba Model KXC18-2, Tokyo, Japan) operated at 180 kVp and 20 mA at a dose of 2.8 Gy/min, which is determined using an X-ray ion chamber survey meter (Aloka, Model ICS-301, Tokyo, Japan).
4. After treatment or irradiation, harvest the cells by trypsinizing with 0.05% trypsin/PBS solution containing 0.05% EDTA, and wash several times in PBS by centrifuging at 200g for 5 min.

3.2. Pulsed-Field Gel Electrophoresis (PFGE)

1. Harvest T-24 cells (2×10^6) treated with H_2O_2 or irradiated with X-ray and suspend the cells in 50 µL of phosphate-buffered saline (PBS). Mix the cell suspension with 50 µL of 1% low melt agarose in PBS kept at 45°C, immediately pour the mixture into a well ($2 \times 5 \times 10$ mm, 100 µL) of a blockformer, and allow the blockformer to stand on ice for 20 min to make hardened agarose blocks containing cell samples.
2. Tap the agarose block out from the blockformer, transfer it to a sample tube (1.5–2 mL) containing 250 µL of ESP and incubate the agarose block in the sample tube with gentle shaking for 24–48 h at 50°C. After incubation, the agarose block should be virtually clear and have sunk.
3. Put the agarose block on Parafilm (American National Can, Chicago, IL) or a slide glass, cut it into four equal-size pieces with a knife and load one piece (*see* **Note 4**) to a well of 1.2% standard separating agarose gel ($15 \times 15 \times 0.44$ cm, 100 mL) on the separating gel-support tray in 0.5X TBE.
4. Set the gel support tray in tank with 2.5 L of 0.5X TBE, and also set a hexagonal electrode unit. PFGE is performed at 14°C for the indicated pulse and run time using a control unit and a power supply.
5. After electrophoresis, remove the standard agarose gel from the gel support tray, stain with ethidium bromide (0.5 µg/mL) in 0.5X TBE for 30–60 min (*see* **Note 5**), visualize on a UV transilluminator at 302 nm after de-staining with distilled wateror 0.5X TBE for a few hours, and take a photograph using #667 Polaroid film or a gel-documentation system (*see* **Note 6**).
6. Chromosomal DNA from *Saccharomyces cerevisiae* and *Saccharomyces pomb* and a mixture of λ DNA, its concatemers and *Hind* III-digested λ DNA can be used as DNA size markers.

3.3. Conventional AGE

1. To prepare small DNA fragments (~50 kbp) containing internucleosomal DNA fragments, collect the digested cell components in ESP solution (250 μL) released from the agarose block prepared for the PFGE experiment (*see* **Subheading 3.2.**).

2. Add the same volume of distilled water (250 μL) to the ESP solution. Extract the solution (500 μL) twice with the same volume (500 μL) of (1:1) phenol-chloroform (1:1) and isoamyl alcohol (24:1, v/v) under mild shaking for 10 min, remove phenol remained by extracting with 500 μL of chloroform for 10 min and collect the DNA in the aqueous phase.

3. Precipitate DNAs by addition of 20 μL of 5M NaCl and 1167 μL of 99.5% ethanol (final concentration 70%) and by keeping at –20°C for 30 min, according to the method described by Sambrook et al. *(19)* (*see* **Note 7**). Collect the precipitates by centrifugation at 12,000*g* for 15 min, wash with 70% ethanol/water (v/v), dry by centrifuging under vacuum, dissolve in 20 μL of TE buffer, and treat with 1.6 μg (4 μL) of RNase A for 60 min at 37°C (*see* **Note 8**).

4. Load a quarter (6 μL) of the aliquot by mixing with 1 μL of loading buffer consisting 0.25% bromo phenol blue in 30% glycerin into a well of 1.5% (w/v) agarose gel for AGE.

5. Electrophoresis is performed with Mupid type II AGE apparatus in 0.5X TBE for 90 min at 15 volts/cm at room temperature.

6. After electrophoresis, stain DNA in the gels with ethidium bromide (0.5 μg/mL) in 0.5X TBE for 30 min and after destaining with 0.5X TBE, take a photograph of the gel on the transilluminator under UV light (302 nm) using #667 Polaroid film.

3.4. Results

3.4.1. Principle of PFGE

The PFGE analysis is a sophisticated method developed by Schwartz et al. *(20)* and Carles's group *(21)* to separate extra-large DNA molecules up to 10 Mbp such as chromosomal DNA. The time it takes to change configuration depends strongly on the size of the molecules. Larger molecules will migrate more slowly through an agarose gel when subjected to changing field directions (**Fig. 1**) (*see* **Note 8**). PFGE are available for analysis of DNAs from not only eucaryotes of animal and yeast cells but also procaryotes of bacteria, phage, and protozoa. Two types of electrode systems, Point and Hexagonal, for Pharmacia LKB 2015 Pulsaphor electrophoresis units have been used for PFGE in this article.

3.4.2. Experimental Considerations

Table 1 shows the parameters for a typical experiment for PFGE and provides a few examples of DNA separation when using 2.5 L of 0.5X TBE

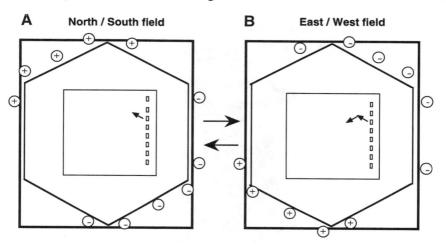

Fig. 1. Two different fields are applied using the hexagonal electrode.

Table 1
General Run Parameters[a]

	DNA size (Mb)				
	1–50 kb	<0.1	0.1–2	2–6	6–12
% agarose	1.2	1.2	1.2	1.2	0.6
Voltage/V (HEX)[b]	450	300	165–200	40–100	25–50
(Point)[b]	450	370	300–330	60–100	30
Puse time	0.3–1 s	1–10 s	10–120 s	3–75 min	4–100 min
Run times	1–4 h	1–6 h	17–24 h	24 h–3 d	3–6 d

[a]In 0.5X TBE buffer, 12–14°C *(22)*.
[b]HEX and Point indicate hexagonal and point type of electrode, respectively.

buffer. These parameters will give good overall separation for 40 kb-12 Mb. For separation of very large DNA (>2000 kb) it seems to be important that the voltage be lowered. For molecules below 100 kb, it appears that higher voltages can be used without problems resulting. In addition to changing the pulse times within 1 s and 100 min, the degree of separation can also be effected by changes in the run time. Generally, separation of large fragments requires run times longer than 2 d. For maximum resolution below 2000 kb, about 24 h is needed. Larger fragments require considerably longer times.

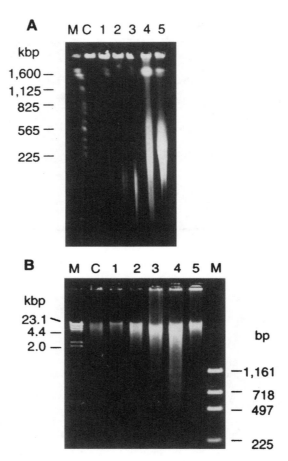

Fig. 2. PFGE **(A)** and AGE **(B)** analysis of double-strand DNA fragments from T-24 cells treated with hydrogen peroxide. The cells were treated with 0.05 mM (lane 1), 0.5 mM (lane 2), 1 mM (lane 3), 5 mM (lane 4), and 10 mM (lane 5) H_2O_2 or without H_2O_2 (lane C) for 18 h at 37°C. Lane M indicates DNA size marker. Electrophoresis was carried out for 28 h at pulse time/run time of 90 s/20 h and 120 s/8 h at 175 volts. AGE was carried out according to the method described in the Methods section.

3.4.3. DNA Fragmentation by H_2O_2 or X-ray Irradiation in T-24 Cells

1. Giant DNA fragments of 1–2 Mbp in size, high molecular-weight DNA fragments with 100–800 kbp, and <100 kbp were observed in T-24 cells treated with various doses of H_2O_2 (**Fig. 2A**) for 18 h. The features of the DNA from the cells are as follows:

   a. The amount of degraded DNA appeared to be proportional to the dose of H_2O_2.

  b. Smaller DNA fragments of less than 100 kbp can be only observed in the cells treated with 0.5, 1, and 5 mM H$_2$O$_2$ (**Fig. 2B**). The amount of DNA fragmentation is proportional to that of the ladder-like internucleosomal DNA fragments but not treated with high concentration of H$_2$O$_2$ such as 10 mM.

  c. The amount of 100–800 kbp fragments produced in the cells treated with 1 mM H$_2$O$_2$ was much less than that in the cells treated with 5 or 10 mM H$_2$O$_2$.

2. Giant DNA fragments of 1–2 Mbp were also produced by X-ray irradiation at 100 Gy. Thereafter the amount of 1–2 Mbp DNA fragments decrease with time (**Fig. 3A**). Marked 100–800 kbp DNA fragment at 10 h are not observed in the irradiated cells (**Fig. 3A**). The 1–2 Mb giant DNA fragments from X-ray irradiated T-24 cells consisted of three bands in the PFGE analysis under different longer electrophoresis conditions of pulse time and run time (**Fig. 3B**). An increase in the amount of ladder-like DNA fragments was observed 10 h and 24 h after the 100 Gy dose (**Fig. 3C**). The amount of 1–2 Mbp giant DNA fragments appeared to be inversely related to the quantity of ladder-like DNA fragments during incubation after X-ray irradiation.

4. Notes

1. There are some instruments including the newest version of the following, such as CHEF (clamped homogeneous electric fields) from Bio-Rad and GENOFIELD (biased sinusoidal field gel electrophoresis system) from Atto Co, Tokyo, Japan available for the PFGE system, other than the Pharmacia Pulsaphor electrophoresis system (LKB 2015, from Amersham Pharmacia Biotech, Bucks, UK) with hexagonal electrode introduced in this article. These DNA separating system are essentially similar in their procedures and performance.

2. Conventional and regular submarine type of AGE apparatus are generally used for separation of internucleosomal DNA fragments.

3. Grade or quality of agarose is critical to the success of isolation of large DNA fragments such as 1–10 Mb in PFGE and great variations have been observed between different manufacturers.

4. Between 0.5 and 1.0 µg of DNA is usually loaded into each well. The remaining DNA sample in the agarose block prepared in ESP can be stored for a few month at 4°C and for longer months if replaced by TE-10 (10 mM EDTA) buffer (without proteinase K and the detergent).

5. The "old" buffer from the tank may now be reused for staining and the destaining with ethidium bromide. The DNA samples on the gel after PFGE can be transferred to a nylon membrane according to the conventional blotting method and be used for Southern-hybridization analysis.

6. The amount of the DNA band could be measured according to the method using a analytical software of gel-documentation system supplied by some corporations such as:

  a. Epi-Light UV FA 1100 from Aishin Cosumosu, Tokyo, Japan;

  b. Bio-Profile with software Bio-1D from Vilber Lourmat, France; and

  c. Gel Doc 1000 with Molecular Analyst software from Bio-Rad Laboratories.

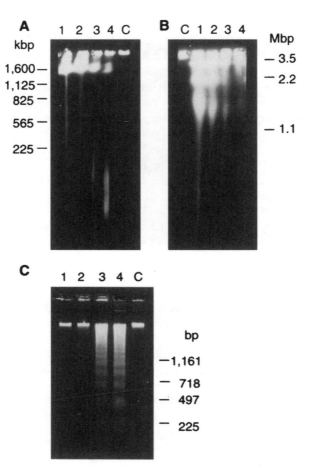

Fig. 3. PFGE (**A** and **B**) and AGE (**C**) analysis of double-strand DNA fragments from T-24 cells irradiated with X-ray. The cells were irradiated with 100 Gy X-ray at room temperature and then incubated with at 37°C for 0 min (lane 1), 3 h (lane 2), 10 h (lane 3), and 24 h (lane 4). Lane C shows nonirradiated and nonincubated samples. In (A), PFGE was carried out for 28 h at pulse time/run time of 90 s/20 h and 120 s/8 h at 175 volts. In (B), PFGE was carried out for 96 h at pulse time/run time 30 min/96 h, at 70 volts and after that as described for (A). AGE was carried out according to the method described in **Subheading 3.**

7. When excess amount of white coprecipitates of EDTA with nucleic acid emerge in the sample solution, they should be collected by centrifugation or remove by conventional spin-column chromatography. Redissolve the precipitates in an appropriate volume of 1X TE buffer and then repeat the DNA preparing procedure by the ethanol precipitating method in **Subheading 3.3., step 1**.

8. During preparation of low molecular-weight DNA fragments, RNase A treatment of DNA preparation is necessary to avoid interference of RNA with DNA not only in the migration in electrophoresis but also in staining with ethidium bromide.

References

1. Bielski, B. H. J. and Gebichi, J. M. (1977) Application of radiation chemistry to biology. *Free Rad. Biol.* **3,** 1–51.
2. Halliwell, B. and Aruoma, O. I. (1991) DNA damage by oxygen-derived species: its mechanism and measurement in mammalian systems. *FEBS Lett.* **281,** 9–19.
3. Ward, J. F. (1988) DNA damage produced by ionizing radiation in mammalian cells. *Prog. Nucleic Acid Res. Mol. Biol.* **35,** 95–125.
4. Wlodek, D. and Hittelman, W. N. (1987) The repair of double-strand breaks correlates with radiosensitivity of L-5178Y-S and L5178Y-R cells. *Radiat. Res.* **112,** 146–155.
5. Yamada, T. and Ohyama, H. (1988) Radiation-induced interphase death of rat thymocytes is internally programmed (apoptosis). *Int. J. Radiat. Biol.* **53,** 65–75.
6. Nevaldine, B., Longo, J. A., King, G. A., Vilenchik, M., Sagerman, R. H., and Hahn, P. J. (1993) Induction and repair of DNA double-strand breaks. *Radiation Res.* **133,** 370–374.
7. Lobrich, M., Ikpeme, S., and Kiefer, J. (1994) DNA double-strand break measurement in mammalian cells by pulsed-field gel electrophoresis: an approach using restriction enzymes and gene probing. *Int. J. Radiat. Biol.* **65,** 623–630.
8. Bryant, P. E. (1984) Enzymatic restriction of mammalian cell DNA using Pvu II and Bam HI; evidence for the double strand break origin of chromosomal aberrations. *Int. J. Radiation Biol.* **46,** 57–65.
9. Oberhammer, F., Wilson, J. W., Dive, C., Morris, I. D., Hickman, J. A., Walkeling, A. E., et al. (1993) Apoptotic death in epithelial cells: cleavage of DNA to 300 and/or 50 kb fragments prior to or in the absence of internucleosomal fragmentation. *EMBO J.* **12,** 3679–3684.
10. Lagarkova, M. A., Iarovaia, O. V., and Razin, S. V. (1995) Large-scale fragmentation of mammalian DNA in the course of apoptosis proceeds via excision of chromosomal DNA loops and their oligomers. *J. Biol. Chem.* **270,** 20,239–20,241.
11. Slater, A., Nobel, S., and Orrenius, S. (1995) The role of intracellular oxidants in apoptosis. *Biochim. Biophys. Acta* **1271,** 59–62.
12. Clutton, S. (1997) The importance of oxidative stress in apoptosis. *Br. Med. Bull.* **53,** 662–668.
13. Stamato, T. D. and Denko, N. (1990) Asymmetric field inversion gel electrophoresis: a new method for detecting DNA double-strand breaks in mammalian cells. *Radiat. Res.* **121,** 196–205.
14. Erixon, K. and Cedervall, B. (1995) Linear induction of DNA double-strand breakage with X-ray dose, as determined from DNA fragment size distribution. *Radiat. Res.* **142,** 153–162.

15. Matsukawa, S. and Higuchi, Y. (1991) The nature of giant DNA molecules produced from nuclear chromosome DNA by active oxygen producing agents, in *Oxidative Damage and Repair* (Davies, K. T. A., ed.), Pergamon Press, pp. 197–201.
16. Higuchi, Y. and Matsukawa, S. (1997) Appearance of 1–2 Mb giant DNA fragments as an early common response leading to cell death induced by various substances which cause oxidative stress. *Free Rad. Biol. Med.* **23,** 90–99.
17. Higuchi, Y. and Matsukawa, S. (1998) Active oxygen-mediated chromosomal 1-2 Mbp giant DNA fragmentation into internucleosomal DNA fragmentation in apoptosis of glioma cells induced by glutamate. *Free Rad. Biol. Med.* **24,** 418–426.
18. Higuchi, Y. and Matsukawa, S. (1999) Glutathione depletion induces chromosomal giant DNA and high molecular weight DNA fragmentation associated with apoptosis through lipid peroxidation and protein kinase C activation in C6 glioma cells. *Arch. Biochem. Biophys.* **393,** 33–42.
19. Sambrook, J., Fritsch, E. F., and Maniatis, T. (1989) *Molecular Cloning*, 2nd ed. Cold Spring Harbor Laboratory Press, Cold Spring Harbor, NY, pp. E10–E15.
20. Schwartz, D. C., Saffran, W., Welsh, J., Haas, R., Goldberg, M., and Cantor, C. R. (1983) New techniques for purifying large DNAs and studying their properties and packaging. *Cold Spring Harb. Quant. Biol.* **47,** 189–195.
21. Carles, G. F., Frank, M., and Olson, M. V. (1986) Electrophoretic separations of large DNA molecules by periodic inversion of the electric field. *Science* **232,** 65–68.
22. Pharmacia LKB Biotechnology. Instruction manual Pulsaphor system 80-1251-98. Pharmacia LKB Biotechnology, S-751.82 Upsala, Sweden. *(New version, from Amersham Pharmacia Biotech, Bucks, UK.)*

19

Evaluation of Antibodies Against Oxygen Free Radical-Modified DNA by ELISA

Rashid Ali and Khurshid Alam

1. Introduction

Oxygen free radicals (OFR) also known as reactive oxygen species (ROS) are formed as a consequence of normal oxidative metabolism. Their generation goes up substantially under chronic inflammatory conditions and ischemia. These free radicals have multifarious effects that include oxidative damage to DNA leading to various human degenerative diseases. Out of these species, hydroxyl radical is highly reactive and can virtually react with any cell macromolecule, the DNA being the most susceptible. Oxidative damage to DNA includes strand breaks, sister chromatid exchange, and subsequent generation of clastogenetic factors, alterations in basic structure and conformational changes *(1,2)*.

It has been established that native DNA in B-conformation is nonimmunogenic *(3)*. Nevertheless, various modified forms of DNA, complexes of DNA-proteins are immunogenic and induce antibodies that recognize native DNA to varying degree *(4–6)*. However, in Systemic Lupus Erythematosus (SLE), a prototype autoimmune disorder of unknown etiology, the presence of circulating antibodies against nDNA is a characteristic feature that serves as a diagnostic marker for the disease *(7)*. The origin and nature of antigens that induce anti-double-stranded DNA antibodies are not known. Many studies suggest modified forms of DNA and polynucleotides, including those by OFR, as the antigenic stimulus for the formation of antibodies having SLE-like auto-antibody characteristics *(8,9)*.

Here we describe a method for the preparation of ROS-modified DNA and detection of antibodies against native and modified polymer by direct binding and competition enzyme-linked immunosorbent assay (ELISA).

From: *Methods in Molecular Biology, vol. 186: Oxidative Stress Biomarkers and Antioxidant Protocols*
Edited by: D. Armstrong © Humana Press Inc., Totowa, NJ

2. Materials

2.1. Equipment

1. UV-visible recording spectrophotometer with temperature controller and programmer as additional accessories.
2. ELISA reader.
3. A refrigerated ultracentrifuge.

2.2. Reagents

1. The following chemicals are obtained from Sigma Chemical Company (St. Louis, MO): highly polymerized calf thymus DNA, nuclease S1, micrococcal nuclease, p-aminosalicylate, BSA, Tween-20.
2. Protein-A-Sepharose CL-4B and Sepharose 4B are from Pharmacia Fine Chemicals (Sweden). Hydrogen peroxide is from E. Merck (India).

3. Methods

3.1. Purification of Calf Thymus DNA

1. Purify commercially available calf thymus DNA free of proteins and single stranded regions as described *(10)*. Allow DNA (2 mg/mL) to swell overnight at 4°C in 0.1X SSC (15 mM sodium citrate and 150 mM sodium chloride, pH 7.4). Add equal volume of chloroform-isoamyl alcohol (24:1, v/v) and transfer the mixture to a stoppered cylinder and swirl end-to-end for 1 h. Separate the aqueous layer containing DNA from the organic layer and re-extract aqueous layer once again with chloroform-isoamyl mixture.
2. Precipitate DNA from the aqueous layer with two volumes of cold absolute ethanol in a beaker. Stir the beaker contents gently with a clean glass rod and allow DNA to collect on it. Press the rod against the wall of the beaker to get rid of excess alcohol and air-dry DNA.
3. Dissolve the purified DNA in acetate buffer (30 mM sodium acetate containing 30 mM zinc chloride, pH 5.0) and treat with nuclease S1 (150 U/mg DNA) at 37°C for 30 min to remove single-stranded regions. Stop the reaction by adding one-tenth volume of 200 mM ethylenediaminetetraacetic acid (EDTA), pH 8.0. Extract the nuclease S1-treated DNA twice with chloroform-isoamyl alcohol as before, precipitate with two volumes of cold absolute ethanol, collect on a glass rod, and air-dry. Dissolve the purified DNA in phosphate-buffered saline (PBS) (10 mM sodium phosphate, 150 mM sodium chloride, pH 7.4). Aliquot in small volumes and freeze.

3.2. Fragmentation of Purified DNA

1. Digest purified DNA by micrococcal nuclease to obtain smaller fragments *(10)*. Dissolve DNA (2 mg/mL) in 6 mM Tris-HCl, 100 mM NaCl, and 2 mM CaCl$_2$, pH 8.0, and treat with nuclease (40 U/mg DNA) at 37°C for exactly 2.5 min.

Incubate DNA and nuclease separately and after few minutes of equilibration, add the desired amount of nuclease to DNA solution. Stop the reaction by adding one-tenth volume of pre-incubated 200 mM EDTA, pH 8.0. Extract the nuclease-digested DNA twice with chloroform-isoamyl alcohol and precipitate DNA fragments with two volumes of cold absolute ethanol. Keep the fragments in cold for a few hours (can be kept overnight) for complete precipitation and collect the precipitate by centrifugation in cold.

2. Remove single-stranded regions by digestion with nuclease S1 as described under DNA purification. Dissolve purified DNA fragments in TBS (10 mM Tris-HCl, 150 mM NaCl, pH 7.4). Separate the DNA fragments on the basis of size by gel filtration on a Sepharose 4B column (46.0 × 1.2 cm) equilibrated with TBS. Elute the column with TBS at a flow rate of 18 mL/h. Collect fractions of 4.0 mL and record absorbance of each fraction at 260 nm. Plot the data as fraction number vs absorbance. Determine the size of DNA in each fraction by 7.5% acrylamide gel electrophoresis using a suitable DNA marker.

3.3. Modification of DNA by Hydroxyl Radical

1. Purified DNA or its fragments of known size can be modified with hydroxyl radical. Make all solutions in 0.01 M sodium phosphate buffer, pH 7.4, unless mentioned otherwise. Chelate the metal ions associated with DNA solution by pre-treatment with EDTA at a final concentration of 1.0 mM/L. Keep the mixture at room temperature for 3 h *(11)*.
2. Prepare assay tubes containing mixtures of DNA (1.515 mM bp), ferrous sulphate (0.525 mM), and hydrogen peroxide (3.03 mM) and incubate for 30 min at room temperature. Solutions of deionized DNA-hydrogen peroxide and DNA-ferrous sulphate will be the controls. At the end of incubation, dialyze the samples extensively against PBS and record UV absorption spectra of all samples, including that of untreated DNA dialyzed simultaneously. Account for any change in sample volume after dialysis.

3.4. Alkaline Sucrose Density Gradient Centrifugation

1. Assess the formation of single-strand breaks in hydroxyl radical modified DNA by alkaline sucrose density centrifugation. Run samples of native and modified DNA simultaneously having identical 260 nm absorbance. Take 0.2 mL of DNA samples, add equal volume of 0.2 N NaOH, and incubate for 10 min to allow denaturation.
2. Carefully layer the samples (0.4 mL each) separately on top of a 4.6 mL linear 5–20% sucrose gradient containing 0.8 M NaCl, 0.2 N NaOH, 0.01 M EDTA, and 0.015 M p-aminosalicylate, pH 12.5 *(12)*. Centrifuge the samples at 30,000 rpm for 1 h at 20°C in an SW 50.1 swinging rotor of Beckman ultracentrifuge or a similar equipment. After centrifugation, carefully pierce the bottom of the tubes and collect fractions of 0.7 mL. Record samples absorbance at 260 nm and plot the data.

3.5. Absorption-Temperature Scan

Carry out the thermal denaturation of DNA samples to ascertain the modification and to determine the melting temperature (Tm). Subject native and modified DNA samples of almost equal 260 nm absorbance (0.5–0.8) to heat denaturation in a UV spectrophotometer having a temperature programmer/controller assembly unit *(13)*. Start samples melting from 30–95°C at an increment of 1.5°C/min after 10 min equilibration at 30°C. Record change in absorbance at 260 nm with increasing temperature. Calculate percent denaturation.

$$\text{Percent denaturation} = \frac{A_T - A_{30}}{A_{95} - A_{30}} \times 100$$

Where A_T represent absorbance at T°C, A_{30} the initial absorbance at 30°C and A_{95} is the final absorbance at 95°C. Plot percent denaturation vs temperature and determine Tm, the temperature for 50% denaturation.

3.6. Isolation of IgG

1. Isolate serum IgG by affinity chromatography on Protein A-Sepharose CL-4B. Dilute serum (0.5 mL) with equal volume of PBS, pH 7.4, and apply to the column (0.9 × 15 cm) previously eqilibrated with PBS. Recycle the wash-through eluent 2–3 times and wash the column extensively with PBS to remove unbound material. Elute bound IgG with 0.58% acetic acid in 0.85% sodium chloride *(14)*. Collect fractions of 3.0 ml in tubes containing 1.0 ml of 1.0 M Tris-HCl, pH 8.5.
2. Record absorbance at 278 nm and determine the concentration of IgG taking 1.4 absorbance at 278 nm as equivalent to 1.0 mg IgG/mL. Extensively dialyze the isolated IgG against several changes of PBS and store in small aliquots at –20°C by adding few drops of 0.1% sodium azide. Ascertain the purity of IgG by nondenaturing PAGE.

3.7. Enzyme-Linked Immunosorbent Assay

1. Coat polystyrene microtiter plate (NUNC, Denmark) with 100 µL of nucleic acid antigen (2.5 µg/mL) in TBS (10 mM Tris-HCl, 150 mM NaCl, pH 7.4) for 2 h at room temperature and overnight at 4°C. Coat half plate with antigen; other half will serve as control. Wash the plate three times with TBS-T (20 mM Tris-HCl, 144 mM NaCl, 2.68 mM KCl, pH 7.4, containing 500 µL/L Tween 20). Shake the plate during each wash. Block the unoccupied sites with 125 µL of 1.5% bovine serum albumin (BSA) for 2 h at room temperature both in antigen coated and control wells. Wash the plate once with TBS-T. Coat the full plate with 100 µL of antibody (1 : 100 diluted serum or appropriate amount of IgG) diluted in PBS. Incubate the plate for 2 h at room temperature. Wash the plate thoroughly three times with TBS-T.

2. Add 100 µL of diluted anti-immunoglobulin alkaline phosphatase conjugate (use the dilution recommended by the manufacturer). Incubate the plate for 2 h at room temperature. Finally wash the plate three times with TBS-T. Add 100 µL of p-nitrophenyl phosphate (500 µg/mL) in carbonate-bicarbonate buffer (15 mM sodium carbonate, 35 mM sodium bicarbonate, pH 9.6, containing 2 mM magnesium chloride). Watch for the development of color and read the plate after proper color development in an ELISA reader at 410 nm. Coat each sample in duplicate. Express results as a mean of A_{test} - $A_{control}$.

3.8. Competition ELISA

Ascertain the antigenic specificity of antigen-antibody interaction by competition inhibition experiments *(9)*. Mix varying amounts of inhibitors to be tested (0–20 µg/mL) with a constant amount of antiserum (generally 1 : 100 dilution) or isolated IgG. Incubate the mixture at room temperature for 2 h and overnight at 4°C. Coat the mixture both in antigen-coated and control wells instead of serum or IgG. Perform the remaining steps as in case of direct-binding ELISA. Express the results as percent inhibition of antigen-antibody interaction.

$$\text{Percent inhibition} = 1 - \frac{A_{inhibited}}{A_{uninhibited}} \times 100$$

Draw the semilog plot of percent inhibition vs inhibitor concentration. Compute the concentration of inhibitor required for 50 inhibition of antigen-antibody interaction. Evaluate the relative affinity of various inhibitors taking the affinity of uninhibited sample as 100.

3.9. Results

1. A representative profile of calf thymus DNA fragments is shown in **Fig. 1**. Purified DNA was digested with micrococcal nuclease (40 U/mg DNA) for 2.5 min followed by nuclease S1 digestion. A *Hae*III restriction nuclease digest of φ × 174 RF DNA was the standard marker. The UV absorption spectra of ROS-modified DNA show hypochromicity at 260 nm compared to native DNA (**Fig. 2**). Alkaline sucrose gradient ultracentrifugation of native DNA demonstrate a single distinct species, whereas ROS-DNA indicate a diffuse sedimentation profile. The profile indicate a random distribution of single-strand breaks that appear as fragments and sediment in a diffused manner in contrast to nDNA (**Fig. 3**). Calf-thymus DNA fragments of 300 bp in length (average size) on ROS modification show decrease in melting temperature to the extent of 8°C, indicating a partial destruction of the secondary structure of DNA. Native 300 bp DNA show a Tm of 55°C (**Fig. 4**).
2. The binding of three randomly selected SLE anti-DNA antibody positive sera to ROS-modified DNA by direct-binding ELISA is shown in **Fig. 5**. Similar results

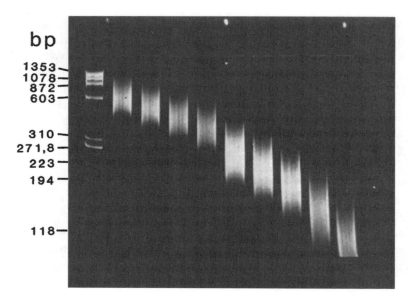

Fig. 1. Polyacrylamide gel electrophoresis of calf-thymus DNA fragments. The fragments were separated on a Sepharose 4B column. Electrophoresis was on a 5% polyacrylamide slab gel at 30 mA for 2.5 h. Gel was stained with ethidium bromide (1 μg/ml). Reproduced with permission from **ref. *10***.

are obtained when IgG is used instead of serum. Increasing concentration of IgG (0–20 μg/mL) result in increase binding to immobilized DNA. Normal human sera/IgG show negligible binding. Competition inhibition ELISA employing modified and native DNA fragments as inhibitors indicate that the modified DNA is a better inhibitor than native fragments (**Fig. 6**). The microtiter plate was coated with unmodified DNA (2.5 μg/mL). By this technique the antigenic specificity of a variety of nucleic acid polymers and their modified forms can be evaluated. **Table 1** demonstrates the antigenic specificity of an anti-ROS-DNA monoclonal antibody (MAb) by this technique with a variety of inhibitors.

4. Notes

1. Sodium azide from samples (DNA or IgG) must be removed by extensive dialysis before UV measurements as it absorbs strongly at 260 and 278 nm.
2. All glass containers must be oven sterilized at 120°C overnight to avoid contamination by nucleases.
3. Wear sterile gloves during DNA sample processing to arrest nucleic acid hydrolysis.
4. Digestion of DNA by micrococcal nuclease requires care. Variable hydrolytic activity of the enzyme has been found depending on the manufacturer and batch

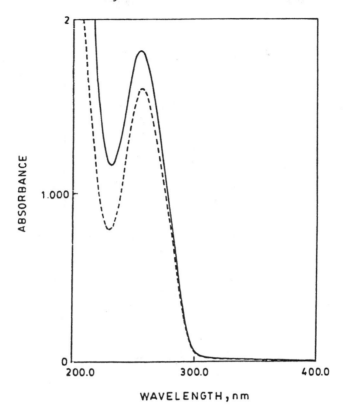

Fig. 2. Ultraviolet absorption profile of native (——) and hydroxyl-radical modified DNA (------).

number. Initially digest a sample of DNA with constant enzyme units but for variable time period. Agarose gel electrophoresis of each sample could be used to select optimum period of incubation. By varying the incubation period and enzyme units, DNA of almost any desire size can be obtained.

5. DNA fragments must be collected by centrifugation, not on a glass rod, as the latter method might result in the loss of smaller fragments.

6. Use deionized water for making buffer solutions and reagents required in the ROS-modification of DNA or its fragments.

7. In ELISA and competition-inhibition assay, proper washing of the ELISA plate is the crucial step. Shaking of the plate in an ELISA shaker or shaking by tapping the side of the plate with fingers, keeping the plate at room temperature for 2–3 min during each wash, and blotting of the ELISA plate to remove the maximum wash solution after each washing. It helps to minimize the unspecific binding and lower the optical density (OD) in control wells.

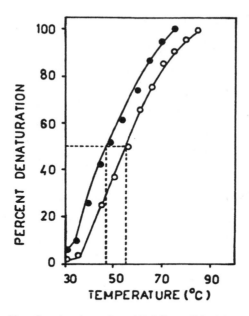

Fig. 3. Melting profile of native (–○–) and ROS-modified (–●–) DNA of 300 bp.

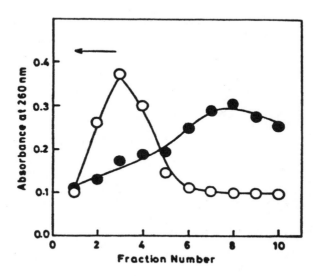

Fig. 4. Sedimentation profile of native (–○–) and ROS-modified DNA (–●–) through alkaline sucrose density gradient. The arrow on top shows the direction of sedimentation.

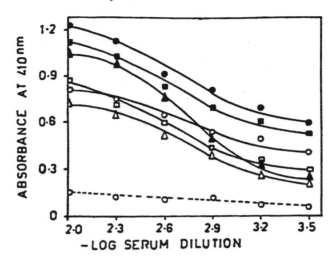

Fig. 5. Binding of SLE anti-DNA antibodies to native (—○—, —□—, —△—) and ROS-modified (—●—, —■—, —▲—) DNA fragments of 300 bp in length. Normal human serum pool with either of the antigens (--○---). The ELISA plates were coated with respective antigens.

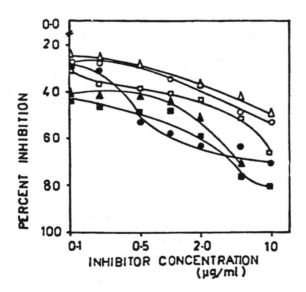

Fig. 6. Inhibition of SLE anti-DNA antibodies binding to native (—○—, —□—, —△—) and ROS-modified (—●—, —■—, —▲—) DNA of 300 bp. The ELISA plates were coated with native DNA fragments.

Table 1
Antigenic Specificity of Anti-ROS-DNA Monoclonal Antibody

Inhibitor	Maximum percent inhibition at 20 μg/mL	Concentration for 50% inhibition	Percent relative affinity
400 bp ROS-DNA	76.5	4.4	100
ROS-DNA	84.0	3.3	146.6
200 bp ROS-DNA	59.7	10	44
Native DNA	65.1	11	40
400 bp UV-DNA	35.9	—	—
Thymine	31.9	—	—
ROS-thymine	66.5	9.0	48.8
Thymidine	27.3	—	—
ROS-thymidine	57.2	9.0	48.8
Poly(dT)	34.8	—	—
ROS-poly(dT)	68.2	3.6	122.2
Guanine	27.3	—	—
ROS-guanine	46.8	—	—
Poly(G)	27.0	—	—
ROS-poly(G)	47.2	—	—

Cytosine, ROS-cytosine, poly(dA), ROS-poly(dA), and poly(dG) showed negligible inhibition. The ELISA plate was coated with 400 bp ROS-DNA.

References

1. Ashok, B. T. and Ali, R. (1999) Mini-review: the aging paradox- free radical theory of aging. *Exp. Gerontol.* **34**, 293–303.
2. Dizdaroglu, M. (*1993*) *DNA and Free Radicals.* Ellis Horwood, Chichester, UK, pp. 19–39.
3. Stollar, B. D. (1989) Immunochemistry of DNA. *Int. Rev. Immunol.* **5**, 1–22.
4. Ara, J. and Ali, R. (1993) Polynucleotide specificity of anti-reactive oxygen species (ROS) DNA antibodies. *Clin. Exp. Immunol.* **94**, 134–139.
5. Arif, Z. and Ali, R. (1996) Antigenicity of poly(dA-dT).poly(dA-dT) photocross-linked with 8-methoxypsoralen. *Arch. Biochem. Biophys.* **329**, 191–198.
6. Alam, K., Ali, A., and Ali, R. (1993) The effect of hydroxyl radical on the antigenicity of native DNA. *FEBS Lett.* **319**, 66–70.
7. Pisetsky, D. S. (1996) The immunologic properties of DNA. *J. Immunol.* **156**, 421–423.
8. Ashok, B. T. and Ali, R. (1998) Binding of circulating antibodies to reactive oxygen species modified-DNA and detecting DNA damage by a monoclonal antibody probe. *Mech. Ageing Dev.* **103**, 69–80.

9. Hasan, R., Ali, A., and Ali, R. (1991) Antibodies against DNA-psoralen crosslink recognize unique conformation. *Biochim. Biophys. Acta* **1073,** 509–513.
10. Ali, R., DerSimonian, H., and Stollar, B. D. (1985) Binding of monoclonal anti-native DNA autoantibodies to DNA of varying size and conformation. *Mol. Immunol.* **22,** 1415–1422.
11. Blakely, W. F., Fuciarelli, A. F., Wegher, B. J., and Dizdaroglu, M. (1990) Hydrogen peroxide-induced base damage in deoxyribonucleic acid. *Radiat. Res.* **121,** 338–343.
12. Fujiwara, Y., Tatasumi, M., and Sasaki, M. S. (1977) Cross-link repair in human cells and its possible defect in Fanconi's anemia cells. *J. Mol. Biol.* **113,** 635–649.
13. Hasan, R. and Ali, R. (1990) Antibody recognition of common epitopes on Z-DNA and native DNA brominated under high salt. *Biochem. Int.* **20,** 1077–1088.
14. Goding, J. W. (1978) Use of Staphylococcal protein-A as an immunological reagent. *J. Immunol. Methods* **20,** 241–253.

II

TECHNIQUES FOR ANTIOXIDANT BIOMARKERS

20

Simultaneous Analysis of Multiple Redox-Active Metabolites from Biological Matrices

Bruce S. Kristal, Karen Vigneau-Callahan, and Wayne R. Matson

1. Introduction

Studies in many areas of biology are hampered by the complexity of the system being studied. This suggests that such areas of study could benefit from the development and application of new and more powerful analytical tools. Traditionally, investigators have chosen analytical methods that offer sensitivity and specificity, ultimately fitting collected results into an overall global picture or an evolving theory. This approach has been effective for the study of disorders manifested by a single etiology or disorders where prior knowledge indicates a clear or probable mechanism. The approach has proven less satisfying, however, when applied to complex systems with many unknown influences. Factors such as the time and cost involved in sorting through myriad possibilities as well as biological constraints (e.g., sample-size constraints) have limited analysis carried out in such complex systems. A related, although distinct, problem lies in the inability to simultaneously measure a series of related metabolites even after the relevance of the metabolic system has been identified.

A high-performance liquid chromatography (HPLC)-based approach has been developed for studying these types of complex systems at the level of small molecule metabolites (1–8). The system complements other approaches currently used to study this problem at different levels. In the area of gene expression, the problem of dealing with complex systems has been addressed through the use of gene arrays. Gene arrays are hybridization chips that are capable of the simultaneous analysis of the expression of literally thousands

From: *Methods in Molecular Biology, vol. 186: Oxidative Stress Biomarkers and Antioxidant Protocols*
Edited by: D. Armstrong © Humana Press Inc., Totowa, NJ

of genes. In the area of proteonomics, complexity has been addressed through the use of 2D-gel electrophoresis coupled with mass spectrometry. The HPLC-based methodology presented here allows simultaneous detection and both qualitative and quantitative analysis of as many as 1200 redox-active low molecular-weight molecules in biological samples.

The CoulArray system combines the increased resolution and sensitivity of modern HPLC systems with an array of coulometric electrode detectors. Coulometric array detectors approach 100% efficiency. Therefore at the appropriate potential compounds react completely. Setting the potentials in a series array (typically 16 channels) at sequentially higher potentials allows resolution and detection of co-eluting compounds of different electro-chemical characteristics. This combination enables simultaneous separation of compounds in a biological sample by both chromatographic retention time and oxidation potentials. This added dimension enhances selectivity without diminishing the sensitivity of the coulometric electrodes (~1–10 pg on column). The approach described is not possible with conventional thin-layer amperometric detectors.

The two major disadvantages of the CoulArray system are the initial cost and the overall sensitivity of the system. Initial cost is high relative to other HPLC systems (currently ~$80,000 for the system described, proportionally lower for other options, such as 4- or 8-channel arrays). Thus, the CoulArray system is likely inappropriate for researchers interested in analysis of only a few metabolites in less complex biological matrices. Such analyses can be equally well-completed on other, less expensive systems, such as one utilizing a dual coulometric electrode. We are unaware of other systems capable of analyses similar to those made possible using the CoulArray. The closest detector systems are probably those based on diode-array detectors, which have approx 1–3 orders of magnitude lower sensitivity. The second disadvantage is that the overall sensitivity for detection is, as must be expected, also reflected in a requirement for tighter analytical controls to fully implement the advantages of the CoulArray. For example, impurities in buffers used in upstream experimental procedures can complicate analyses.

The major advantages of the CoulArray system include the sensitivity of the coulometric electrodes and the greater specificity and selectivity made possible by the coulometric array detectors. This system makes possible simultaneous analysis of greater numbers of analytes (e.g., entire pathways) relative to other systems. This advantage can help conserve scarce biological resources. The array also simplifies method development by allowing more complete separation of co-eluting peaks.

2. Materials

2.1. Equipment and Supplies

1. CoulArray System (ESA, Inc., Chelmsford, MA). The CoulArray system used includes two model 580 pumps, one gradient mixer/pulse dampner, one PEEK pulse dampner, a model 540 autoinjector, column heater, and a CoulArray detection system with 4 coulometric electrode detector cell modules (16 total channels).
2. Two TosoHaas TSK-GEL ODS-80$_{TM}$ columns in series (4.6 × 250 mm, 5μ C18)^2.
3. A computer and color ink-jet printer.
4. 1 mL and 200 μL adjustable pipets.
5. Polypropylene autosampler vials (250 μL) and caps.
6. Glass autosampler vials (1.5 mL) and caps.
7. Crimper for HPLC sample tube.
8. Vial trays.
9. Microcentrifuge evaporator with cold trap.
10. High-speed centrifuge.
11. Fume hood.

2.2. Reagents

1. Acetic acid (0.4%) in acetonitrile.
2. Methanol.
3. Isopropanol.
4. Acetonitrile.
5. Lithium acetate.
6. Glacial acetic acid.
7. Pentane sulfonic acid.
8. 0.85% saline solution.
9. Distilled, de-ionized water.

3. Methods

3.1. Sample Processing

3.1.1. Plasma or Serum

1. Plasma or sera samples (250 μL) are mixed with 1 mL of 0.4% acetic acid in acetonitrile in a 1.5-mL microcentrifuge tube. The samples are vortexed for 20 s at top speed, then centrifuged for 15 min at 12,000 rpm (11,000g). One mL of supernatant is taken off to a polypropylene autosampler vial and evaporated to dryness in a microcentrifuge evaporator. The dried residue is dissolved in 200 μL of mobile phase A (*see* **Subheading 3.2.1.**). Aliquots of 50 μL are injected into the Coularray system. This protocol conserves reactive species such as ascorbate, homogentistic acid, and 6-OH dopamine at 1 ng/mL concentrations.

2. Blood is typically drawn into a vacutainer, chilled to 4°C on ice, and centrifuged at $1800 \times g$ at 4°C for 15 min to obtain the plasma (in presence of anticoagulants) or serum (in the absence of an anticoagulant). Plasma is typically aspirated, avoiding the buffy coat since platelets and leukocytes can affect results for certain pathways, e.g., serotonin and metabolites. Blood samples may be collected with no anticoagulant or in a variety of different anticoagulants with certain restrictions. The use of EDTA as anticoagulant will lead to highly unstable ascorbic acid values unless the samples are processed immediately, i.e., treated with the acetic acid/acetonitrile precipitating agent within ca. 30 min after centrifugation. Heparin typically contains peaks that can interfere with indoles in the later times of the gradient assay. Oxalate contains peaks that can interfere with MHPG (3-methoxy-4-hydroxy phenyl glycol).

3. Serum values for serotonergic metabolites are less consistent than plasma values in ethylenediaminetetraacetic acid (EDTA) or heparin. Essentially any collection protocol must be evaluated for its specific effects and possible artifacts on the patterns as a whole.

3.1.2. Mitochondria

1. Mitochondria are isolated by standard differential centrifugation techniques. Following isolation *(7,8)*, mitochondrial samples are pelleted in a microcentrifuge and the supernatant is removed. As described *(7)*, we have found that washing the mitochondria in 160 m*M* KCl removes most of the organic buffers usually used in the isolation procedure. In most cases, this wash step is recommended, as the buffers used, e.g., HEPES, are highly electrochemically active and otherwise obscure portions of the chromatogram.

2. For analysis, mitochondrial pellets (typically containing ~2–10 mg protein) are suspended in 100 μL of ddH$_2$O in their original tube. The samples are vortexed for 20 s at top speed. One mL of 0.4% acetic acid in acetonitrile is added and the samples vortexed for another 20 s at top speed. The samples are placed in a cold pack (MeOH-containing tray prechilled at –80°C) and sonicated for 10 s at a setting of 4 (sonicator/cell disrupter, Model W-220F, Heat Systems-UltraSonics, Inc.). The samples are then centrifuged for 15 min at 12,000 rpm (11,000*g*). One mL of supernatant is taken off to a polypropylene autosampler vial and evaporated to dryness in a microcentrifuge evaporator. The dried residue is dissolved in 200 μL of mobile phase A (*see* **Subheading 3.2.1.**). An aliquot of 50 μL is injected into the Coularray system.

3.1.3. Cerebral Spinal Fluid

Samples are centrifuged for 5 min at 12,000 rpm (11,000*g*). An aliquot of 50 μL of the supernatant is injected into the Coularray system.

3.1.4. Urine

Urine samples are diluted 1:10 with ddH$_2$O, and aliquots of 50 μL of the diluted specimen are then injected directly into the array.

3.2. HPLC Mobile Phases

3.2.1. Mobile Phase A

1. Mobile Phase A stock preparation: Weigh out 399.98 g of pentane sulfonic acid. Add 1300 mL of ddH$_2$O, filter through GF/F Whatman filter paper. Filter through 0.2 μM nylon filter. Add 200 mL glacial acetic acid. Bring to 2 L with ddH$_2$O. The concentrated sulfonic acid stock solution, which is inherently contaminated, is cleaned by electrolyzing the final preparation over pyrolytic graphite at 1000 mV vs (αPd(H)) for 12–24 h.
2. Working Mobile Phase A preparation: Dilute 50 mL mobile phase A stock to a final volume of 1 L. Add 1 mg/L citric acid. Filter through 0.2 μm nylon filter. The pH of the final mobile phase is 3.02.

3.2.2. Mobile Phase B

1. Working Mobile Phase B preparation: Mix 8 L methanol, 1 L isopropanol, 1 L acetonitrile, 100 mL 4 M lithium acetate, pH 4.1, 200 mL glacial acetic acid, 10 mg citric acid/L. Filter through 0.2 μm nylon filter.
2. Lithium acetate (4 M): Add 672 g of lithium hydroxide to 1760 mL of glacial acetic acid, mix well over an ice bath. Add 880 mL of glacial acetic acid. Add 400 ml of ddH$_2$O. Adjust the pH with glacial acidic acid or 2 M LiOH such that a 1:20 dilution with ddH$_2$O is 4.10 (typically 200 μL acetic acid or 1 mL 2 M LiOH/0.05 pH units). Bring to a total volume of 4 L with ddH$_2$O. Filter with GF/F Whatman Filter paper.

3.3. HPLC Standards

3.3.1. Preparation and Storage

1. In practice, we use different sets of standards (~40–80 compounds each) *(5,6)* for different experiments. For example, one set of standards might be useful for measurements related to oxidative stress, and might include markers such as *o*-, *m*-, *p*-, *N*-tyrosine, glutathione, and glutathione disulfide, and so on, whereas markers for neurologic studies might include markers such as dopamine, kynurenine, kynurenic acid, homovanillic acid, and so on. These different standard sets can be used either alone or in combination, as appropriate. Under optimized conditions the CoulArray can resolve all 40–80 compounds in any given standard in a single chromatographic run.
2. Individual stock standards are currently prepared as follows: 10–100 mg of each standard to be used is placed into an appropriately labeled 100 mL amber bottle with a Teflon lined cap (Wheaton). Each standard is dissolved in 100 mL of either 20% MeOH solution or 0.85% saline solution, depending on the solubility of the standard being used. Appropriate dilutions of these standards (based on the samples to be analyzed) are then made into a final volume of 1 L using 0.85% saline as a diluent. We generally use between 20 μL and 1 mL of each stock standard. Individual standards and aliquots of mixed standards are stored in autosampler

Table 1
Gradient Method

Step	Time	Comment	%Mobile Phase B	Flow Rate (mL/min)
01	0.00	Flow	0	1.00
02	0.10	Autozero on		
03	0.56	Autosampler inject (1 s)		
04	1.00	File Start		
05	30.00	Flow	12	1.00
06	35.00	Flow	20	1.00
07	55.00	Flow	48	0.70
08	90.00	Flow	100	0.99
09	95.00	Flow	100	1.20
10	100.00	Flow	100	1.20
11	100.10	Flow	0	1.20
12	104.00	Flow	0	1.20
13	107.00	Flow	0	1.00
14	110.00	File Stop		
15	110.00	Clean Cell On		
16	114.00	Flow	0	1.00
17	114.50	Clean Cell Off		
18	120.00	Flow	0	1.00

vials at −80°C. Standards appear stable under these conditions for >7 yr. Vials of mixed standards are thawed to 4°C and mixed thoroughly before using.

3.3.2. HPLC Separations and Coulometric Array Analysis

1. The CoulArray system allows analysis under either isocratic or gradient conditions. The basic gradient method that we use (**Table 1**) has been published (*5–8*). Briefly, samples are sequentially eluted over 120 min as the mobile-phase gradient is increased from 0–100 %B. The last ~10 min incorporate a high-potential cell-cleaning step and restore the column condition to 100% mobile phase A. The detergent action of the sulfonic acids in the A mobile phase and the high organic solvent levels in the B mobile phase keep the column clean of residual proteins and lipids from the preparative protocols. The mobile phase selection and repetitive cell-cleaning enables continuous stable operation over 3–6 mo periods. Flow rates are adjusted to compensate for azeotropic viscosity. Analyte detection is accomplished using a 16-channel coulometric array detector as described in the Introduction and in the legend to **Fig. 1**, which shows chromatograms generated from studies of rat sera and rat liver mitochondria.
2. An important capability in generating multicomponent patterns is that chromatographic profiles can be easily modified to suit a specific individual application. Such modifications may include shortening the gradient profile (to reduce run

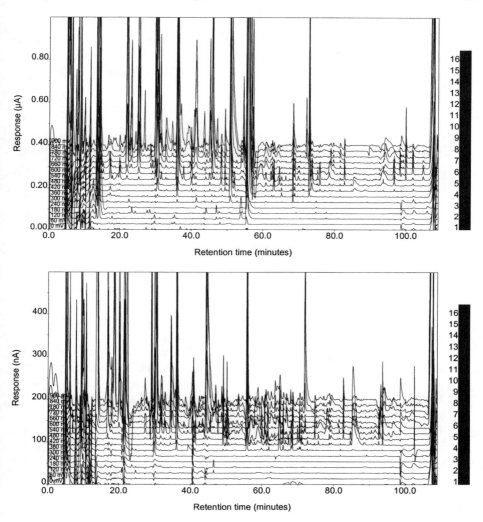

Fig. 1. Sample chromatograms. Analysis of sera collected from a 6-mo-old male Fischer x Brown Norway F1 rat **(top panel)**. Analysis of a liver mitochondrial sample collected from a 4-mo-old male Fischer 344 rat **(bottom panel)**. The specific mitochondrial sample shown was isolated by differential centrifugation using sucrose mannitol-based buffers. The mitochondria were processed as described in the text, including the 160 mM KCl wash. Full scale on the top and bottom chromatograms is 1 µA and 500 nA, respectively. In each case the chromatographic profile was obtained at 120 min.

time) when the analytes of interest are more hydrophilic, or lengthening portions in order to separate peaks which co-elute. For example, we have observed co-elutions between methionine and guanosine in some samples. If these peaks were important for a specific study, the chromatographic profile would be altered to accommodate these desired changes.

3.3.3. Data Analysis

1. HPLC analysis on the coulometric array can be used to generate databases of all of the redox active compounds in a sample. In the case of the chromatographic parameters presented here, all redox active molecules with hydrophilicities between ascorbate and tocopherol and redox potentials from 0–900 mV vs (αPd(H)) can be readily detected. Databases can be analyzed either for specific compounds of interest (e.g., dopamine), related compounds (e.g., the lipoates, hydroxylated phenylalanine byproducts), entire pathways (e.g., purine catabolites [uric acid, xanthine, hypoxanthine, xanthosine, guanosine, and guanine]), or combinations of these. Alternatively, metabolic patterns may be addressed using multivariate analysis techniques (e.g., cluster analysis, pattern recognition, etc.). Initial analyses in all cases are conducted using software supplied as a component of the CoulArray package.

2. The basic CoulArray for Windows 32 package (ESA, Inc.) is capable of carrying out all of the analysis described below (**Subheading 3.3.3.–3.4.**). This includes the qualitative analysis of peak identity as well as the quantitation of the peaks relative to either absolute or reference standards. Proprietary algorithms within the software automatically subtract backgrounds resulting from gradient drift. The software includes wizards designed to help individuals through most basic analyses.

3. The dynamic range enables analysis of analytes ranging from ~1 nA (~1–10 pg on column) upwards as much as five or more orders of magnitude (1–3 μg on column). In practice, compounds at upper limits of detection may display altered chromatographic behavior and/or electrochemical response on the sensors. Typically at very high levels there is "spill over" to following sensors and a resultant change in the response ratios. These changes may be controlled by selection of appropriate concentration levels in the control standards and are typically less of a factor in control of precision than recoveries during preparative procedures (for samples requiring such procedures).

4. Analysis of a series of analytes in a complex mixture is automated by first generating a compound table from a standard or from a pooled sample comprised of aliquots from most or all of the members that comprise the sample set.

5. In practice, initial quantitation requires manual oversight to confirm that the software has correctly identified peaks of interest. An earlier version of the software was, however, capable of >95% success in peak analysis after three training runs (we have not yet tested the Windows version under equivalent conditions). Peak values can be directly transferred to Windows applications (e.g., Lotus, Excel)

3.3.4. Qualitative Analysis

Basic qualitative analyses of data generated using coulometric arrays are generally carried out on the basis of three criteria: retention time, dominant channel, and the ratio of reactivity on the dominant channel to reactivity on subdominant channels, as has been explained previously *(1–6)*. The majority of these analyses are handled automatically by the CoulArray software, but the user can alter the parameters as appropriate, for example to relax or tighten standards to resolve potential conflicts. Comparisons are made to standards run in parallel or to appropriate peaks in the pool. Peak identity can be further investigated by spiking the sample(s) with the analytes of interest.

3.3.5. Quantitative Analysis

1. Quantitative analysis can be carried out in either of two ways. Absolute quantitation of specific, known analytes of interest can be carried out by direct comparison to an analytical standard of known analyte(s) of known concentration(s). The analytical standard can be run alone, or if desired, spiked into a duplicate sample.
2. Alternatively, currently unidentified peaks can be quantitated relative to a standard pooled sample. In this case, all analytes in the pool are assigned a specific arbitrary value (e.g., 100). Peaks in individual samples are then quantitated relative to this standard.

3.4. Results

1. Sera: Analyses of sera carried out at ESA, Inc. suggest that chromatographic retention times, monitored using authentic standards, do not vary by more than ~1% over a 30-d period. Absolute qualitative channel ratio responses do not vary more than ±20% and are controlled for by inclusion of authentic standards to within ±5%.
2. Mitochondria: Chromatographic parameters for mitochondria have been published *(7,8)* and variability observed was only slightly worse than that observed using sera samples. For our initial mitochondrial study, which was run on a CEAS (earlier generation of the CoulArray) criteria for qualitative acceptance of peaks was set at a retention time match of ±1.2% and a ratio accuracy of ±16% vs authentic standards. In this study, over a 1-mo time period, retention time of standards was held to within 1.7% based on raw data and to 1.1% when referenced to tyrosine. Mean C.V.% of retention times of analytes measured was 0.35%, mean C.V.% of the ratio of the dominant to the subdominant reactivity was ~11%. Note that, as described, some compounds are not included in this analysis.
3. Examples of the application of this chromatographic approach to sera and to mitochondria are shown in **Fig. 1**.
   In many cases, peaks having maximum amplitudes of 1 nA can be successfully visualized and examined. Thus peaks having amplitudes of ~0.1–0.2% of the full scale shown can generally be studied.

In both panels, the array was set from 0–900 mV in even increments of 60 mV. The temperature of cells and columns was maintained at 35°C. The exact chromatographic method used in the two chromatograms is shown in **Table 1**.

4. Notes

1. Other necessary equipment/supplies includes standard lab safety equipment (gloves, eye protection, lab coat), glassware (including funnels), microcentrifuge, and 10 mL polypropylene tubes. Also, –80°C or liquid-nitrogen storage will be required if the samples are not processed and run immediately.
2. Other columns have been successfully used in prior work (e.g., META 250, 4.6 × 250 mm, 5μ C18; 2 series MCM, 4.6 × 250 mm, 5μ C18).

Acknowledgments

We thank our colleagues for their comments on this manuscript, and Alex Shestopalov for preparing the mitochondrial sample shown in **Fig. 1**. This research was supported by NIA R01-AG15354 (BSK), NIH P01-AG14390 (J.P. Blass, P.I., BSK P.I., subproject 2), SBIR 2R44MH53810 (WRM), and internal funds from both the Burke Medical Research Foundation and ESA, Inc.

References

1. Matson, W. R., Langiais, P., Volicer, L., Gamache, P. H., Bird, E. D., and Mark, K. A. (1984) n-electrode three dimensional liquid chromatography with electrochemical detection for determination of neurotransmitters. *Clin. Chem* **30,** 1477–1488.
2. Svendsen, C. N. (1993) Multi-electrode array detectors in high-performance liquid chromatography: a new dimension in electrochemical analysis. *Analyst* **118,** 123–129.
3. Acworth, I. N. and Gamache, P. H. (1996) The coulometric electrode array for use in HPLC analysis, Part 1. Theory. *Am. Lab.* **5,** 33–38.
4. Acworth, I. N., Boweus, M. (1997) An introduction to HPLC-based electrochemical detection: From single-electrode to multi-electrode arrays. In: Coulometric Electrode Array Detections for HPLC, Progress in HPLC-HPCE, (Acworth, I.N., Naoi, M., Parvez, S., and Parvez, H., eds.) VSP Publications, Utrecht, Holland, vol. 6, pp. 1–48.
5. Milbury, P. E. (1997) CEAS generation of large multiparameter databases for determining categorical process involvement of biomolecules, in: Progress in HPLC, VSP International Science Publication, Utrecht, Holland. *Coulometric Electrode Array Detectors for HPLC*, vol. 6, pp. 125–141.
6. Milbury, P. E., Vaughan, M. R., Farley, S., Matula, G. J., Jr., Convertino, V. A., and Matson, W. R. A comparative bear model for immobility osteopenia. *Ursus* **10,** 507–520.
7. Kristal, B. S., Vigneau-Callahan, K. E., and Matson, W. R. (1998) Simultaneous analysis of the majority of low-molecular weight, redox-active compounds from mitochondria. *Anal. Biochem.* **263,** 18–25.
8. Kristal, B. S., Vigneau-Callahan, K. E., and Matson, W. R. (1999) Purine Catabolism: Links to mitochondrial respiration and antioxidant defenses? *Arch. Biochem. Biophys.* **370,** 22–33.

21

Determination of Uric Acid in Urine by Fast-Scan Voltammetry (FSV) Using a Highly Activated Carbon Fiber Electrode

Roberto Bravo, Dawn M. Stickle, and Anna Brajter-Toth

1. Introduction

Uric acid (UA), (7,9-dihydro-1H-purine-2, 6, 8 (3H)-trione), is the principal end product of purine metabolism (1); therefore, determinations of UA in biological samples can serve as a marker in detection of disorders associated with purine metabolism, such as gout, Lesch-Nyhan syndrome (2,3), and others (4–6). Diagnosis is confirmed by monitoring UA serum, or urinary levels. Normal UA serum levels range from 4.1–8.8 (mg dl^{-1}), and urinary excretion is typically 250–750 mg/d^{-1} (7). In addition recent studies have identified UA as a strong determinant of antioxidant capacity in serum (8) and the antioxidant capacity of UA has been determined by a total oxidant scavenging capacity (TOSC) assay (9). Consequently, it is essential to develop a simple and rapid method for the determination of UA, for routine clinical analysis.

UA has been determined in the clinical laboratory by colorimetric, enzymatic, and electrochemical methods. The colorimetric determinations are the most popular of all the methods, and involve oxidation of UA to allantoin and CO_2, by phosphotungstic acid, which is reduced to a tungsten blue chromophore compound. The absorbance of the chromphore, between 660 and 720 nm, is proportional to the concentration of UA (10,11). Numerous modifications of the colorimetric procedure have been developed (12–14), but the tendency of UA to co-precipitate with plasma proteins, the possible formation of turbidity in the final colored solution, the nonlinearity in relation to the color yield, and concentration of UA, over the range of commonly encountered concentrations

From: *Methods in Molecular Biology, vol. 186: Oxidative Stress Biomarkers and Antioxidant Protocols*
Edited by: D. Armstrong © Humana Press Inc., Totowa, NJ

of UA, and interferences, such as ascorbic acid (AA) and other reducing agents, can restrict the analytical applications of this method *(15)*.

Many enzymatic methods, based on uricase, which catalyzes the oxidation of UA to allantoin and hydrogen peroxide, have been reported. These methods are more sensitive and selective than the colorimetric methods, but sample preparation involves many steps, and is time-consuming *(16,17)*.

Direct electrochemical determinations of UA have also been reported *(18)*. Unfortunately, one of the major problems in the determinations of UA in biological samples are the electrochemical interferences, such as AA. At graphite electrodes, commonly used for the electrochemical determinations in biological samples, AA and UA have similar oxidation potentials, $E_{1/2}{\sim}200$ mV vs saturated calomel electrode (SCE), and since AA can be present at high concentrations in the biological samples, the large signal of AA interferes with the signal of UA. New electrochemical methods, based on surface-modified electrodes, have also been used, but these methods either require complex sample preparation, or have detection limits that need to be improved *(19,20)*.

Previous results from this group have demonstrated good selectivity and sensitivity in the electrochemical determinations of low concentrations of UA, in the presence of high concentrations of AA *(21,22)*. By using fast-scan voltammetry (FSV) with electrochemically pretreated carbon-fiber electrodes (CFEs), and by pushing voltammetric scan rates above 500 V s^{-1}, voltammograms of UA and AA have been resolved without a need for a permselective film on the surface of the CFEs. This is because the electrochemical kinetics of the two anions are sufficiently different in FSV, at the electrochemically pretreated CFEs, that the electrode reaction of UA can be detected without the interference from AA. Simultaneously, in the determinations of UA an improvement in the signal-to noise (S/N) ratio is achieved. The S/N is improved because of the high temporal resolution of FSV, which allows acquisition of a large number of scans, that can be signal averaged in a short period of time *(22)*.

FSV is a background subtraction technique, and analytical determinations by FSV are limited by background interferences, such as from a large background current, which accompanies all FSV measurements. The background current has at least two components: from the double-layer capacitance, associated with a layer of ions, which forms at the charged electrode/solution-interface, and from the oxidation and reduction of the electrode surface-bound species.

The sensitivity of FSV determinations depends on the stability of the background current *(23)*. Previously we have established experimentally that after the electrochemical pretreatment of the CFE, by continuous cycling of the electrode potential, in the potential window of 1.5 to –1.0 V vs SCE, for 30 min, at a scan rate of 10 Vs^{-1}, a stable background current was obtained at the pretreated CFE.

The electrochemical determination of UA in urine by the method of standard additions and FSV, at the electrochemically pretreated CFE, with signal averaging, is illustrated here.

2. Materials

2.1. Instrumentation

1. A Bioanalytical Systems Electrochemical Analyzer (BAS-100, West Lafayette, IN), coupled to a home-made pre-amplifier *(24)*, was used in all voltammetric determinations, at scan rates lower than 1 Vs^{-1}. The electrochemical data were downloaded to an IBM PS/2 model 50 computer and analyzed using Origin 5.0 (Microcal Software, Inc., Northampton, MA).
2. The instrumental set up for FSV has been previously described *(24)*. Briefly, a function generator (Universal Programmer, Model 175 EG & G, Princeton, NJ) was used to apply a triangular waveform to the SCE, in a two-electrode configuration potentiostat. The current at the carbon-fiber working electrode was converted to voltage, amplified, and was recorded by a digital oscilloscope (LeCroy Model 9310, Chestnut Ridge, NY). The stored waveforms were transferred from the oscilloscope to a computer for plotting and processing of the data. A copper-mesh Faraday cage was used to minimize the environmental noise.

2.2. Chemicals

1. A 5×10^{-3} M potasssium ferricyanide solution was prepared in 0.5 M KCl, pH 6.0.
2. UA has a low solubility in water, 6.5 mg/100 mL *(25)*. A 3×10^{-4} M standard stock solution of uric acid (Sigma, Chemical Co., St. Louis, MO) was prepared in 7×10^{-2} M potassium phosphate buffer, pH 7.4. Buffer pH was adjusted with HCl or NaOH.

2.3. Electrodes

An SCE was used as a reference electrode. Carbon fiber (7 µm diameter; Textron specialty materials, Lowell, MA) was used as the working electrode. Fabrication of the CFE has been previously described *(24,26)*. Briefly, a single carbon fiber was connected to a copper wire with silver epoxy (EPO-TEK 410 E, Epoxy Technology Inc., Billerica, MA). After the silver epoxy dried (ca. 24 h, at room temperature), the fiber was inserted into a micropipet tip and sealed with Shell epoxy. Shell epoxy was made by mixing the resin (Shell Epon 828, Miller-Stephenson Chemical Co., Danbury, CT), and hardener (12% by weight, metaphenylenediamine; Miller-Stephenson Chemical Co., Danbury, CT). The epoxy mixture was heated, in a water bath, until the epoxy became transparent and water-like. The micropipet tip was filled with the liquid epoxy, the electrode was left overnight at room temperature, and was next cured in an oven for 1 h at 150°C. After curing, the excess epoxy was removed from the electrode surface, by polishing on a 600-grit silicon carbide paper (Mark V laboratory, East Granby, CT), for 1 min, using a polishing wheel (Ecomet I, Buehler Laboratory, Evanston, IL).

The response of the polished electrodes was tested by cyclic voltammetry in 5×10^{-3} M $Fe(CN)_6^{3-}$ in 0.5 M KCl, pH 6.0, at a scan rate of 50 mVs^{-1}. The electrodes that did not respond were repolished; the electrodes that did not respond after the repolishing were discarded. In addition, electrodes with an initial high background capacitance, which was indicated by a large background current in cyclic voltammetry in the response of ferricyanide, were discarded.

Finally, the electrodes were gently polished for 1 min on an alpha A polishing cloth (Mark V laboratory, East Grandy, CT), with gamma alumina suspension of 0.1 µm particle size (Gamal, Fisher Scientific Co., Pittsburgh, PA), using a polishing wheel (Ecomet I, Buehler Laboratory, Evanston, IL). Immediately, after the last polishing step, the electrodes were dipped in 2-propanol for 10–15 min, and were then sonicated in H_2O for 5 min.

3. Methods

3.1. Sample Preparation

Urine was collected on the morning of the analysis. Exactly 100 µL was volumetrically diluted to 100 mL, with 7×10^{-2} M potassium phosphate buffer, pH 7.4, to give a dilution factor of a thousand. The diluted urine samples were mixed and were used the same day in FSV determinations of UA. All determinations were performed at room temperature.

3.2. Selection of CFE for Electrochemical Pretreatment (ECP)

1. The cyclic voltammetric (27) response of 5×10^{-3} M ferricyanide, $Fe(CN)_6^{3-}$, in 0.5 M KCl, pH 6.0, at 50 mVs^{-1}, was obtained in a potential window from 500 to -100 mV, vs SCE, as illustrated in **Fig. 1**, and was used in the initial selection of CFEs for the electrochemical pretreatment. The CFEs that were selected for the electrochemical pretreatment produced, before the pretreatment, a limiting current of ferricyanide (at -100 mV vs SCE) of $(5.0 \pm 0.5 \times 10^{-9}$ A, and a half-wave potential of $E_{1/2} \approx E^{\circ\prime}$, of 135 ± 10 mV vs SCE, as shown in **Fig. 1**. From the limiting current of ferricyanide of $(5.0 \pm 0.5) \times 10^{-9}$ A, an electrode radius of $(3.4 \pm 0.3) \times 10^{-6}$ mm is calculated, assuming a disk geometric area of the electrode (28).

2. The CFEs that produced the limiting current and the $E_{1/2}$ values summarized in **step 1** were electrochemically pretreated (ECP), in 7×10^{-2} M potassium phosphate buffer pH 7.4, for 30 min, by continuous cycling of the electrode potential, at a scan rate of 10 Vs^{-1}, in a potential window from -1.0–1.5 V vs SCE (22).

3. The apparent electrode capacitance of the CFE, which was measured before and after the ECP, C_{obs}, was used to verify that the ECP occurred. The C_{obs} was additionally used to verify the reproducibility of the pretreatment procedure.

4. The C_{obs} was determined from the voltammetric background current, that was measured in 7×10^{-2} M potassium phosphate buffer, pH 7.4, at a scan rate of

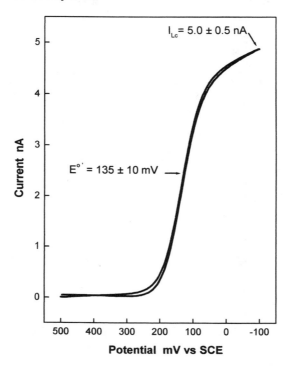

Fig. 1. Cyclic voltammetry at carbon fiber disk electrode (ca. 7 μm diameter) of 5×10^{-3} M Fe(CN)$_6^{3-}$ in 0.5 M KCl (pH 6.0) before ECP. Scan rate 50 mV s^{-1}.

10 Vs^{-1}, at 0.75 V vs SCE, as illustrated in **Fig. 2**. The background current was recorded in the same experimental potential window that was used in the ECP, 1.5 to −1.0 V, in the absence of the analyte in the buffer solution. It was assumed that at CFE, at 0.75 V, the background current due to surface faradaic reactions of graphite is low, and the voltammetric background current, which is measured by cyclic voltammetry, results only from double-layer charging. The C_{obs} was normalized by the area of the CFE, using the radius determined from cyclic voltammetry in ferricyanide solution, as described in **step 1**. The C_{obs} was calculated from the experimental background current, obtained by cyclic voltammetry in buffer *(29)*, as shown in **Fig. 2**.

Equation 1 was used to calculate the capacitance.

$$C_{obs} = \Delta I/2 \ vA \qquad (1)$$

In **Eq. 1** ΔI is the background current at 0.75 V, in A, measured from the separation between the anodic and cathodic sections of the voltammogram, as shown in **Fig. 2**, A is the area of the CFE in (cm^2), and v is the scan rate in Vs^{-1}.

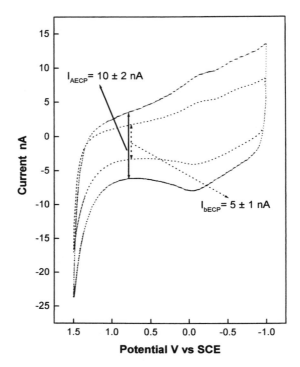

Fig. 2. Background current in cyclic voltammtery, in 7×10^{-2} M potassium phosphate buffer, pH 7.4, at a voltammetric scan rate of 10 V s^{-1}, before and after continuous potential cycling in 7×10^{-2} M potassium phosphate buffer, pH 7.4, in the potential window of -1.0 to 1.5 V vs SCE. Cyclic voltammograms obtained: (dashed curve) with a freshly polished carbon fiber electrode (ca. 7 µm diameter); (solid line) after 30 min of continuous voltammetric potential cycling, at a scan rate of 10 V s^{-1}. The voltammograms were signal averaged 50 ×.

3.3. Electrochemical Determination of UA by the Method of Standard Addition

1. Three 25-mL aliquots of a diluted urine sample were delivered into three 50-mL Erlenmeyer flasks. The Erlenmeyers were marked as sample, addition one, and addition two.
2. An 170-µL aliquot of a 3×10^{-4} M UA standard stock solution was delivered into the Erlenmeyer, marked as addition one. The solution was mixed, and the mixed solution was ready to be analyzed by FSV.
3. An 340-µL aliquot of a 3×10^{-4} M UA standard stock solution was delivered into the erlenmeyer, marked as addition two. The solution was mixed and was ready to be analyzed by FSV.

4. FSV is a technique which uses background subtraction. A FSV background was obtained first, by recording a voltammogram of 7×10^{-2} M potassium phosphate buffer, pH 7.4, at an experimental scan rate of 500 Vs^{-1}, using two hundred fifty scans (cycles), in the experimental potential window of 1.5 to -1.0 V. The background current, obtained in each cycle by FSV, was added and averaged, and was stored to be used for digital background subtraction. Next, 250 scans were recorded by FSV, in the same potential window and at the same scan rate, in the solution of UA, using the solution of UA from the Erlenmeyer marked as sample. The current, that was obtained in each cycle by FSV in UA solution, was added, averaged, stored, and used for digital data processing.

In the processing of the data, the averaged stored FSV background current was subtracted from the averaged FSV current of UA. For convenience of performing the background subtraction, the UA analyte and the phosphate buffer solutions, were each injected into the electrochemical cell (which has a volume of 80 μL) with a syringe. The injections of the solutions allowed the solutions to be pumped into the cell without moving the electrodes (24).

Each determination of UA was repeated at least four times, and the average of the readings is reported. The current of UA was measured by FSV at 0.5 V vs SCE. **Figure 3A** illustrates the voltammogram of UA obtained by FSV. The current of UA was measured at the oxidation peak, at 0.5 V vs SCE.

5. The same procedure as in **step 4**, above, was repeated in the determination of UA in urine, using the samples from the Erlenmeyers marked as addition one and addition two.

3.4. Results

3.4.1. Selection of CFE for the ECP

1. The limiting current and the $E_{1/2}$ of the voltammetric wave of ferricyanide, $Fe(CN)_6^{3-}$, in slow scan cyclic voltammetry was used to evaluate the quality of the CFEs before and after the ECP, and during the determination of UA. The apparent radius of the electrode was determined from the limiting current (22). Voltammetric response of ferricyanide as the electroactive probe was used in the evaluation of the electrodes because of the absence of adsorptive behavior of ferricyanide at CFEs and the well-known redox properties of ferricyanide (30).

2. Cyclic voltammetry at low scan rates, 50 mVs^{-1} was used to find the apparent geometric radius of the CFE, by considering a disk equivalent area of the electrode (28). The apparent radius of the CFEs that were used in the FSV determinations of UA is $(3.4 \pm 0.3) \times 10^{-6}$ mm. The radius was calculated from the limiting current of ferricyanide, from the voltammograms, such as that shown in **Fig. 1**, at -100 mV vs SCE, using the diffusion coefficient of $Fe(CN)_6^{3-}$, $D_o = 7.7 \times 10^{-6}$ (cm^2 s^{-1}) (31).

3. In the additional characterization of the CFEs, the kinetics of $Fe(CN)_6^{3-}$ at the CFE were determined from the plots of the data, such as those in **Fig. 1**, of potential vs log $((I_l - I_a) / I_a)$ of ferricyanide (27). A plot, before the ECP,

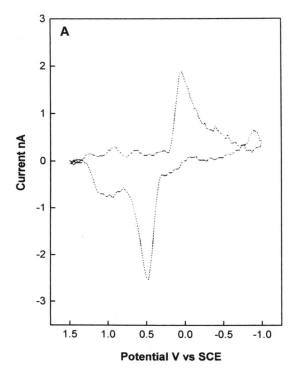

Fig. 3A. Fast scan cyclic voltammetry (FSV) of uric acid at carbon fiber electrodes (ca. 7 μm diameter) after the electrochemical pretreatment of the electroder. Figure **(A)** is the response of 3×10^{-6} *M* uric acid (UA) standard solution in 7×10^{-2} *M* potassium phosphate buffer, pH 7.4.

gave a slope of 100 ± 2 mV, which is indicative of an irreversible wave for a one-electron electrode redox reaction of ferricyanide, as expected for ferricyanide before the activation of the CFE by ECP *(32)*.

3.4.2. Electrochemical Activation of CFE

1. Background current observed in FSV, in a blank solution of 7×10^{-2} *M* potassium phosphate buffer, pH 7.4, is due to several processes, including double-layer charging, redox reactions of the surface functional groups, such as quinones, and the redox reactions of impurities in the electrolyte solution *(32)*. The amplitude of the background current may be more than two orders of magnitude greater than that of the current of the analyte. Consequently, FSV requires a stable background current to produce a reliable background, that can be used in background subtraction. Oxides *(32)* can form at the electrode surface, and the roughness of the electrode surface can change during the voltammetric determination, and these changes can produce significant modifications of the electrode background current.

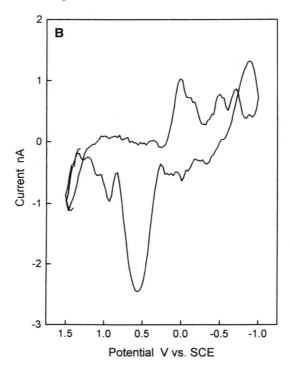

Fig. 3B. *(continued)* **(B)** is the response of UA in a urine sample (diluted solution) at a FSV scan rate of 500 V s^{-1}. 7×10^{-2} M phosphate buffer, pH 7.4, was used for dilution and for obtaining the background which was used in background subtraction. The cyclic voltammograms are background subtracted. 250 cycles for the background and UA were used for data processing and signal averaging.

2. Previously, we have established that a stable background current can be obtained at a CFE after 30 min of continuous cycling of the electrode potential, in the potential window from –1.0–1.5 (V) vs SCE, in 7×10^{-2} M potassium phosphate buffer, pH 7.4, at a scan rate of 10 Vs^{-1}. In addition, voltammetric results have shown that by using this form of the ECP of the CFE, it is possible to obtain high sensitivity and stability of the CFE in the FSV determination of UA.

3. A comparison of two voltammograms in **Fig. 2**, the voltammogram shown by a dashed line (before the ECP), and the voltammogram shown by a solid line (after the ECP), illustrates the significant changes at the CFE surface, which occur after 30 min of continuous cycling of the electrode potential in the potential window from – 1.0–1.5 V vs SCE, at a scan rate of 10 Vs^{-1}, as reflected by the large increase in the background current. As shown in **Fig. 2**, as a result of the ECP as specified in **step 2,** charging current increases in the entire potential window. The increase in the background current is directly proportional to the increase in the capacitance of the electrode, as shown by **Eq. 1**.

4. The increase in the capacitance of the CFE after the ECP has been associated with the formation of nanostructurated carbon-fiber surface as a result of the reactions of graphite, which occur at the CFE surface during the ECP, under the conditions specified in **step 2**. After the ECP, the high electrode capacitance, which is reflected by the large background current, indicates that the surface structure of the CFE has a high density of carbon defects, with a relatively low density of nonconducting surface oxides, as indicated by the relatively low current at ca.1.5 V. The formation of some defect oxides at the CFE surface after the ECP has been proposed *(22,26,33)*.

5. The voltammetric background current values, measured in 7×10^{-2} *M* potassium phosphate buffer, pH 7.4, at 0.75 V vs SCE, at a scan rate of 10 Vs^{-1}, before (5 ± 1 nA) and after (10 ± 2 nA) the electrochemical treatment of the CFE surface, were therefore used as the diagnostic values in the determination of the quality of the CFE surface before and after the ECP, and in the determination of the stability of the CFE after the electrochemical determination of UA. Only, the electrodes that had the background current values specified above were used in the analytical determinations of UA.

3.4.3. Electrochemical Determination of UA

1. In previous work *(22)*, linear calibration curves of UA were obtained in the concentration range of 1.0–20 µ*M*, in 7×10^{-2} *M* potassium phosphate buffer, pH 7.4, with a slope (sensitivity) of 0.44 ± 0.02 nA µ*M*$^{-1}$, a correlation factor of 0.998, and a limit of detection (LOD) of 0.5 µ*M* (S/N $\geq$ 3).

2. It is well-known that ascorbic acid coexists with uric acid in many biological samples, such as urine; therefore, the interference by AA has been previously investigated in detail *(21,22)*. It was found that the high sensitivity and selectivity of FSV at (500 Vs^{-1}) at a CFE, after the ECP of the electrode surface according to the procedure described in **step 2** of **Subheading 3.4.2.**, allows the determination of low concentration of UA in the presence of high concentrations of ascorbic acid, without a permselective film at the surface of the CFE *(22)*.

3. **Figure 3A** shows the FSV that was obtained in a standard solution of UA, in phosphate buffer, and in **Fig. 3B** a FSV of UA in urine is illustrated for comparison. The voltammogram in B was obtained before the addition of the standard stock solution of UA. A clear oxidation peak of UA is obtained by FSV at 0.5 V vs SCE, at a CFE after the ECP, and the oxidation peak of UA is well-developed in both voltammograms. The concentration of UA in urine was determined by FSV, using the method of standard additions, and the plot of the data is illustrated in **Fig. 4**.

4. The concentration of UA in the diluted urine sample was calculated from the plot of the data in **Fig. 4**, by using the following equation:

$$C_{Diluted} = (- V_s' C_s) / V_t \tag{2}$$

V_s' was determined form the plot, at I* value equal to zero, as shown in **Fig. 4**, C_s was the concentration of the standard stock solution, and V_t was the total

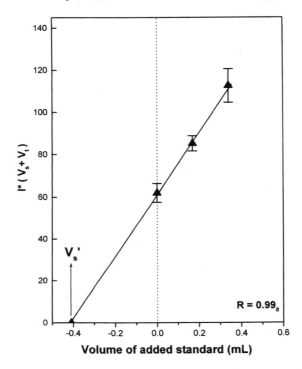

Fig. 4. Standard addition plot for the determination of UA in urine by FSV at a voltammetric scan rate of 500 V s^{-1}, at an electrochemically pretareted carbon fiber electrode (ca. 7 μm diameter). Data obtained in urine sample after 1000-fold dilution with 7×10^{-2} *M* potassium phosphate buffer, pH 7.4. 7×10^{-2} *M* potassium phosphate buffer, pH 7.4, was used to prepare standard solutions of UA used in standard addition.

volume of the sample. The total UA concentration was calculated by multiplying the concentration, determined for the diluted sample from the plot in **Fig. 4**, by a thousand, which was the dilution factor.

5. **Table 1** summarizes some of the analytical results of the determination of UA in urine sample, using the method of standard addition, and FSV at the CFE. The results in **Table 1** indicate that the concentrations of UA, that were determined in urine samples by FSV at 500 Vs^{-1}, are similar to the values reported in the literature, which are between 2–5 m*M* (*34–36*).

6. The linearity of the standard addition plots (R = 0.99$_8$) suggests that the high activity of the CFE in FSV, which is achieved after the ECP of the CFE surface, can be maintained during the determination of UA in urine samples. Consequently, freedom from poisoning of the CFE surface during the FSV determinations, by compounds present in urine, is observed in the determinations of UA in urine samples that were prepared by dilution, by methods described in the Procedures section.

Table 1
Determination of UA in Urine Samples at Active CFE Surface
by FSV at 500 Vs^{-1a}

Sample	Diluted concentration $\times 10^{-6}$ M	Total concentration[b]	
		$\times 10^{-3}$ M	$\times 10^{-8}$ mg dl^{-1}
1	3.0 ± 0.7	3.0 ± 0.7	5 ± 1
2	4.3 ± 0.5	4.3 ± 0.5	7 ± 1
3[c]	2.3 ± 0.4	2.3 ± 0.4	3.9 ± 0.6

[a]Results were background subtracted. 250 cycles for background and analyte were stored, averaged and used for processing of the data. Electrode diameter = 7 μm. 7×10^{-2} M phosphate buffer, pH 7.4, was used as background.

[b]Total value obtained by multiplying the diluted concentration value with the dilution factor of 1000.

[c]Sample belongs to a person who was diagnosed with UA elimination problems.

7. An additional advantage of FSV at the CFE, as used here, is the simplicity of the method and the short time required to complete the determination of UA, to obtain a voltammogram that is used to measure the current of UA, and subsequently the concentration of UA in an urine sample. This facilitates signal averaging, and thus S/N ratios are improved, as is the sensitivity in the determination of UA. We need to point out that the determination UA was achieved without deoxygenation of the urine samples.

8. Utility of FSV at CFEs, in the determination of UA, after the ECP of the CFE surface is a result of the method of the electrochemical treatment of the CFE. Potential cycling of CFE at 10 Vs^{-1} in 7×10^{-2} M potassium phosphate buffer pH 7.4, in a potential window from −1.0–1.5 V vs SCE, before the determination, and the acquisition of the background current between each determination of UA, facilitates background subtraction, and produces a CFE surface that is characterized by a large area. High density of surface defects contributes to the high area and these, in turn, contribute to the high activity of the electrode surface in FSV determinations of UA. Such CFE surfaces allow sensitive and reproducible determinations of UA in urine samples, as demonstrated here.

9. The high sensitivity and the low LOD of this method of UA determinations allows the use of a large dilution factor of 1000 for urine samples, which produced two advantages.

   a. First, the matrix effects d not affect the simple and direct determination of UA.

   b. Second, no sample preparation, other than dilution with a physiological buffer, was required for the analysis, which makes the determinations very rapid.

10. The high activity of the CFE surface in FSV determinations of UA in urine samples is reflected by the large slope of the calibration curves (sensitivity) and the low LOD (S/N ≥ 3).

References

1. Dryhurst, G. (1977) *Electrochemistry of Biological Molecules*. Academic Press, New York, pp. 71–180.
2. Star, L. V. (1995) Gout: options for its therapy and prevention. *Hosp. Med.* **31,** 25–37.
3. Harper, H. A. (1977) *Review of Physiological Chemistry, 16th ed.* Lange Medical Publications, CA, p. 406.
4. Heptinstall, R. H. (1966) *Gout. Pathology of the Kidney*, 2nd ed. Little Brown, pp. 495–496.
5. Krakoff, I. H. (1965) Studies of uric acid biosynthesis in the chronic leukemias. *Arthritis Rheum.* **8,** 772.
6. Puig, J. G. and Ruilope, L. M. (1999) Uric acid as a cardiovascular risk factor in arterial hypertension. *J. Hypertent.* **16,** 40.
7. Kissinger, P. T., Pachla, L. A., Reynolds, L. D., and Wright, S. (1987) Analytical methods for measuring uric acid in biological samples and food products. *J. Assoc. Off. Anal. Chem.* **70,** 1–14.
8. Rosell, M., Regnsrtom, J., Kallner, A., and Hellenius, M. L. (1999) Serum urate determines antioxidant capacity in middle-aged men- a controlled, randomized diet and exercise intervention study. *J. Intern. Med.* **246,** 219.
9. Regoli, F. and Winston, G. W. (1999) Quantification of total oxidant scavenging capacity of antioxidants for peroxynitrite, peroxyradicals, and hydroxyl radicals. *Toxicol. Appl. Pharmacol.* **156,** 96.
10. Folin, O. and Denis W. (1912–1913) A new (colorimetric) method for the determination of uric acid in blood. *J. Biol. Chem.* **12,** 469.
11. Pilleggi, J. V., DiGiorgio, J., and Wybenga, R. D. (1972) A one tube serum uric acid method using phosphotungstic acid as protein precipitant and color reagent. *Clin. Chim. Acta.* **37,** 141.
12. Benedict, R. (1930) An inmproved method for the determination of uric acid in blood. *J. Biol. Chem.* **86,** 179.
13. Jung, H. D. and Parekh, C. A. (1970) An imporved reagent system for the measurement of serum uric acid. *Clin. Chem.* **16,** 247.
14. Tietz, N. W. (1987) *Fundamentals of Clinical Chemistry*, 3rd ed. W. B. Saunders, Philadelphia, PA., pp. 686–687.
15. Wyngaarden, B. J. and Kelley, N. W. (1976) *Gout and Hyperuricemia*. Grune & Stratton, pp. 60–63.
16. Dilena, B. A., Peake, M. J., Pardue, H. L., and Skorg, J. W. (1986) Direct ultraviolet method for enzymatic determination of uric acid with equilibrium and kinetic data-processing options. *Clin. Chem.* **32,** 486.
17. Haeckel, H. (1978) Improved enzymatic uric acid determination. *Clin. Chim. Acta* **24,** 1846.
18. Mueller, K., Palmour, R., Andrews, C. D., and Knott, P. J. (1985) In voltammteric evidence of production of uric acid by rate caudate. *Brain Res.* **335,** 231.
19. Markas, A., Gilmartin, T. A. M., Hart, P. J., and Birch, J. B. (1992) Voltammetric and amperometric behaviour of uric acid at bare and surface-modified screen-printed electrodes: studies toward a disposable uric acid sensor. *Analyst* **117,** 1299.

20. Markas, A., Gilmartin, T., and Hart, J. (1994) Development of amperometric sensor for uric acid based on chemically modified graphite-epoxy resin and screen-printed electrodes containing coblat phtalocyanine. *Analyst* **119,** 243.

21. Hsueh, C. C., Bravo, R., Jaramillo, A., and Brajter-Toth, A. (1997) Enhancement of selectivity and sensitivity analysis with FSV at scan rates above 1,000 V/s. *Anal. Chim. Acta* **349,** 67.

22. Bravo, R., Hsueh, C. C., Jaramillo, A., and Brajter-Toth, A. (1997) Possibilities and limitations in miniaturized sensor design for uric acid. *Analyst* **123,** 1625.

23. Stamford, J. (1989) In vivo voltammetry: prospects for the next decade. *Trends Neurosci.* **12(10),** 407.

24. Hsueh, C. and Brajter-Toth, A. (1994) A simple current transducer for ultramicroelectrode measurements at a wide range of time scales. *Anal. Chim. Acta* **349,** 67.

25. Wyngaarden, B. J. and Kelley, N. W. (1976) Physiology and biochemistry of purine compounds, in *Gout and Hyperuricemia*, Grune & Stratton, New York, pp. 53–67.

26. Hsueh, C. and Brajter-Toth, A. (1993) Fast scan voltammetry in aqueous solution at CFUME's with on-line IR compensation. *Anal. Chem.* **65,** 1570.

27. Bard, A. J. and Faulkner, R. L. (1980) *Electrochemical Methods. Fundamentals and Applications.* Wiley, New York.

28. Kelly, R. and Wightman, M. R. (1986) Bevelled carbon fiber ultramicroelectrodes. *Anal. Chim. Acta* **187,** 79.

29. Anjo, M. D., Karh, M., Khodabakhsh, M. M., Nowinski, S., and Wanger, M. (1989) Electrochemical activation of carbon electrodes in base: minimization of dopamine adsorption and electrode capacitance. *Anal. Chem.* **61,** 2603.

30. Kawiak, J., Kulesza, J. P., and Galus, Z. (1987) A search for conditions permitting model behavior of the $Fe(CN)_6^{3-}$ system. *J. Electroanal. Chem.* **226,** 305.

31. Stackelberg, M., Pilgram, M., and Toome, V. Z. (1953) Bestimmung von Diffusionskoeffizienten einiger Ionen in WaBriger Losung in Gegenwart von Fremdelektrolyten. *Zeit. Electrochem* **57,** 342.

32. McCreery, L. R. (1991) Carbon electrodes: structural effects on electron transfer kinetics, in *Electroanalytical Chemistry, vol. 17.* (Bard, A. J., ed.), Marcel Dekker, New York, p. 221.

33. Bravo, R. and Brajter-Toth, A. (1999) A highly active carbon fiber electrode surface for the determination of uric acid in physiological buffers. *Chem. Anal. (Warsaw)* **44,** 423.

34. Cai, X., Kalcher, K., Neuhold, C., and Oborevc, B. (1994) An improved voltammetric method for the determination of trace amounts of uric acid with electrochemically pretreated carbon paste electrodes. *Talanta* **41,** 407.

35. Zen, M. J. and Chen, P. J. (1997) A selective voltammetric method for uric acid and dopamine detection using clay-modified electrodes. *Anal. Chem.* **69,** 5087.

36. Zen, M. J. and Ilangovan, G. (1998) Selective voltammetric method for uric acid detection using pre-anodized Nafion-coated glassy carbon electrodes. *Analyst* **123,** 1345.

22

Measurement of α-Tocopherol Turnover in Plasma and in Lipoproteins Using Stable Isotopes and Gas Chromatography/Mass Spectrometry

Elizabeth J. Parks

1. Introduction

As the primary fat-soluble antioxidant in human tissues, α-tocopherol absorption and metabolism have been the focus of active investigation, as reviewed recently *(1,2)*. Although the concentration of any metabolite measured in plasma or tissue at a given time can be used as an indicator of the status of that pool, the true biological activity of molecules is better represented by their turnover. The use of stable isotopes in biological research has expanded steadily since the 1950s. The availability of methods has also increased as a result of improvements in instrument sensitivity and automation. Vitamin E studies utilizing gas chromatography-mass spectrometry (GC-MS) have aided in the elucidation of vitamin E absorption and transport, as well as the identification of vitamin E oxidation byproducts *(3)*. Human vitamin E studies using stable isotopes of α-tocopherol (deuterated molecules) have evaluated plasma α-tocopherol levels in response to increasing vitamin E doses *(4)* and the selective secretion of *RRR*-α-tocopherol from the liver out in nascent very low-density lipoproteins (VLDL) *(5,6)*.

A number of excellent methodology papers exist for the quantitation of vitamin E by high-performance liquid chromatography (HPLC) *(7–11)* and methods for measurement by mass spectrometry are also available *(12–15)*. The stable isotope method presented here is a modification of that developed by Burton and Daroszewska *(16)*, while focusing on plasma and lipoprotein concentrations and including information on the measurement of vitamin E turnover. When combined with lipoprotein and tissue subfractionation proce-

From: *Methods in Molecular Biology, vol. 186: Oxidative Stress Biomarkers and Antioxidant Protocols*
Edited by: D. Armstrong © Humana Press Inc., Totowa, NJ

dures, these techniques can be used to study the secretion of vitamin E by the liver and its clearance from the blood to increase our understanding of the role of vitamin E in human health and disease.

The measurement of vitamin E concentration in plasma and lipoprotein fractions and the calculation of its turnover in these pools is a relatively simple multistep process that encompasses: 1) isolation of lipoprotein fractions, 2) extraction of vitamin E, 3) measurement of the enrichment of d_6-*RRR*-α-tocopherol by GC-MS, 4) calculation of vitamin E concentration and turnover.

2. Materials

2.1. Equipment/Supplies

1. Purple-top vacutainers [Beckton Dickinson #366457, 10 mL containing K_3 ethylenediaminetetraacetic acid (EDTA)].
2. Ultracentrifuge (Beckman L6).
3. 50.3 Ti rotor (Beckman).
4. Gas chromatography-mass spectrometer with electron-impact ionization in the selected-ion monitoring mode (e.g., Hewlett-Packard 6890 GC and 5973 MS, Palo Alto, CA).
5. DB-17HT column: 30 m, 0.25 i.d., 0.15-micron film thickness (J&W Scientific, Folsom, CA).
6. Vortexor.
7. Low-speed refrigerated, table-top centrifuge.
8. Nitrogen evaporator.
9. GC vials, with Teflon or PFTE caps.

2.2. Isotopes

1. $2R$ $4'R$ $8'R$-α-$(5,7-(C^2H_3)_2)$ tocopheryl acetate (d_6-*RRR*-α-tocopheryl acetate) from the Henkel Corporation (Lagrange, IL).
2. $2R$ $4'R$ $8'R$-α-$(5,7,8-(C^2H_3)_3)$-α-tocopherol (d_9-α-tocopherol) is used as an internal standard. Standards are diluted in *n*-decane, which has a relatively high boiling point (174°C) (*see* **Note 1**).

2.3. Solvents/Chemicals

1. Heptane (HPLC grade).
2. Methylene chloride.
3. Ethanol (absolute).
4. Acetonitrile.
5. Pyridine: stored in a stoppered bottle to revent water saturation (*see* **Note 2**).
6. d_0-α-tocopherol.
7. N,O-(bistrimethylsilyl)trifluoroacetamide:trimethylchlorosilane (BSTFA:TMCS).
8. Butylated hydroxytoluene (BHT).

3. Methods

3.1. Administration of Isotope with a Meal

A capsule containing (d_6-*RRR*-α-tocopheryl acetate) is administered to the subject with a meal in which the energy derived from fat is at least 15% (*see* **Note 3**). At repeated timepoints after administration of the capsule (time denoted T_{zero}) plasma is then drawn into EDTA-containing vacutainers (1 mg/mL blood) and BHT is added to a final concentration of 0.01%.

3.2. Isolation of Lipoprotein Fractions

1. Plasma is separated by centrifugation of the samples at 2,500 rpm for 25 min in an ultracentrifuge rotor (50.3) at 5°C (1.6×10^6g) to separate chylomicrons *(17)*.
2. Lipoprotein samples of interest are isolated by sequential density ultracentrifugation *(18)*: VLDL (d < 1.006 g/mL), LDL (d = 1.006–1.063), and/or HDL (d > 1.063).
3. Each lipoprotein fraction is collected in the top 1 mL of density solution and transferred to amber borosilicate glass tubes.
4. Samples in tubes are nitrogen-topped, capped (teflon or PTFE, polytetrafluoroethylene caps) and refrigerated until extraction.

3.3. Vitamin E Extraction and Preparation for GC/MS

Different quantities of internal standard (d_9-α-tocopherol) are added to plasma and the various lipoprotein fractions because the concentrations of the vitamin E in plasma and lipoproteins are quite different. The quantity of internal standard should approximate the total amount of d_0-α-tocopherol present in the fraction. Typical quantities of lipoproteins present in normolipidemic plasma are shown in **Table 1**, along with the average quantity of α-tocopherol in these fractions. Subjects with elevated cholesterol levels should have higher concentrations of tocopherol in their blood and the internal standard should be adjusted accordingly.

1. Add a known quantity of internal standard (d_9-α-tocopherol) to the 1 mL of sample.
2. Add 1 mL 100% ethanol and 1 mL heptane.
3. Vortex 1 min.
4. Centrifuge at 2,000g for 5 min in a refrigerated table-top centrifuge to achieve phase separation.
5. Transfer organic layer (top) to a GC vial, and dry under N_2.
6. Re-extract the original sample vial with a second aliquot of heptane (1 mL), add to GC vial and re-dry under nitrogen.
7. Add 200 μL methylene chloride (azeotrope) to the GC vials.
8. Vortex once lightly, then re-evaporate under nitrogen.

Table 1
Concentrations of Cholesterol and Vitamin E in Plasma and Lipoprotein Fractions

| | Concentration of cholesterol | | Quantity of vitamin E | | | |
| | FC | CE | Men | | Women | |
Fraction[a]	mmol/L	mmol/L	nmol/ml	%[b]	nmol/ml	%[b]
Plasma	1.296	3.75	25		25	
VLDL	0.104	0.207	2	8	3	12
LDL	0.959	2.506	12	59	10	42
HDL	0.233	1.033	6	33	12	59

[a]Abbreviations: FC, free cholesterol; CE, cholesterol ester; VLDL, very low-density lipoprotein; LDL low-density lipoprotein; HDL, high-density lipoprotein.
[b]Percentage of total plasma vitamin E in this lipoprotein fraction.

9. Add 100 µL pyridine and 50 µL BSTFA:TMCS (10:1, v/v).
10. Nitrogen top vials, cap, and vortex sample briefly.
11. Heat at 65°C for 20 min.
12. Following heating, the sample is directly injected into the GC-MS column.

3.4. Gas Chromatography-Mass Spectrometric Analysis

The d_0-, d_6-, and d_9 enrichments are assessed by GC-MS using a DB-17HT column with acetonitrile as a syringe rinse solvent. The temperature program is as follows: initial temperature: 85°C, then increasing at a rate of 40°C/min to 310°C, then an increase of 3°C/min to 317°C, which is held for 8 min. The injector temperature is 300°C and the detector 300°C. Selected ion monitoring is carried out at m/z 502, 508, and 511. A sample electron ionization mass spectra of the derivatives is shown in **Fig. 1** (*see* **Note 4**).

3.5. Calculations

3.5.1. d_6-α-Tocopherol Enrichments and Concentrations of d_6- and d_0-α-Tocopherol

The amount of d_9 internal standard added to the samples depends on the estimated amount of d_0 present in the sample (*1*). The absolute concentrations of unlabeled (d_0-) and labeled (d_6-) α-tocopherol in plasma and lipoprotein are calculated using two standard curves, the preparation of which is described next.

3.5.1.1. PREPARATION OF D6 AND D9 STANDARD CURVES

1. The d_6/d_0-α-tocopherol standard curve is made by mixing known amounts of the two isotopes. The points of the d_6 standard curve, from 1–20% (% d_6-enrichement)

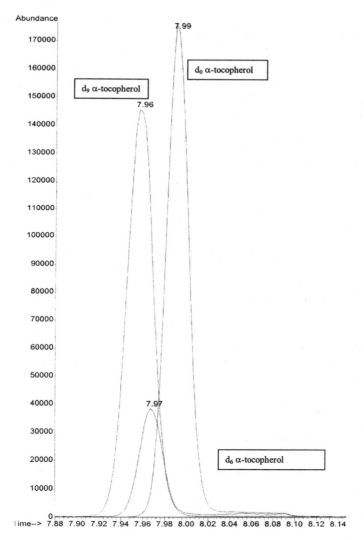

Fig. 1. Electron ionization mass spectra of silyl derivatives of d_0, d_6, and d_9 α-tocopherol (m/z 502, 508, and 511).

are based on the expected ratio of d_6-α-tocopherol to total ($d_6 + d_0$) α-tocopherol. We use a five-point standard curve typically including 1, 2, 5, 10, and 20% d_6 (*see* **Note 5**).

$$\%d_6 = d_6 \text{ conc} / (d_6 \text{ conc} + d_0 \text{ conc}) \qquad (1)$$
or, for example, the 10% $d_6 = 4$ ng d_6 / (4 ng d_6 + 36 ng d_0)

As shown below, the convention of the sum of isotopes ($d_6 + d_0$) in the denominator must be carried throughout the final calculations, otherwise the amount of d_6 in the sample will be underestimated (this point is often missed).

2. Second, a similar d_9/d_0-α-tocopherol standard curve is constructed where the designated percentage of d_9 equals the absolute quantity of d_9 divided by the sum of the known amounts of d_9 and d_0. Our d_9/d_0-α-tocopherol standard curve ranges from 5–30% and includes 5, 10, 15, 20, 25, 30%.

3. A GC-MS run was constructed in the following order: The %d_6 standard curve was run, followed by the %d_9 standard curve, followed by the samples (unknowns) run in duplicate, followed by a repeat of the %d_6 standard curve and the %d_9 standard curve. During this run, the same MS method is used for all standards and samples. This MS method obtains the spectrum for all ions, d_0, d_6, d_9 (m/z 502, 508, 511) but not all of these data are used in every calculation. The reason all ions are collected for each run is to standardize the amount of detector time given for the acquisition of both samples and standards.

4. The %d_6 (1, 2, 5, 10, 20% d_6) standard curve points are plotted against the ratio of the areas under the curve as: AUC d_6 / AUC d_0 + AUC d_6 and a regression equation is calculated. Again, note that although d_9 was acquired during the run, none of the d_9 areas are used in the %d_6 standard curve calculation. Confidence levels of the least-squares regression equation describing the standard curve should be within acceptable ranges (ca. 99.99%).

5. Using both of the %d_6 standard curves (before and after the samples), the %d_6 in the sample is calculated twice (once with each standard curve) and these numbers are averaged. To perform these calculations, the ratio of unknown's chromatographic area is manipulated as follows:
unknown data: AUC d_6/AUC d_0 + AUC d_6. This ratio is plugged into the regression equation as described above to back calculate the %d_6 of the sample.

6. The %d_9 standard curve is constructed the same way as the %d_6. That is, for each of the standard curve points, the %d_9 (10, 15, 20, 25, or 30%) is plotted against the measured AUC d_9 / AUC d_0 + AUC d_9 for that point. The least-square regression equation is calculated. The %d_9 of the unknown is estimated by plotting the unknown d_9 AUC d_9/ AUC d_0 + AUC d_9 and calculating the %d_9 using the standard curve-regression equation. Now that the %d_9 of the sample is known, the concentration of d_0-α-tocopherol in the sample can be calculated.

3.5.1.2. Calculating the concentration of D_0 in the sample

1. The concentration (conc) of d_0 is calculated using **Eq. 2**.

$$\%d_9 = d_9 \text{ conc} / d_9 \text{ conc} + d_0 \text{ conc} \qquad (2)$$

This formula rearranges to:

$$d_0 \text{ conc} = (d_9 \text{ conc} / \%d_9) + \text{conc } d_9 \qquad (3)$$

2. As an example, if the sample was measured by GC-MS to have 10% d_9, and the absolute amount of d_9 internal standard added to the original sample was 4 ng, then the unknown amount of d_0 in the sample is 36.0 ng.

3.5.1.3. CALCULATING THE CONCENTRATION OF D_6 IN THE SAMPLE

1. The concentration of d_6 can similarly be calculated using the measured concentration of d_0 and the calculated %d_6 obtained from the regression equation (*see* **Note 6**).

$$\%d_6 = d_6 \text{ conc } / \ (d_0 \text{ conc} + d_6 \text{ conc}) \tag{4}$$

This formula rearranges to:

$$d_6 \text{ conc} = \frac{(\%d_6(d_0 \text{conc})}{(1- \%d_6)} \tag{5}$$

2. As an example, suppose the percentage of d_6 in plasma was calculated to be 26% and the concentration of d_0 (from **Eq. 3**) was calculated to be 36.0 ng. The concentration of d_6 would be 12.65 ng (total α-tocopherol in plasma is 48.65 ng).

3.5.2. Turnover of α-Tocopherol

1. The transport rate of α-tocopherol into and out of a compartment is determined by plotting the %d_6 data against time (**Fig. 2**). Using standard curve-fitting software (Deltagraph, Sigmaplot, etc.), these data are then fit to the equation:

$$y = A\infty \bullet (1 - e^{-ks(t-c)}) \tag{6}$$

2. Within equation 6, y = lipoprotein- or plasma-α-tocopherol enrichment, A∞ = the plateau or asymptote value of α-tocopherol enrichment within the pool, t = time in hours, and c = lag period before isotope incorporation into the pool. An example of the replacement rate for the VLDL fraction is *VLDL-α-tocopherol transport rate (nmol • kg FFM⁻¹h⁻¹) = VLDL-α-tocopherol concentration (1.5 nmol/mL) • Plasma volume (3482 mL) • k_s (0.266 h⁻¹) / Fat free mass (64.9 kg)*. The transport rate of VLDL-α-tocopherol was calculated to be 21.4 nmol • kg FFM⁻¹h⁻¹. Formulas for estimating plasma volume in humans are readily available *(19,20)*.

3.5.3. α-Tocopherol Clearance Rate

1. The clearance rate of a substance is equivalent to the amount of plasma completely cleared of the substance/min and represents its efficiency of removal *(21)*. At steady-state, clearance is calculated as the ratio of the substance's transport (or turnover rate) to its concentration. For example, to calculate the clearance rate of VLDL α-tocopherol (mL/min), the transport rate (nmol/min) is divided by the plasma concentration (nmol/mL).

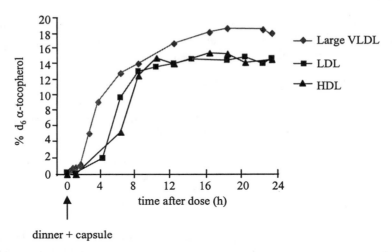

Fig. 2. Incorporation of labeled vitamin E into lipoprotein fractions of a single subject.

2. As an example, one of our research subjects had a replacement rate (K_s) of 0.00332 min^{-1}, a total plasma volume of 3482 mL, and a d_6 α-tocopherol concentration of 3.28 nmol/mL. Therefore, 0.00332/min • 3.28 nmol/mL • 3482 mL equals a transport rate (R_a) of 37.83 nmol/min. Divide this by the d_6 concentration of 3.28 nmol/mL to obtain a clearance rate of 11.53 mL/min.

4. Notes

1. All isotopes of vitamin E should be nitrogen topped, and stored in a freezer in amber-glass containers until use. The vitamin E content of lipoproteins should be performed on freshly isolated lipoproteins that have been protected from light and heat to reduce oxidation.
2. Water competes with α-tocopherol during and after the derivatization procedure. One solution to avoid water saturation of pyridine, involves keeping it nitrogen-topped in a stoppered bottle. When an aliquot is desired, a long needle attached to a syringe containing N_2 is inserted through the stopper (syringe #1) so it rests just above the pyridine in the bottle. Then, a second empty syringe (with long needle attached) is also inserted through the stopper and the needle is pushed into the pyridine. As the N_2 is injected into the bottle through syringe #1, the pyridine is withdrawn into syringe #2 under positive pressure. This technique limits the amount of contact the pyridine has with room air.
3. As little as 15% of the energy of the meal can be fat for efficient vitamin E absorption *(23,24)*. From experience, however, one can say that the capsule taken on an empty stomach or with a glass of whole milk does not provide sufficient stimulus for bile acid secretion and/or fat in the intestine to support measurable vitamin E absorption.

4. As can be seen in **Fig. 1**, the presence of nine deuterium in the internal standard changes the interaction of this isotope with the GC column such that it comes off ahead of the d_0- and d_6-α tocopherol.
5. Following the 50 mg dose of d_6-α-tocopherol, the enrichment of d_6-α-tocopherol will not usually exceed 40%. Thus, a five-point standard curve of d_6/d_0-α-tocopherol ranging from 1–30% should be sufficient to analyze most plasma and lipoprotein samples. The d_9/d_0-α-tocopherol standard curve ranges from 10–50%. This is based on the expected concentration of d_0-α-tocopherol in plasma of 25 nmol/mL and the typical distribution of vitamin E among the lipoprotein fractions as shown in **Table 1**.

Summary

Burton and Daroszewska (*16*) have presented an excellent method for quantifying α-tocopherol in human and animal tissues. The present paper expands that method by including the theory and calculations for α-tocopherol turnover in human plasma and lipoprotein fractions. Recent advances in mathematical modeling in experimental nutrition (*22*) have been aided by the increased availability of labeled isotopes and sensitive analytical methods. Applied to the study of α-tocopherol, these techniques will allow the characterization of the kinetic behavior of this micronutrient in vivo and expand the understanding of this key nutrient's role in preventing disease.

References

1. Traber, M. G. (1999) Vitamin E, in *Modern Nutrition in Health and Disease* (Shils, M. E., Olson, J. A., Shike, M., Ross, and A. C., eds.), Williams and Wilkins, Baltimore, MD, pp. 347–362.
2. Pryor, W. A. (2000) Vitamin E and heart disease: basic science to clinical intervention trials. *Free Rad. Biol. Med.* **28,** 141–164.
3. Burton, G. W., Ingold, T. U., Cheeseman, K. H., and Slater, T. F. (1990) Application of deuterated α-tocopherols to the biokinetics and bioavailability of vitamin E. *Free Rad. Res. Commun.* **11,** 99–107.
4. Traber, M. G., Tader, D., Acuff, R. V., Ramakrishnan, R., Brewer, H. B., and Kayden, H. J. (1998) Vitamin E dose-response studies in humans with use of deuterated *RRR*-α-tocopherol. *Am. J. Clin. Nutr.* **68,** 847–853.
5. Traber, M. G., Burton, G. W., Hughes, L., et al. (1992) Discrimination between forms of vitamin E by humans with and without genetic abnormalities of lipoprotein metabolism. *J. Lipid Res.* **33,** 1171–1182.
6. Traber, M. G., Rudel, L. L., Burton, G. W., Hughes, L., Ingold, K. U., and Kayden, H. J. (1990) Nascent VLDL from liver perfusions of cynomolgus monkeys are preferentially enriched in *RRR*- compared with *SRR*-α-tocopherol: studies using deuterated tocopherols. *J. Lipid Res.* **31,** 687–694.

7. Kramer, J. K., Blais, L., Fouchard, R. C., Melnyk, R. A., and Kallury, K. M. (1997) A rapid method for the determination of vitamin E forms in tissues and diet by high-performance liquid chromatography using a normal-phase diol column. *Lipids* **32,** 323–330.

8. Kramer, J. K. G., Fouchard, R. C., and Kallury, K. M. R. (1999) Determination of vitamin E forms in tissues and diets by high-performance liquid chromatography using normal-phase diol column, in *Methods in Enzymology, Oxidants and Antioxidants, Part A* (Packer, L., ed.), Academic Press, San Diego, pp. 318–329.

9. Bieri, G., Toliver, T. J., and Catignani, G. L. (1979) Simultaneous determination of α-tocopherol and retinol in plasma and red cells by high pressure liquid chromatography. *Am. J. Clin. Nutr.* **32,** 2143–2149.

10. Podda, M., Weber, C., Traber, M. G., and Packer, L. (1996) Simultaneous determination of tissue tocopherols, tocotrienols, ubiquinols, and ubiquinones. *J. Lipid Res.* **37,** 893–901.

11. Podda, M., Weber, C., Traber, M., Milbradt, R., and Packer, L. (1999) Sensitive high-performance liquid chromatography techniques for simultaneous determination of tocopherols, tocotrienols, ubiquinols, and ubiquinones in biological samples, in *Methods in Enzymology, Oxidants and Antioxidants, Part A* (Packer, L., ed.), Academic Press, San Diego, pp. 330–341.

12. Rentel, C., Strohschein, S., Albert, K., and Bayer, E. (1998) Silver-plated vitamins: a method of detecting tocopherols and carotenoids in LC/ESI-MS coupling. *Anal. Chem.* **70,** 4394–4400.

13. Lauridsen, C., Leonard, S. W., Griffin, D. A. et al. (2001) Quantitative analysis by liquid chromatography-tandem mass spectrometry of deuterium-labeled and unlabeled vitamin E in biological samples. *Analytical Biochemistry.* **289,** 89–95.

14. Liebler, D. C., Burr, J. A., Philips, L., and Ham, A. J. (1996) Gas chromatography-mass spectrometry analysis of vitamin E and its oxidation products. *Anal. Biochem.* **236,** 27–34.

15. Liebler, D. C., Burr, J. A., Philips, L., and Ham, A. J. L. (1999) Gas chromatography-mass spectrometry analysis of vitamin E and its oxidation products, in *Methods in Enzymology, Oxidants and Antioxidants, Part A* (Packer, L., ed.), Academic Press, San Diego, pp. 309–319.

16. Burton, G. W. and Daroszewska, M. (1996) Deuterated vitamin E: measurement in tissues and body fluids, in *Free Radicals: A Practical Approach* (Punchard, N. A. and Kelly, F. J., eds.), Oxford University Press, Oxford, pp. 257–270.

17. Hatch, F. T. and Lees, R. S. (1968) Practical methods for plasma lipoprotein analysis, in *Advances in Lipid Research* (Paoletti, R. and Kritchevsky, D., eds.), Academic Press, New York, pp. 2–68.

18. Lindgren, F. T., Jensen, L. C., and Hatch, F. T. (1972) The isolation and quantitative analysis of serum lipoproteins, in *Blood Lipids and Lipoproteins: Quantification, Composition, and Metabolism* (Nelson, G. S., ed.), John Wiley and Sons, New York, NY, pp. 181–274.

19. Ganong, W. F. (1987) *Review of Medical Physiology 13.* Appleton & Lange, Norwalk, CT.

20. Moore, F. D., Oleson, K. H., McMurrey, J. D, Parker, H. V., Ball, M. R., and Boyden, C. M. (1963) The body cell mass and its supporting environment, in *Body Composition in Health and Disease*, W.B. Saunders, Philadelphia, p. 67.
21. Ryan, W. G. and Schwartz, T. B. (1965) Dynamics of plasma triglyceride turnover in man. *Metabolism* **14,** 1243–1254.
22. Clifford, A. J. and Muller, H-G. (1998) in *Mathematical Modeling in Experimental Nutrition.* 1. Plenum Press, New York.
23. Parks, E. J. and Traber, M. G. (2000) Mechanisms of vitamin E in aging. *Antioxid. Redox Signal.*, **2,** 405–412.
24. Parks, E. J., Dare, D., Frazier, K., Hughes, E., Hellerstein, M. K., Neese, R. A., and Traber, M. G. (2000) Dependence of plasma α-tocopherol flux on very low density lipoprotein triglyceride clearance in humans. *Free Rad. Biol. Med.*, **29,** 1151–1159.

23

Analysis of Tocotrienols in Different Sample Matrixes by HPLC

Kalyana Sundram and Rosnah Md. Nor

1. Introduction

Vitamin E is comprised of two homologous series of tocochromanols, termed "tocopherols" and "tocotrienols." They are structurally related, having a common chromanol ring, but distinguished by their side chains. Tocopherols have a saturated phytyl tail, whereas the tocotrienols possess an unsaturated isoprenoid side chain. Four homologs of each type, characterized by their Greek prefixes (α, β, γ, δ), are known to exist in nature *(1)*.

Tocopherols are present in a variety of foods and most edible oils are rich sources. Tocotrienols on the other hand are concentrated mostly in cereals (barley, wheat germ, oat, and rye) and palm and rice bran oils. Modern technology has made it possible to extract and concentrate both tocopherols and tocotrienols from by-products of the edible oil-refining industry *(2)*.

The tocols have potent antioxidant properties that are well-recognized *(3)*. In addition, tocotrienols have been reported to have a number of physiological properties that are not apparent in the tocopherols. These include possible plasma cholesterol-lowering effects mediated through the regulation of HMG-CoA reductase activity *(4)*. They also play a protective role in reducing a number of lipid and nonlipid related cardiovascular risk factors *(5)*. Tocotrienols have been reported to have possible antitumor activities in different cancer systems *(6)*. These and other therapeutic properties of tocotrienols have recently been reviewed by Theriault et al. *(7)*.

In line with this growing interest in tocotrienols, we describe here an isocratic high-performance liquid chromatography (HPLC) method that allows the simultaneous detection and quantification of both tocopherols and tocotrienols.

From: *Methods in Molecular Biology, vol. 186: Oxidative Stress Biomarkers and Antioxidant Protocols*
Edited by: D. Armstrong © Humana Press Inc., Totowa, NJ

A number of extraction procedures are also detailed since the use of unsuitable lipid-extraction procedures will result in false values for the different tocotrienol isomers. This is especially true when handling biological samples of different origin.

2. Materials

2.1. Equipment

1. Rotary vacuum dryer.
2. Centrifuge.
3. Vortex mixer.
4. Analytical HPLC Instrumentation (Jasco, Hachioji, Japan).
   a. Jasco PU-980 HPLC Pump.
   b. Jasco 821-FP Fluoresence Detector.
   c. Jasco 851-A Autosampler.
   d. Borwin Version 1.2 Chromatography Data Acquisition System (LeFontanil, France).
   e. Columns: FinePak SIL-5, Jasco, Hachioji, Japan.
5. Autosampler vials and inserts.

2.2. Reagents

1. The following solvents are HPLC grade and obtained from Merck (Darmstadt, Germany): hexane, isopropyl alcohol.
2. The following solvents are analytical-grade and obtained from Merck: chloroform, ethanol, petroleum ether (60–80°C).
3. Nitrogen for drying samples (preferably passed through a moisture and oxygen trap).
4. Tocopherol standards (Merck).

2.3. HPLC Procedure

2.3.1. Pump and Mobile Phase

The pump is a Jasco PU-980 (Hachioji, Japan) unit, operated under constant flow rate of 2.0 mL/min. The mobile phase is an isocratic mixture of hexane: isopropyl alcohol (99:1) prepared by adding 40 mL isopropyl alcohol to 4.0 L hexane. The mobile phase should be degassed and filtered through a 0.45-μm nylon membrane using a suitable filtration unit.

2.3.2. Columns

Two normal-phase silica columns (length: 250 mm; internal diameter: 4.6 mm; silica particle size 5-μm and pore size: 120°A) are fitted in series to enhance separation of all tocol isomers (*see* **Note 2**).

2.3.3. Detector

A fluorescence detector is required for this analysis and we routinely use a Jasco 821-FP model. Excitation and emission wavelengths of 295 and 330 nm, respectively, are specific for all tocols. The most suitable detector parameters for optimum sample detection and quantification, prepared as above (*see* **Subheadings 2.3.1.** and **2.3.2.**) are: response–standard; attenuation–32, and gain–X100 (*see* **Note 3**).

2.3.4. Tocol Standards

a. Single-component tocopherol standards (α, β, γ, δ), obtained from either Sigma (St. Louis, MO) or Merck and conveniently packed in 50 mg sealed vials, are mostly > 95% pure. Transfer the contents of the vial into a weighed 25 mL volumetric flask with HPLC-grade hexane. Ensure complete transfer by repeated washes.

   Evaporate the solvent from the volumetric flask using nitrogen and weigh the content of tocol isomer. Then, make up volume to exactly 25 mL in the same volumetric flask. This is the stock standard solution.

b. The working solution of each isomer is prepared by diluting the stock solution serially such that an injection volume contains from 0.2–1.0 µg. Additionally, equal weights of each isomer are mixed together in a vial and analyzed by HPLC.

c. Run pure standard preparation on HPLC to determine the purity of the standard. Correct standard concentration by multiplying the concentration by the percent of the total area generated by the standard under optimal HPLC conditions.

3. Methods

3.1. Oils and Fats

1. Melt the oil or fat at a low temperature (40–60°C) and homogenize the sample using a vortex mixer.
2. Weigh accurately 0.2–0.5 g into a 5 mL volumetric flask. Make up the volume with redistilled hexane and mix by vortexing.
3. The sample, usually a clear solution, is ready for HPLC injection (*see* **Note 1**).

3.2. Food and Diet Samples Using Soxhlet Extraction

1. Food or diet samples intended for vitamin E analysis should first be mixed and homogenized to ensure complete homogeneity of the sample. (If the samples contain much moisture, they should be oven-dried at 60–80°C for 2 h to remove their water content. Dry the sample to a state when it can be transferred into a Soxhlet thimble. Higher temperatures should be avoided because they can affect vitamin E content).

2. Weigh accurately 3.0–5.0 g of the dried sample into a Soxhlet thimble and extract the fat by standard Soxhlet procedures using petroleum ether (60–80°C) as the solvent.
3. Recover and accurately quantify the fat content of the sample by rotor-drying under vacuum.
4. Weigh 0.2–0.5 g of the fat into a 5 mL volumetric flask, make up volume with hexane, and vortex to mix contents. The sample is ready for HPLC analysis.

3.3. Food and Diet Samples By Cold Extraction with Chloroform: Ethanol (1:1)

1. Dry food and diet samples may also be cold extracted with chloroform:ethanol (1:1) mix. This method is also suitable for vitamin E analysis in egg yolks.
2. Weigh accurately about 2.0 g sample into a 100 mL Erlenmeyer flask and slowly add 25 mL ethanol while vortexing followed by 25 mL chloroform. Allow to stand overnight. Filter using a Whatman 1PS filter paper and recover the clear (often yellowish-tinted) filtrate. Wash and rinse the flask and filter paper with 5 mL solvent mix to ensure complete recovery. Evaporate the solvent under vacuum or by rotor-drying and obtain accurate weight of lipids recovered.
3. Transfer known weight (about 0.2 g) of the lipid into a 5 mL volumetric flask using hexane, make up volume with hexane, mix, and ready for HPLC injection.

3.4. Biological Samples: Plasma or Serum

1. Collect whole blood into a Vacutainer-type tube and prepare plasma or serum as required.
2. Pipet 0.5 mL plasma or serum into a capped centrifuge tube. Add 2 mL ethanol, vortex for 5 min and inspect that proteins have precipitated. Add 4 mL hexane, cap, and vortex tube for 2 min. Centrifuge tubes for 2 min to separate phases and remove upper layer. Repeat extraction of remaining lower layer with an additional 4 mL hexane; pool the extracts.
3. Evaporate extracts under nitrogen. Reconstitute in 100 µL hexane and sonicate for 60 s. Transfer to sample vials for HPLC analysis.

3.5. Biological Samples: Tissue Samples (e.g., Adipose, Liver)

1. Pulverize samples (approx 200 mg) under liquid nitrogen and rapidly transfer into a weighed, capped, centrifuge tube. Record weight of pulverized sample transferred. Add 4 mL ethanol and vortex vigorously followed by 4 mL chloroform and vortex for 5 min. Stand for 15 min and filter through a Whatman 1PS filter paper.
2. Remove solvent into a preweighed vial. Repeat extraction **Subheading 3.5., step 1** twice with the addition of 3 mL CHCL$_3$:EtOH (1:1) mix each time. Pool the solvent layers, evaporate under nitrogen or by rotor drying, record weight of lipids recovered, and reconstitute in 200 µL hexane for HPLC injection.

3.6. HPLC Analysis

A JASCO 851-A Autosampler controls automated analysis. Standards or samples are loaded into 1.5-mL vials fitted with a siliconized polytetra-fluoroethylene (PTFE) septum. For smaller volumes, 250-µL conical glass inserts with a centering support are placed inside the autosampler vials. The autosampler can be programmed to inject sample volumes ranging from 5–100 µL depending on the sample concentrations. 20 µL injections are routine and are eluted isocratically using the mobile phase described in **Subheading 2.3.1.** at a constant flow rate of 2.0 mL/min. For two columns in series, this produces a typical pressure of approx 105 kg/cm². Fluorescence absorption data is collected at intervals of 2 pts/s by the PC-controlled Borwin (LeFontanil, France) chromatography data-acquisition system. The data is automatically stored in the PC and this allows postanalysis processing of data and chromatogram outputs. Each automated run is programmed for 25 min and interjected by a wash and rinse cycle to ensure no carry-over effects from previous runs.

The HPLC system is also fitted with a Rheodyne (Cotati, CA, USA) 7725i valve with a 20-µL loop. Manual injections are carried out using a 100 µL Hamilton syringe fitted with a blunt-tipped needle.

3.7. Results

3.7.1. Chromatographic Separation: Oils, Fats, Foods, and Diets

A representative chromatogram from a standard mixture of tocopherol and tocotrienols and palm vitamin E (tocotrienol rich fraction, TRF) is shown in **Fig. 1A** and **B**, respectively. The advantage of this procedure is the clear separation of the tocopherol and tocotrienol isomers. The procedure also allows the quantification of vitamin E isomers in any given sample matrix, provided the extraction procedures previously detailed are adhered to.

3.7.2. Chromatographic Separation: Biological Samples

Figure 1C shows a representative chromatogram from a serum extract of a rat fed a palm-oil diet enriched with TRF. The vitamin E content of the rat serum extract reveals the presence of both tocopherols and tocotrienols suggesting accumulation of tocotrienols in the serum from dietary palm oil enriched with TRF. Similar chromatographic separations are also evident in other biological tissues including adipose, colon, liver, and mammary fatty tissue.

A number of attempts have been made to quantify tocotrienols in fasted human plasma or serum especially following dietary intake of palm oil or palm vitamin E. However, tocotrienols are hardly detected in fasted human plasma and serum. Instead increases in the content of α-tocopherol, the major vitamin E

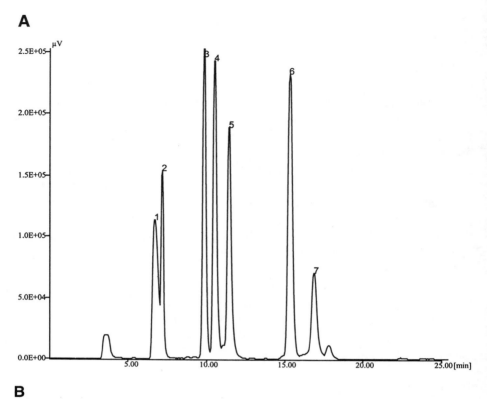

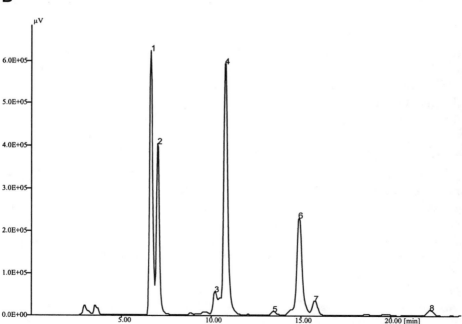

Fig. 1.

C

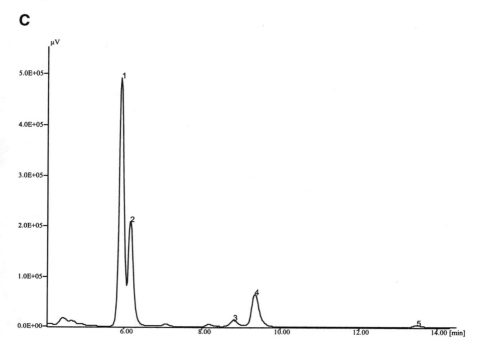

Fig. 1 *(continued)* **(A)** Standard mixture of tocopherols and tocotrienols analyzed by HPLC. 1, α-tocopherol; 2, α-tocotrienol; 3, β-tocopherol; 4, γ-tocopherol; 5, γ-tocotrienol; 6, δ-tocopherol; 7, δ-tocotrienol. **(B)** Tocol composition of palm tocotrienol-rich fraction (TRF). 1, α-tocopherol; 2, α-tocotrienol; 3, γ-tocopherol; 4, γ-tocotrienol; 5, δ-tocopherol; 6, δ-tocotrienol; 7, unknown; 8, didesmethyl tocotrienol. **(C)** Serum tocotrienol profile of a rat fed TRF. 1, α-tocopherol; 2, α-tocotrienol; 3, γ-tocopherol; 4, γ-tocotrienol; 5, δ-tocotrienol.

isomer in human plasma is evident. The current detection system, however, is not viewed as a problem for this sample matrix. Spiking plasma samples with tocotrienols and extracting them for analysis yields good recovery rates using this method. This suggests that tocotrienols metabolize in the human circulatory system in a manner that is currently unknown.

3.7.3. Calculation of Results

It is usually the practice to use a single tocol isomer (usually α-tocopherol acetate) as a standard (either internal or external) for calculating total vitamin E content and distribution of various tocol isomers in the sample. Since our system detects both tocopherols and tocotrienols, we prefer to use the four common tocopherol isomers (α, β, γ, and δ) as external standards to quantify not only tocopherols but also the related tocotrienol analogues.

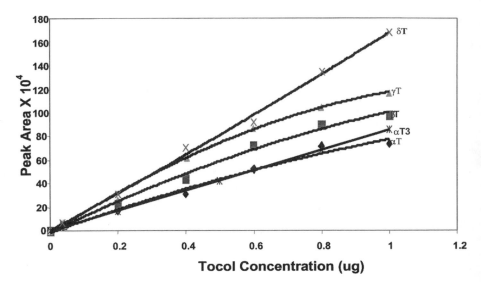

Fig. 2. Effect of tocol concentration on HPLC peak area response. Pure (>97% purity) tocopherol and tocotrienol isomers were prepared at various concentrations ranging from 0.05–1.20 μg and analyzed by HPLC as described in the Methods. The peak area response of each vitamin E isomer was plotted against concentration.

Determine the concentration of each isomer in the sample by applying the following calculations:

$$\text{Isomer Conc. } (\mu g/g) = \frac{\text{Peak area of isomer}}{\text{Peak area 1 } \mu g \text{ Std}} \times \frac{\text{Total sample volume}}{\text{Volume injected}} \times \frac{1}{\text{Sample wt (g)}}$$

The sum of all tocol isomers thus calculated gives the total vitamin E content.

Since the above is based on the weight of the lipid or fat extracted from the sample, total vitamin E content of the sample is determined by multiplying with the % of lipid or fat.

Pure tocopherol isomers and α-tocotrienol at various concentrations were analyzed by the aforementioned method to establish the validity of the HPLC technique and calculation of results. From experience, we have realized that using a single standard could result in inaccurate vitamin E quantification. For example, using α-tocopherol as the standard would overestimate the concentration of δ-tocopherol, whereas δ-tocopherol would underestimate α-tocopherol. This is clearly demonstrated in **Fig. 2**. In our laboratory, the tocotrienol isomers are calculated based on the standard peak area of its related tocopherol analog. This increases accuracy of the determinations (*see* **Note 4**).

3.7.4. Observed Ranges of Tocotrienols in Oils and Fats

The vitamin E content of a number of oils and fats analyzed by this HPLC procedure is shown in **Table 1**. Palm oil, its fractions (palm olein and palm stearin), and rice-bran oil are rich in tocotrienols, whereas other oils contain tocopherols. In these oils and fats, all tocol isomers are clearly separated and quantified.

3.7.5. Quantification of Tocotrienols in Adipose Tissue of Guinea Pigs Fed Palm Vitamin E

The ability of this HPLC procedure to detect tocotrienols in biological samples is shown in **Table 2**. Various tocotrienol isomers contained in palm vitamin E were detected and quantified in the adipose tissue of guinea pigs fed diets containing palm vitamin E (TRF) at three different concentrations.

The HPLC analytical technique previously described has been demonstrated to resolve the various vitamin E isomers that are present in different sample matrixes. The use of the correct extraction techniques for different sample types cannot be overemphasized since the exact quantification of their vitamin E content is highly dependent on these procedures.

4. Notes

1. Refined oils and fats usually appear as clear solutions. For crude oils or fats containing impurities, following hexane addition in the volumetric flask, centrifuge samples at $2000g$ for 10–15 min. The clear upper layer should be used for analysis. Alternatively, the hexane solution may be filtered using a syringe fitted with a 0.45 μM nylon or PTFE pore filter.
2. It is possible to obtain relatively good separation of the tocols using a single silica-gel column (5-μM particle size). However, this often results in overlapping peaks representing the α-tocopherol and α-tocotrienol isomers. Such overlaps will also compromise quantification of these individual tocol isomers. Separation is improved by the addition of tetrahydrofusant (THF) (1%) to the mobile phase. However THF is unpleasant on the operator. It does not keep well and the mobile phase must be prepared fresh at regular intervals. All these problems are simply resolved by connecting two silica columns in series.
3. Detector settings have been found optimum for the HPLC system and separation conditions that we currently operate. When using other HPLC systems, these settings should be individually optimized in your own laboratory.
4. Unfortunately, commercial tocotrienol standards are unsuitable because they do not give single suitable peaks as desired. For identification of tocotrienols, we routinely use an in-house preparation of palm vitamin E (purity > 90%), rich in various known toctotrienols (previously identified by mass spectrometry [MS]). When desired, these individual tocotrienol isomers are purified by thin-layer and column chromatography.

Table 1
Tocopherol and Tocotrienol Content (μg/g) of Common Refined Edible Oils

Tocol Isomers	Soybean oil	Corn oil	Olive oil	Sunflower oil	Milk fat (ghee)	Wheat germ oil	Rice bran oil	Palm oil	Palm olein	Palm stearin
α-Tocopherol	117.2	248.9	151.4	485.2	32.7	218.9	64	188.2	179.0	50.0
β-Tocopherol	19.8	10.1	13.3	3.0	n.d.	33.2	10.6	n.d.	n.d.	n.d.
γ-Tocopherol	560.7	464.1	10.9	51.0	n.d.	84.7	n.d.	n.d.	17.6	n.d.
δ-Tocopherol	178.2	58.2	n.d.	n.d.	33.8	n.d.	187	n.d.	n.d.	n.d.
α-Tocotrienol	n.d.	n.d.	n.d.	n.d.	n.d.	n.d.	31.4	198.1	219.9	47.4
β-Tocotrienol	20.2	n.d.	n.d.	n.d.	n.d.	347.5	83.2	10.0	8.1	9.0
γ-Tocotrienol	6.2	n.d.	n.d.	8.3	n.d.	n.d.	783.2	198.8	332.7	134.9
δ-Tocotrienol	n.d.	n.d.	n.d.	n.d.	n.d.	18.4	38.6	98.4	67.0	31.4
Total	902.2	781.4	175.6	547.5	66.5	702.7	1198	693.5	824.3	272.8

[a]Oils and fats were prepared (*see* **Subheading 3.1.**) and analyzed by HPLC (*see* **Subheading 2.3.2.**) as described in Methods.
[b]n.d. – not detected.

Table 2
Occurrence of Tocotrienols in Adipose Tissue of Guinea Pigs Fed TRF-Enriched Diets

	αT	αT3	βT3	γT3	δT3	Total tocols
			(mg/g concentrate)			
TRF	290	140	20	280	140	890
			Tocols (ug/g tissue)			
TRF added to diet (mg/kg diet)						
1000	23.93	26.80	3.75	10.45	4.23	69.16
250	20.00	14.58	1.57	5.78	2.38	44.31
125	13.56	7.38	0.9	2.22	1.23	25.53

[a]Guinea pig diets were supplemented with tocotrienol-rich fraction (TRF) from palm oil (89% purity) at three different levels (1000, 250, 125 mg/kg diet). Animals were maintained on these diets for 8 wk prior to sacrifice. Lipids were extracted from adipose tissue (see **Subheading 2.3.2.**) and analyzed by HPLC (see **Subheading 3.5.**) as described in Methods.
[b]T, tocopherol; T3, tocotrienol.

References

1. Kamal-Eldin, A. and Appelqvist, L. A. (1996) The chemistry and antioxidant properties of tocopherols and tocotrienols. *Lipids* **31,** 671–701.
2. Sundram, K. and Gapor, A. (1992) Vitamin E from palm oil: its extraction and nutritional properties. *Lipid Technol.* **4,** 137–141.
3. Serbinova, E. A., Tsuchyia, M., Goth, S., Kagan, V. E., and Packer, L. (1993) Antioxidant action of α-tocopherol and α-tocotrienol in membranes, in *Vitamin E in Health and Disease* (Packer, L. and Fuchs, J., eds.), Marcel Dekker, New York, pp 235–243.
4. Qureshi, A. A., Burger, W. C., Peterson, D. M., and Elson, C. E. (1986) The structure of an inhibitor of cholesterol biosynthesis isolated from barley. *J. Biol. Chem.* **261,** 10,544–10,550.
5. Tomeo, A. C., Geller, M., Watkins, T. R., Gapor, A., and Bierenbaum, M. L. (1995) Antioxidant effects of tocotrienols in patients with hyperlipidemia and carotid stenosis. *Lipids* **30,** 1179–1183.
6. Komiyama, A. K., Iizuka, K., Yamaoka, M., Watanabe, H., Tsuchiya, N., and Umezawa, I. (1989) Studies on the biological activity of tocotrienols. *Chem. Pharm. Bull.* **37,** 1369–1371.
7. Theriault, A., Jun-Tzu, C., Wang, Q., Gapor, A., and Adeli, K. (1999) Tocotrienol: a review of its therapeutic potential. *Clin. Biochem.* **32,** 309–319.

24

Measurement of β-Carotene 15,15′-Dioxygenase Activity by Reverse-Phase HPLC

Alexandrine During, Akihiko Nagao, and James Cecil Smith, Jr.

1. Introduction

β-Carotene is the major precursor of vitamin A (retinol) and is found mainly in fruits and vegetables. In humans, 30–40% of β-carotene is absorbed via the intestine as an intact molecule and then incorporated into chylomicrons; however, the majority of β-carotene (60–70%) is metabolized to retinyl esters and transported by the lymph *(1)*. Indeed, in the enterocyte of the small intestine, β-carotene 15,15′-dioxygenase (EC 1.13.11.21), a cytosolic enzyme, cleaves the central double bond (15-15′) of the β-carotene molecule to produce two molecules of retinal in the presence of oxygen *(2,3)*. Retinal is a direct precursor of both: 1) retinol by reduction through short-chain dehydrogenase/ reductase activity, and 2) retinoic acid by irreversible oxidation through aldehyde dehydrogenase activity or cytochrome P450 activity *(4)*. Thus, in mammals, β-carotene cleavage yielding retinal is the first and requisite step to produce retinoids, which are involved in many essential biological functions, including vision, reproduction, cell differentiation, gene expression, and general maintenance *(5)*. Numerous methods have been used to determine β-carotene 15,15′-dioxygenase activity in mammalian tissues, these techniques employed radioactive β-carotene *(2,3,6,7)*, thin-layer chromatography (TLC) *(2)*, alumina column chromatography *(3)*, and more recently high-performance liquid chromatography (HPLC) *(6,7)*. However, these procedures, which often required vigorous large volume solvent extractions followed by evaporations, could result in both β-carotene and retinal being oxidized and/or isomerized. Therefore, we developed a simplified β-carotene 15,15′-dioxygenase assay that eliminates several complicated and time-consuming steps; thus, this

From: *Methods in Molecular Biology, vol. 186: Oxidative Stress Biomarkers and Antioxidant Protocols*
Edited by: D. Armstrong © Humana Press Inc., Totowa, NJ

methodology will be useful for further investigations of β-carotene metabolism *(8)*. This enzyme assay is described here.

2. Materials

2.1. Equipment

1. Vortex.
2. Circulating water bath.
3. Microcentrifuge with refrigeration.
4. Analytical HPLC instrumentation:
   a. Shimadzu (Kyoto, Japan) LC10AS pump or Beckman 114M pump (Beckman Instruments, Inc., CA).
   b. Shimadzu SPD-10A UV-VIS detector or UV-970 UV-VIS detector (Jasco, Tokyo, Japan).
   c. Shimadzu C-R3A integrator (or Beckman System Gold chromatography software).
   d. TosoHaas (Montgomeryville, PA) ODS-80Ts C18 HPLC column (4.6 × 150 mm, 5 μm particle size, 80Å pore).
   e. Supelco (Bellefonte, PA) Pelliguard LC-18 guard column (2 × 20 mm)
   f. Rheodyne (Cotati, CA) 200 μL-loop sample.

2.2. Reagents

All chemicals are obtained from Sigma Chemical Company (St. Louis, MO), except those specified. All solvents are HPLC-grade, filtered through 0.22-μm nylon membranes, and de-gassed prior to HPLC analyses.

1. Acetonitrile.
2. Formaldehyde at 37% (w/w).
3. Ammonium acetate.
4. Tween 40 at 4.5% in acetone.
5. Dithiothreitol (DTT) solution at 10 mM stored at –20°C (1 mL aliquot).
6. Sodium cholate solution at 40 mM.
7. Nicotinamide solution at 300 mM.
8. *all-trans*-Retinal is purified by HPLC procedure using a TSK GEL Silica 60 column (4.6 × 250 mm) (Tosohaas) and the mobile phase: hexane/ethyl acetate (95:5; v/v). Purified retinal (≥ 99%) in ethanol (≈ 20 μM) containing α-tocopherol (1/10 molar part) is stable for several months at –20°C.
9. *all-trans*-β-Carotene (type IV) is purified by passage through a neutral alumina column (Brockman III) in hexane. Purified β-carotene (≥ 98.8%) in hexane (≤ 80 μM) containing α-tocopherol (1/10 molar part) is stable for several months at –20°C. The final β-carotene concentration is ≤ 80 μM in hexane.
10. d-α-Tocopherol (Eisai Co., Tokyo, Japan or Henkel Co., LaGrange, IL) is solubilized in ethanol (20 mM) and stored at –20°C for several months.

2.3. Buffer Solutions

1. Solution A: 50 mM N-(2-hydroxyethyl)-piperazine-N'-(2-ethansulfonicacid) (HEPES)-KOH buffer, pH 7.4, 1.15 % KCl, 1 mM ethylenediaminetetraacetic acid (EDTA), and 0.1 mM DTT. Stable for 6 mo at 4°C.
2. Solution B: 10 mM HEPES-KOH buffer, pH 7.4, 50 mM KCl, 0.1 mM EDTA, and 0.1 mM DTT. Stable for 6 mo at 4°C.
3. Tricine buffer, pH 8.0: 0.5 M N-tris(hydroxymethyl)methylglycine (Tricine)-KOH buffer, pH 8.0. Stable for 6 mo at room temperature.

3. Methods

3.1. Enzyme preparation

3.1.1. Tissue Homogenate

Organs are harvested immediately after killing of animals, washed with a NaCl solution (0.9%), and weighed. For the small intestine, the mucosa is gently scraped off and then weighed. Finally, organ sample is homogenized in 5 volumes of Solution A followed by a centrifugation at 10,000g for 30 min (4°C).

3.1.2. Gel Filtration

The resultant supernatant (2.5 mL) is applied to a Sephadex G-25M column (1.5×5.5 cm; Pharmacia Biotech, Uppsala, Sweden) equilibrated with Solution B (4°C). The enzyme fraction is eluted with 3.5 mL of Solution B (4°C) (*see* **Note 1**).

3.2. Enzyme Assay

3.2.1. Substrate Solution

α-Tocopherol (0.5 μmol) in ethanol and purified *all-trans*-β-carotene (75 nmol) in hexane are placed in a glass test tube and solvents evaporated under a stream of argon or nitrogen gas. Compounds are redissolved with Tween 40 (7.5 mg) in acetone, which is then evaporated. The final dried mixture is solubilized in 1 mL of deionized water and thoroughly vortexed to obtain a clear micellar solution.

3.2.2. Factor Requirements for the Enzyme Activity (see **Table 1**)

1. a SH-reagent (DTT or GSH) protecting one or more sulfhydryl groups of the protein essential for the enzyme activity,
2. a trihydroxy bile salt (sodium cholate or glycocholate) to improve the dispersion of β-carotene in the incubation medium,
3. A "nicoti"-compound with a CO-group in position 6 (either nicotinamide, nicotinamide N-oxide, or nicotinic acid); its role in the reaction is unclear.

Table 1
Critical Factors for Optimizing β-Carotene 15,15′-Dioxygenase Assay

Compounds	Relative effect on enzyme response[a]
Buffer	Optimum pH = 8.0
Detergent	Tween 40 > Tween 20 = Tween 80 > Triton X-100 at 0.15%[b]
SH-Reagent	Dithiothreitol at 0.5 mM[b] = Gluthathione at 5 mM[b]
Bile salt (sodium)	Cholate = Glycocholate >> Taurocholate = Desoxycholate at 6 mM[b]
"Nicoti"-compound	Nicotinamide = Nicotinamide N-oxide = Nicotinic acid >> Nicotine at 15 mM[b]

[a]Dioxygenase activity equal (=), greater (>), and much greater (>>) in presence of one compound than in presence of the other compounds in the incubation medium.
[b]Final concentration in the incubation medium.

3.2.3. Standard Incubation Reaction

The final incubation medium (0.2 mL) contains 0.1 M Tricine buffer, pH 8.0, 0.5 mM DTT, 4 mM sodium cholate, 15 mM nicotinamide, 15 μM β-carotene, 0.1 mM α-tocopherol, 0.15% Tween 40, and the enzyme preparation (≤ 0.5 mg protein). First, the reaction medium without β-carotene, α-tocopherol, and Tween 40, is placed in a glass tube and preincubated at 37°C for 5 min. Then, the enzyme reaction is initiated by adding 40 μL of the aqueous substrate solution, followed by an incubation at 37°C for 30 min.

3.2.4. Formaldehyde Treatment

The reaction is stopped with 50 μL of formaldehyde and the mixture is incubated at 37°C for 10 min. This step is required since the extraction of retinal from biological samples has been reported to be difficult and thus a formaldehyde treatment has been recommended to enhance the recovery of retinal *(9,10)*.

3.2.5. Acetonitrile and Centrifugation

After adding 500 μL of acetonitrile, the mixture is thoroughly vortexed, transferred into an Eppendorf tube, and kept on ice for 5 min. The mixture is then centrifuged at 10,000*g* for 10 min (4°C). The supernatant is thus ready for HPLC analysis.

3.3. HPLC Procedure

3.3.1. Mobile Phase

An acetonitrile/water mixture (90 : 10, v/v) containing 0.1% of ammonium acetate is used as mobile phase, flow rate of 1.0 mL/min.

3.3.2. HPLC analysis

Two hundred μL of the supernatant obtained in **Subheading 3.2.5.** are injected into the HPLC system equipped with an ODS-80Ts reverse-phase (RP) column and a 200 μL sample loop. Retinal is monitored at 380 nm (*see* **Notes 2** and **3**).

3.3.3. External Standard

The standard curve of retinal is obtained by varying purified *all-trans*-retinal concentration (0.2–50 pmol/200 μL injected) dissolved in the solvent mixture of water/formaldehyde/acetonitrile (200:50:500; v/v/v), identical to that used for sample.

3.4. Characteristics of the Method

3.4.1. Sensitivity

The detection limit of the method is 0.2 pmol of retinal/200 μL of enzyme assay.

3.4.2. Recovery

The recovery of retinal added to the reaction mixture containing all components except β-carotene is decreased with increased: 1) quantity of the enzyme preparation and 2) time of incubation. However, under the standard conditions, the recovery of retinal is greater than 93%.

3.4.3. Reproducibility

Intraday coefficients of variation are ≤ 3% for $n = 3$ assays per enzyme preparation.

3.5. Results

3.5.1. Chromatogram of the Product of the Enzyme Reaction (see **Fig. 1**)

The elution profile obtained for a complete reaction in presence of β-carotene and the enzyme preparation shows a peak at 7.5 min corresponding to the standard retinal, while no peak is detected for the control incubation (zero time control).

3.5.2. Time and Protein Linearity (see **Fig. 2**)

Retinal formation in the incubation mixture containing a saturating level of β-carotene (15 μ*M*) is directly proportional to both: 1) incubation time (5–60 min) and 2) protein concentration (0.05–0.7 mg proteins per assay).

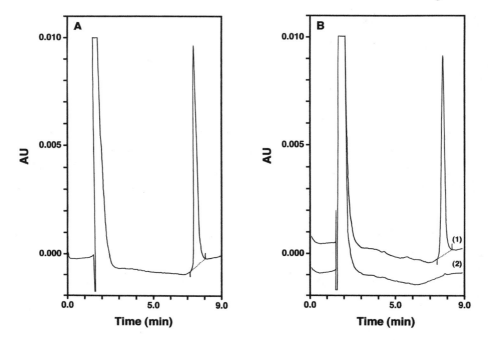

Fig. 1. HPLC elution profiles of standard retinal and cleavage product of β-carotene. **(A)** Purified all-trans-retinal (28 pmol/200 µL injected) in the mixture of water/formaldehyde/acetonitrile (200:50:500; v/v/v) and **(B)** Product of the reaction after incubation of β-carotene with an enzyme preparation of rat intestinal mucosa (0.26 mg protein) under the standard conditions (1) for 60 min and (2) for 0 min (zero-time control).

3.5.3. Inhibitors of the Enzyme

1. In vitro, we found that the activity varied markedly when a plastic test tube (1.5 mL) was used for the reaction; thus some antioxidants or plasticizer eluted from the tube may inhibit the activity. To avoid the possible elution of inhibitors from plastic tubes and pipets, plasticware should be avoided.
2. Moreover, of those tested, BHT, which is frequently used as antioxidant for extraction and storage of oxidation-liable substances, was the most potent inhibitor of the dioxygenase activity. Therefore, such antioxidants should not be added to the reaction medium via either the reagents and/or enzyme preparation. For example, tetrahydrofuran, widely used to solubilize carotenoids, usually contains BHT as a stabilizer. However, α-tocopherol can be used instead of BHT as antioxidant during β-carotene storage and the assay procedure; it does not produce significant inhibition at concentrations of less than 0.5 mM.
3. In addition to antioxidants, the dioxygenase activity can be inhibited by carotenoids, such as α-carotene, β-cryptoxanthin, lutein, lycopene, canthaxanthin,

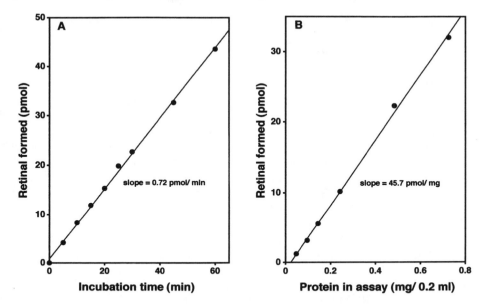

Fig. 2. Effect of incubation time and protein concentration on retinal formation in the incubation mixture (0.2 mL) containing the enzyme preparation of rat liver and β-carotene (3 nmol). **(A)** The enzyme preparation (0.48 mg protein) was incubated at 37°C for different times and **(B)** Different protein concentrations of the enzyme preparation were incubated under the standard conditions for 30 min.

and zeaxanthin *(11–13)*. Hydrophobic substances analogous to carotenoids may inhibit the dioxygenase. Tissue homogenates may contain these types of inhibitors from dietary and/or tissues sources. Indeed, the activity of intestinal homogenates may be markedly and directly influenced by dietary hydrophobic substances.

4. Notes

1. Enzyme preparation is submitted to a gel filtration in order to eliminate compounds at low molecular weight such as cofactors (NADH and NADPH) involved in retinal reduction to retinol or retinal oxidation to retinoic acid *(4)*.
2. Twenty percent of all-*trans*-retinal is isomerized to 13-*cis*-retinal during the incubation reaction containing the enzyme preparation, but these two isomers of retinal are not resolved under the described HPLC conditions. Thus, due to the presence of 13-*cis* retinal, which has a lower extinction coefficient than all-*trans*-retinal (E%, cm = 1250 compared to 1530 at 380 nm, respectively) a factor of correction (1.045) is applied to the final value.
3. Under these HPLC conditions, β-carotene is not eluted. Thus, after many samples, β-carotene accumulates in the column; that will disturb the further analyses by ghost peak, baseline drift, and other anomalies. Therefore, after

daily analyses, the column should be washed with less polar solvents, such as the chloroform/methanol mixture (1:1; v/v) or dichloromethane alone injected through the sample loop several times.

References

1. Goodman, D. S., Blomstrand, R., Werner, B., Huang, H. S., and Shiratori, T. (1966) The intestinal absorption and metabolism of vitamin A and β-carotene in man. *J. Clin. Invest.* **45,** 1615–1623.
2. Olson, J. A. and Hayaishi, O. (1965) The enzymatic cleavage of beta-carotene into vitamin A by soluble enzymes of rat liver and intestine. *Proc. Natl. Acad. Sci. USA* **54,** 1364–1369.
3. Goodman, D. S. and Huang, H. S. (1965) Biosynthesis of vitamin A with rat intestinal enzymes. *Science* **149,** 879–880.
4. Duester, G. (1996) Involvement of alcohol dehydrogenase, short-chain dehydrogenase/reductase, aldehyde dehydrogenase, and cytochrome P450 in the control of retinoid signaling by activation of retinoic acid synthesis. *Biochemistry* **35,** 12,221–12,227.
5. Olson, J. A. (1989) Biological actions of carotenoids. *J. Nutr.* **119,** 94–95.
6. Lakshman, M. R., Mychkovsky, I., and Attlesey, M. (1989) Enzymatic conversion of all-trans-β-carotene to retinal by a cytosolic enzyme from rabbit and rat intestinal mucosa. *Proc. Natl. Acad. Sci. USA* **86,** 9124–9128.
7. Devery, J. and Milborrow, B. V. (1994) β-Carotene-15,15′-dioxygenase (EC 1.13.11.21) isolation reaction mechanism and an improved assay procedure. *Br. J. Nutr.* **72,** 397–414.
8. During, A., Nagao, A., Hoshino, C., and Terao, J. (1996) Assay of β-carotene 15,15′-dioxygenase activity by reverse-phase high-pressure liquid chromatography. *Anal. Biochem.* **241,** 199–205.
9. Suzuki, T., Fujita, Y., Noda, Y., and Miyata, S. (1986) A simple procedure for the extraction of the native chromophore of visual pigments: the formaldehyde treatment. *Vision Res.* **26,** 425–429.
10. Nagao, A. and Olson, J. A. (1994) Enzymatic formation of 9-cis, 13-cis, and all-trans retinals from isomers of beta-carotene. *FASEB J.* **8,** 968–973.
11. Ershov, Yu V., Bykhovsky, V. Y., and Dmitrovskii, A. A. (1994) Stabilization and competitive inhibition of β-carotene 15,15′-dioxygenase by carotenoids. *Biochem. Mol. Biol. Int.* **34,** 755–763.
12. Van Vliet, T., Van Schaik, F., Schreurs, W. H. P., and Van den Berg, H. (1996) In vitro measurement of β-carotene cleavage activity-methodological considerations and the effect of other carotenoids on β-carotene cleavage. *Intl. J. Vit. Nutr. Res.* **66,** 77–85.
13. Grolier, P., Duszka, C., Borel, P., Alexandre-Gouabau, M. C., and Azais-Braesco, V. (1997) In vitro and in vivo inhibition of β-carotene dioxygenase activity by canthaxanthin in rat intestine. *Arch. Biochem. Biophys.* **348,** 233–238.

25

Ubiquinol/Ubiquinone Ratio as a Marker of Oxidative Stress

Yorihiro Yamamoto and Satoshi Yamashita

1. Introduction

Oxidative stress is defined as a disturbance in the prooxidant-antioxidant balance in favor of the former *(1)* and has been suggested be a causative factor in aging and degenerative diseases such as heart attack, diabetes, and cancer. The ubiquinol/ubiquinone ratio should therefore be a good marker of oxidative stress because ubiquinol is very labile in the oxidation of low-density lipoprotein (LDL) or plasma. In fact, a decrease in ubiquinol/ubiquinone ratio has been reported in patients with adult respiratory distress syndrome *(2)* and in patients with hepatitis, cirrhosis, and hepatoma *(3)*. Interestingly, newborn babies are under oxidative stress as judged by plasma ubiquinol/ubiquinone ratio *(4)*. Here we describe a simple and reproducible high-performance liquid chromatography (HPLC) method for the detection of ubiquinol and ubiquinone using an on-line reduction column and an electrochemical detector (ECD), and its application to human plasma *(5)*.

2. Materials
2.1. Equipment

1. Vortex mixer.
2. Centrifuge.
3. Analytical HPLC instrumentation.
   a. Rheodyne (Cotati, CA) 7125 sample injector.
   b. Shimadzu (Kyoto, Japan) LC-10AD HPLC pump.
   c. Shimadzu SPD-10A UV detector.
   d. Irica (Kyoto, Japan) Σ985 ECD.

From: *Methods in Molecular Biology, vol. 186: Oxidative Stress Biomarkers and Antioxidant Protocols*
Edited by: D. Armstrong © Humana Press Inc., Totowa, NJ

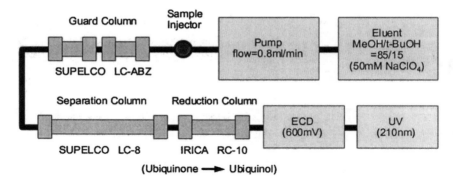

Fig. 1. HPLC system for the detection of plasma ubiquinol-10 and ubiquinone-10.

    e. Shimadzu C-R7A plus integrator.
    f. Supelco (Bellemonte, PA) guard column (4.6 × 20 mm, 5 μm).
    g. Supelco LC-8 column (4.6 × 250 mm, 5 μm).
    h. Irica RC-10-1 reduction column (*see* **Note 1**).
  4. Heparin Vacutainer Tubes.

2.2. Reagents

1. Sodium borohydride and sodium perchlorate monohydrate were obtained from Wako Pure Chemical (Osaka, Japan).
2. Solvents (hexane, methanol, and *tert*-butyl alcohol) used were of analytical-grade from Nacalai Tesque (Kyoto, Japan) (*see* **Note 2**).

3. Methods

3.1 Procedure

1. Collect whole blood into a heparin vacutainer, then centrifuge at 3000*g* for 10 min and analyze immediately or store plasma at –80° C.
2. Human heparinized plasma (50 μL) was mixed vigorously with 250 μL of methanol and 500 μL of hexane in a 1.5 mL polypropylene tube. After centrifugation at 10,000*g* for 3 min at 4°C, 5 μL of hexane layer (corresponding to 0.5 μL of plasma) was injected immediately and directly for HPLC analysis. HPLC system is shown in **Fig. 1**.

3.2. HPLC Mobile Phase

The mobile phase was 50 m*M* sodium perchlorate in methanol/*tert*-butyl alcohol (85/15, v/v) with a flow rate of 0.8 mL/min.

3.3. Standards

1. Ubiquinone-10 was dissolved in hexane. This standard solution was stable at least 1 mo when stored at –20°C.

2. Ubiquinol-10 was prepared by the reduction of ubiquinone-10 with sodium borohydride. Ubiquinone-10 (0.2–0.3 g) was dissolved in 2 mL of hexane in a test tube. After the addition of approx 50 μL of methanol and 20 mg of sodium borohydride, the mixture was stirred for 3 min and allowed to stand at room temperature in the dark for 5 min. Completion of the reduction was judged by the disappearance of the quinone yellow color. After addition of 1 mL of water containing 100 μM ethylenediaminetetraacetic acid (EDTA), the test tube was vigorously shaken and centrifuged at 1500g for 3 min. All of the ubiquinone-10 is quantitatively converted to ubiquinol-10 and this standard solution was stable at least 1 wk when stored at –20°C.

3.4. Results

1. **Figure 2A** shows a typical ECD chromatogram of the hexane extract from plasma of a healthy donor. Peaks 1–5 were identified as tocopherols (mostly α-tocopherol), lycopene, β-carotene, ubiquinol-10, and ubiquinone-10. **Figure 2B** shows a chromatogram of the same sample monitored at 210 nm, indicating that free cholesterol, cholesteryl arachidonate, cholesteryl linoleate, and cholesteryl oleate can be also quantified. The ABZ guard columns are necessary to separate tocopherols and free cholesterol. It also separates ubiquinol-10 and unknown peak 6.
2. The stability of plasma tocopherols (mostly α-tocopherol) and ubiquinol in the hexane extract was determined at room temperature, 0° (on ice), –20, and –78°C (**Fig. 3**). Ubiquinol-10 in the plasma extract was stable only at –78°C. The rate of oxidation of ubiquinol-10 to ubiquinone-10 increased with increasing storage temperature. On the other hand, tocopherols were stable at all conditions, as was also the case for lycopene and β-carotene (data not shown). A decrease in ubiquinol-10 content and a concomitant increase in ubiquinone-10 level were observed after drying 50 μL of the hexane extract under a stream of nitrogen followed by redissolving the residue in 50 μL hexane, although these procedures took only a few minutes. These results clearly indicate that the hexane extract should be analyzed immediately after extraction.
3. **Table 1** summarizes the results of 15 male donors ranging from 23–56 yr old (*see* **Notes 3** and **4**). It was found that the ubiquinol-10/ubiquinone-10 ratio is about 95/5, indicating that plasma coenzyme Q-10 exists mostly in the reduced form.

4. Notes

1. Ubiquinone-10 is usually converted to ubiquinol-10 quantitatively by the reduction column. This can be confirmed by comparing the values of peak areas per injected amount of ubiquinol-10 and ubiquinone-10. If the efficiency of the reduction goes down, the reduction column can be activated by washing with distilled water, aqueous ascorbate (0.1 M), then distilled water for 10 min each at a flow rate of 0.5 mL/min. Washing the whole system with 2-propanol is also useful to remove precipitates, which cause poor resolution, low sensitivity, and low reduction efficiency.

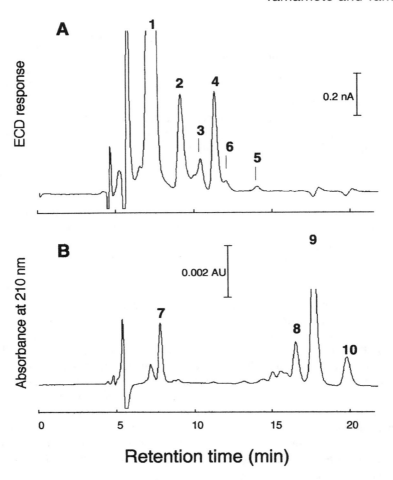

Fig. 2. (**A,B**) Typical HPLC chromatograms of hexane extract from a normal subject. Peaks are identified as tocopherols (1), lycopene (2), β-carotene (3), ubiquinol-10 (4), ubiquinone-10 (5), unknown (6), free cholesterol (7), cholesteryl arachidonate (8), cholesteryl linoleate (9), and cholesteryl oleate (10).

2. We are now using 2-propanol instead of *tert*-butyl alcohol since *tert*-butyl alcohol gives an unstable baseline that we did not see before. The reason for this is unknown.
3. The reproducibility of the assay is excellent *(5)*. However, the coefficient of variation value for ubiquinone-10 was relatively high due to its low concentration in human plasma.
4. Identical ubiquinol-10/ubiquinone ratio was observed in both fresh plasma and twice-frozen and thawed plasma, indicating frozen plasma can be used *(5)*.

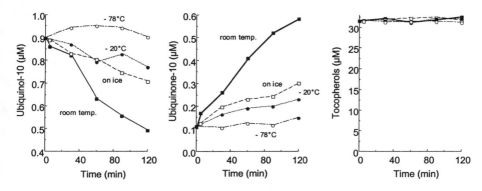

Fig. 3. Effect of storage temperature on the stability of ubiquinol-10, ubiquinone-10, and tocopherols in the hexane extract of human plasma.

Table 1
Plasma Levels of Lipid-Soluble Antioxidants in Healthy Humans

Substance	Healthy humans ($n = 15$) Means ± SD
Tocopherols (μM)	27.7 ± 11.1
Lycopene (μM)	0.65 ± 0.36
β-carotene (μM)	1.82 ± 1.57
Ubiquinol-10 (nM)	927 ± 214
Ubiquinone-10 (nM)	41 ± 12
Total[a]	968 ± 218
Ubiquinol-10/Total[a] (%)	95.6 ± 1.6
Ubiquinone-10/Total[a] (%)	4.4 ± 1.6

[a]Total = Ubiquinol-10 + Ubiquinone-10.

References

1. Sies, H. (1985) Oxidative stress: Introductory remarks, in *Oxidative Stress* (Sies, H., ed.), Academic Press, London, pp. 1–8.
2. Cross, E. C., Forte, T., Stocker, R., Louie, S., Yamamoto, Y., Ames, N. B., and Frei, B. (1990) Oxidative stress and abnormal cholesterol metabolism in patients with adult respiratory distress syndrome. *J. Lab. Clin. Med.* **115**, 396–404.
3. Yamamoto, Y., Yamashita, S., Fujisawa, A., Kokura, S., and Yoshikawa, T. (1998) Oxidative stress in patients with hepatitis, cirrhosis, and hepatoma evaluated by plasma antioxidants. *Biochem. Biophys. Res. Commun.* **247**, 166–170.
4. Hara, K., Yamashita, S., Fujisawa, A., Ogawa, T., and Yamamoto, Y. (1999) Oxidative stress in newborn infants with and without asphyxia as measured by

plasma antioxidants and free fatty acids. *Biochem. Biophys. Res. Commun.* **257,** 244–248.

5. Yamashita, S. and Yamamoto, Y. (1997) Simultaneous detection of ubiquinol and ubiquinone in human plasma as a marker of oxidative stress. *Anal. Biochem.* **250,** 66–73.

26

Catechol- and Pyrogallol-Type Flavonoids

Analysis of Tea Catechins in Plasma

Keizo Umegaki, Mituaki Sano, and Isao Tomita

1. Introduction

Green tea is rich in flavonols, mainly catechins (flavan 3-ols). Recently, catechins, particularly (-)-epigallocatechin-3-*O*-gallate (EGCg), have received attention owing to its various biological effects such as antioxidative, antimutagenic, and anticarcinogenic activities *(1–3)*. To evaluate the biological effect of catechins in vivo, however, it is important to elucidate its metabolic fate by determining the concentration in blood and tissues.

Generally, catechins can be analyzed by high-performance liquid chromatography (HPLC) with ultraviolet (UV) detector. However, the UV detecting system is not sensitive enough to analyze catechins in plasma samples, because the concentrations of EGCg, a major component of green tea catechins, in plasma 1–2 h after giving excess tea extract to human subjects is about 100–600 nM *(4–6)*. Therefore, several sensitive HPLC systems have been developed so far, equipped with either electrochemical detectors *(4,5)* or chemiluminesence detectors *(6)*. On the other hand, in most of the studies, tea catechins in plasma have been extracted by ethyl acetate several times. The samples extracted contains more nonpolar substances than catechins, which are eluted after catechins in analytical chromatogram, and may interfere for the repeated analysis. In addition, it is difficult to handle many samples at one time in the solvent extraction method. This problem is solved by solid-phase extraction method, where more nonpolar substances and polar substances than catechins can be eliminated by selecting an appropriate washing- and eluting-solvent during the solid phase-extraction procedure.

From: *Methods in Molecular Biology, vol. 186: Oxidative Stress Biomarkers and Antioxidant Protocols*
Edited by: D. Armstrong © Humana Press Inc., Totowa, NJ

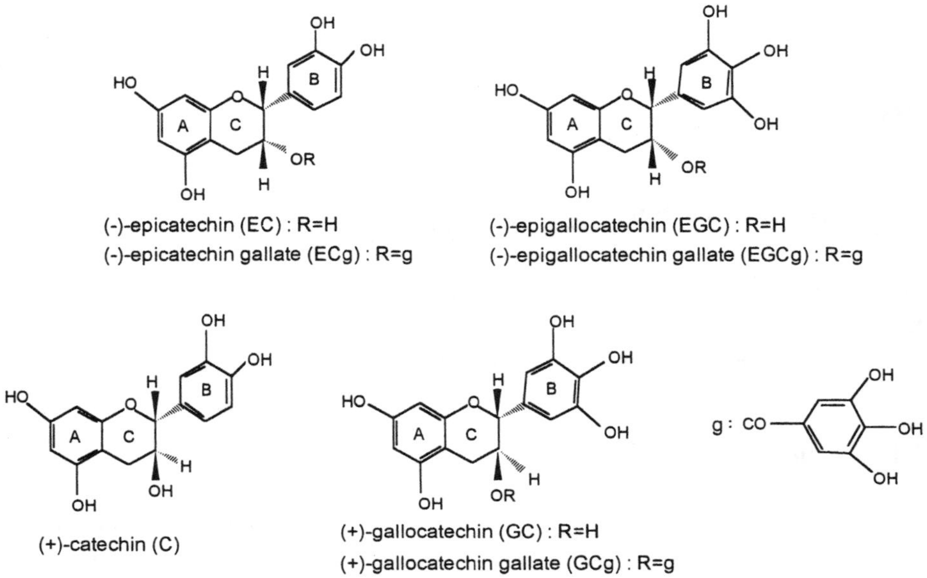

Fig. 1. Chemical structure of tea catechins.

In green tea leaves, there are usually 7 catechins, the chemical structures of which are shown in **Fig. 1**: the major catechins are catechol-type flavonoids such as (-)-epicatechin (EC) and (-)-epicatechin-3-O-gallate (ECg); and pyrogallol-type flavonoids such as (-)-epigallocatechin (EGC), and (-)-epigallocatechin-3-O-gallate (EGCg). In this section, we describe the analytical method of the aforementioned four major catechins in plasma by solid-phase extraction followed by HPLC using an electrochemical detector.

2. Materials

2.1. Equipment

1. Vortex mixer.
2. Centrifuge for microtube.
3. Vacuum centrifuge (Model EC-57, SAKUMA), with Freeze dryer FD-80 (EYELA).
4. Analytical HPLC instrumentation.
   a. HPLC pump (LC-10AD) with a noise filter unit (Shimadzu, Kyoto).
   b. Autoinjector (Shiseido Nanospace SI-1, Tokyo).
   c. Column oven (Shiseido Nanospace SI-1).
   d. Electrochemical detector with guard cell (model 5020) and analytical cell (model 5010) (Coulochem II, ESA Inc., Bedford, MA).

   e. Integrator (Chromatopac CR-7A plus; Shimadzu).

   f. RP-18 C18 HPLC column (4.6 × 250 mm, 5 μm particle) with pre-filter box (No.8210-p) (IRICA Instruments Inc., Kyoto, Japan).

   g. Helium gas (prepurified).

5. Vacuum Manifold for SPE (GL Science, Tokyo).
6. Bond Elut C18 cartridge (500 mg/3 mL, No. 5010-11023, VARIAN, Harbor City, CA).
7. Microtube (2 mL, Eppendorf, Hamburg, Germany).
8. Filter unit (Milli-Cup HV, Millipore Co. Bedford, MA).

2.2. Reagents

1. EGCg (FW458.36), ECg (FW442.36), EC (FW290.26), EGC (FW306.26), GC (FW306.26), GCg (FW458.36), and C (FW290.26), all analytical-grade (Funakoshi Co. Ltd., Tokyo).
2. Ethyl gallate, ascorbic acid, phosphoric acid, disodium ethylenediaminetetraacetic acid (EDTA), acetic acid, sodium hydroxide, methanol, and acetonitrile (Wako Chemical Ind., Osaka). The solvents are all HPLC-grade and the others are reagent-grade, respectively.
3. Beta-glucuronidase (Type H-2, from Helix pomatia) (Sigma Chemical Co., St Louis, MO).
4. Milli-Q water (18 MΩ) by MilliQ system (Millipore Co., Bedford, MA) is used as water.

2.3. Solutions

2.3.1.

1. Beta-glucuronidase solution (beta-glucuronidase 100,000 μ/mL and sulfatase 5000 μ/mL).
2. 0.2 M sodium acetate buffer, pH 5.0 containing 0.5 mM EDTA.
3. 10 μM ethyl gallate solution (internal standard).
4. 50 mM sodium phosphate (pH 3.0).

2.3.2.

1. Methanol
2. 50 mM sodium phosphate buffer, pH 3.0.
3. 50 mM sodium phosphate buffer, pH 3.0, containing 5 % methanol.
4. MilliQ water.
5. 10% acetonitrile-90% methanol.

3. Methods

3.1 Sample Processing

1. Collect whole blood into a tube treated with heparin. Centrifuge at 3000g for 10 min, at 4°C and isolate the plasma.

2. Combine plasma (200 µL) and 10 µL of ascorbic acid (16%) in a microtube, and analyze immediately, or store at –80°C until analysis.

3.2. Preparation of Standard and Internal Standard

Each catechin and internal standard (ethyl gallate) is dissolved in 50 m*M* sodium phosphate buffer (pH 3.0) containing 20% methanol to make a stock solution (about 2 m*M*). Working standard solution, which contains 1 µ*M* of each catechin and internal standard, is prepared by diluting with mobile phase of HPLC described in **Subheading 3.5.1.** As internal standard solution, 10 µ*M* ethyl gallate is prepared by diluting stock solution with 50 m*M* phosphate buffer (pH 3.0).

3.3. Digestion of Conjugates

1. Add 20 uL of β-glucuronidase solution, 100 uL of 0.2 *M* sodium acetate buffer, pH 5.0, containing 0.5 m*M* EDTA, and 20 uL of 10 u*M* ethyl gallate solution (200 pmol/tube) to plasma sample (210 uL) in a microtube, then mix. The final concentrations of beta-glucuronidase and sulfatase in the mixture are 5,700 and 290 u/mL, respectively. When the free forms of catechins, are analyzed addition of β-glucuronidase solution and next incubation step are omitted
2. Incubate the mixture at 37°C, for 45 min to digest the conjugate (*see* **Note 1**).
3. Add 350 uL of acetonitrile to the reaction mixture, and vigorously mix for 5 min in a shaker, then centrifuge at 10,000*g* for 15 min to precipitate protein.
4. Transfer the resulting supernatant (500 uL) to another tube, and apply to an evaporator to reduce acetonitrile concentration in the sample. The acceptable volume of the sample after this evaporation is less than 300 uL.
5. Add 1 mL of 50 m*M* sodium phosphate buffer, pH 3.0, for further decreasing acetonitrile concentration to less than 10%, and apply to the solid-phase extraction described in **Subheading 3.4.** (*see* **Note 2**).

3.4. Solid-Phase Extraction of Catechins

1. Set a Bond Elut cartridge (C18, size 500 mg/3 mL) in Vacuum Manifold for SPE.
2. Rinse cartridge with 3–4 mL of methanol, then with 4–5 mL of water to activate the cartridge. It is important not to allow the solvent to dry.
3. Apply the sample mixture to the preconditioned cartridge. At the time, vacuum of the Manifold is around minus 5 inches Hg.
4. Rinse with 4 mL of 50 m*M* sodium phosphate buffer, pH 3.0), followed with 2 mL of 50 m*M* phosphate buffer, pH 3.0, containing 5% methanol. These two rinses completely eliminate excess ascorbic acid added in the sample (*see* **Note 3**).
5. Set a microtube in the Manifold.
6. Elute catechins with 1 mL of 10% acetonitoril-90% methanol, and collect the elute into the microtube.
7. Apply to a vacuum centrifuge with Freeze dryer.

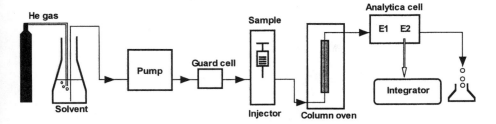

Fig. 2. HPLC system with electrochemical detector.

8. Re-dissolve the residue with 200–500 uL of mobile phase of HPLC as described in **Subheading 3.5.1.** (*see* **Note 4**).

3.5. HPLC Procedure

3.5.1. Mobile Phase

Mobile phase is consisted of 50 m*M* phosphoric acid, 0.05 m*M* EDTA, 14% acetonitrile, pH 2.5. To make the solution, pre-mix solution without acetonitrile is filtered through filter unit (Milli-Cup HV), then mix with acetonitrile. If He gas is not used for bubbling during analysis, the mobile phase should be de-gassed before use.

3.5.2. HPLC analysis

Figure 2 shows schema of HPLC systems. HPLC running conditions are as follows: column temperature, 35°C; flow rate, 1 mL/min; guard cell, +250 mV; analytical cell, –200 mV for electrode 1 and +150 mV for electrode 2 (*see* **Note 5**). The typical pressure of the system is approx 150 kg/cm^2. Electrochemical data from electrode 2 are collected for the analysis, and the typical basal cell current is 2 nA. Samples (50–80 uL) or standard mixture (5–10 uL) are injected into HPLC, and elute isocratically. Sample is identified by the retention time relative to standards. It is better to inject standard mixture solution once every 10 samples to check retention time of each catechin. Running time of one sample is about 45 min. Concentration of each catechin is calculated according to the following formula.

Concentration (nmol/mL) = 200 (pmol) × (A/St) / 0.2 mL / 1000

A: amount of respective catechin (pmol/injection)

St: amount of internal standard (pmol/injection)

3.6. Results

Figure 3A,B show the typical chromatograms. In this condition, EGC, EGCg, EC, Ecg, and GCg peaks are seen as a single peak in plasma sample

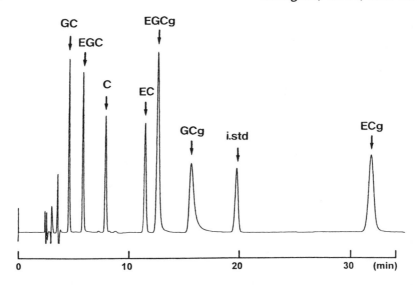

Fig. 3A. Typical chromatogram of standard. Ten uL of standard mixture (1 μ*M*) was injected into HPLC with an electrochemical detector.

0.5–3 h after mixture of catechins (47mg EGC, 49mg EC, 200 mg EGCg, 13 mg ECg) had been given to human subjects. However, it is difficult to detect GC and C as single peaks owing to the interference of unknown peaks in the sample (*see* **Note 6**). The detection limit of EGCg is about 2 nM in plasma. Half-life of catechins in plasma is relatively short, and the peak in the chromatogram has been detected only 1–2 h after taking mixture of catechins. There are two type of electrochemical detectors: coulometric and amperometric. The amperometric detector is also available *(5,7)*, where the optimum applied potential of the detector is +600mV vs Ag/AgCl. However, the sensitivity of the amperometric detector is about 5–8 times lower than that of coulometric detector owing to the structural difference of the detectors.

4. Notes
1. In this condition, the digestion of conjugates is completed with 45-min incubation. Further incubation, especially more than 2 h decreased EGCg, which may be related to the degradation of EGCg.
2. If acetonitrile concentration in the sample is more than 12%, catechins are not trapped in Bond Elute cartridge used (size 500 mg/5 mL). We tried to use other type of cartridge such as Empore DISK cartridge C18, but acetonitrile in the sample markedly prevented absorption of catechins in the cartridge.

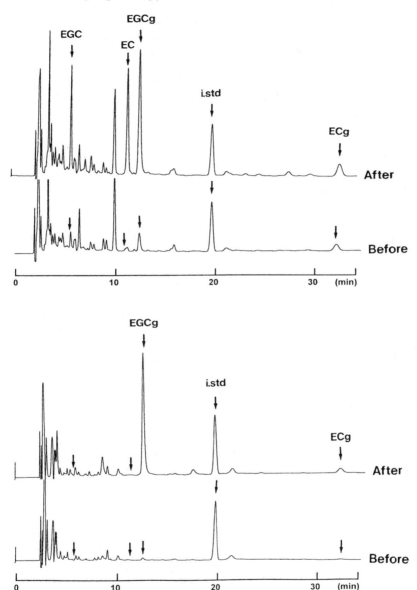

Fig. 3B. Typical chromatogram of human plasma sample treated with and without glucuronidase. Plasma samples treated with *(upper)* and without *(lower)* glucurnonidase as described in the text were injected into HPLC with an electrochemical detector.

3. In HPLC with electrochemical detector, ascorbic acid appears as a big front peak in chromatogram, and interferes the detection of EGC, EC, and EGCg. Therefore, before applying HPLC, it is important to eliminate excess ascorbic acid added to plasma samples.
4. Increased concentration of acetonitrile and methanol in HPLC sample increase peak height and area of EGCg and ECg in HPLC chromatogram, but not those of EGC and EC *(8)*. Thus, it is important to adjust the concentration of methanol and/or acetonitrile in standard and sample the same. Accordingly we prepared or diluted HPLC sample and working standard with mobile phase before applying HPLC.
5. Negative potential of electrode 1 in analytical cell is necessary to detect ECG. The detection limit of EGC increase as increasing the negative potential of electrode 1, and the maximum effect is obtained at −150 to −200mV.
6. For the detection of EC and C, fluorescence detector (ex; 280nm, em; 310nm) may also be applicable, but the detection sensitivity is not enough for the analysis of EC and C in plasma.

References

1. Tomita, I., Sano, M., Watanabe, J., Miura, S., Tomita, T., Yoshino, K., and Nakano, M. (1995) Tea and its components as powerful antioxidants, in *Oxidative Stress and Aging* (Culter, R. G., Packer, L., Bertram, J., and Mori, A., eds.), Brikhauser Verlag, Basel, Swizerland, pp. 355–365.
2. Tomita, I., Sano, M., Sasaki, K., and Miyase, T. (1998) Tea catechin (EGCG) and its metabolites as bio-antioxidants, ACS symposium Series No.701. Oxford University Press, Oxford, UK, pp. 209–216.
3. Yang, C. S. and Wang, Z-Y. (1993) Tea and cancer. *J. Natl. Cancer Inst.* **85,** 1038–1049.
4. Lee, M. J., Wang, Z. Y., Li, H., Chen, L., Sun, Y., Gobbo, S., Balentine, D. A., and Yang, C. S. (1995) Analysis of plasma and urinary tea polyphenols in human subjects. *Cancer Epidemiol. Biomark. Prev.* **4,** 393–399.
5. Unno, T., Kondo, K., Itakura, H., and Takeo, T. (1996) Analysis of (-)-epigallocatechin gallate in human serum obtained after ingesting green tea. *Biosci. Biotechnol. Biochem.* **60,** 2066–2068.
6. Nakagawa, K. and Miyazawa, T. (1997) Chemiluminescence-high-performance liquid chromatographic determination of tea catechin, (-)-epigallocatechin 3-gallate, at picomole levels in rat and human plasma. *Anal. Biochem.* **248,** 41–49.
7. Yoshino, K., Suzuki, M., Sasaki, K., Miyase, T., and Sano, M. (1999) Formation of antioxidants from (-)-epigallocatechin gallate in mild alkaline fluids, such as authentic intestinal juice and mouse plasma. *J. Nutr. Biochem.* **10,** 223–229.
8. Umegaki, K., Esashi, T., Tezuka, M., Ono, A., Sano, M., and Tomita, I. (1996) Determination of tea catechins in food by HPLC with an electrochemical detector. *J. Food Hyg. Soc. Jpn.* (in Japanese) **37,** 77–82.

27

Pyruvate Dehydrogenase Complex as a Marker of Mitochondrial Metabolism

Inhibition by 4-Hydroxy-2-Nonenal

Mulchand S. Patel and Lioubov G. Korotchkina

1. Introduction

Pyruvate dehydrogenase complex (PDC) is a multienzyme complex that plays a crucial role in mitochondrial metabolism of carbohydrates and some amino acids. It catalyzes the oxidative decarboxylation of pyruvic acid with production of carbon dioxide, acetyl-CoA, and NADH, linking the utilization of carbohydrates with the tricarboxylic acid cycle and biosynthetic processes *(1,2)*. Mammalian PDC consists of multiple copies of three catalytic components: pyruvate dehydrogenase (E1), dihydrolipoamide acetyltransferase (E2), dihydrolipoamide dehydrogenase (E3), and one binding protein, the E3-binding protein (E3BP). The E2 and E3BP components have lipoyl moieties covalently linked to specific lysyl residues in the lipoyl-binding domains (*see* **Table 1**). Additionally, there are two regulatory enzymes, E1-kinase (PDK) and phospho-E1-phosphatase present in the cell in several isoforms, that regulate the activity of PDC through a phosphorylation (inactivation)-dephosphorylation (activation) mechanism *(1,2)*. The activities of PDK and phospho-E1-phosphatase in turn are regulated by different metabolic conditions of the cell. Short-term upregulation of PDK activity is achieved by increasing the levels of acetyl-CoA and NADH, products of the PDC reaction, and downregulation is achieved by increasing the levels of ADP and pyruvate *(3,4)*. Activity of phospho-E1-phosphatase depends on the concentrations of Ca^{2+} and NADH. The mitochondrial ratios of ATP/ADP, NAD^+/NADH and acetyl-CoA/CoA, and Ca^{2+} and pyruvate concentrations determine the activities of PDK and phospho-

From: *Methods in Molecular Biology, vol. 186: Oxidative Stress Biomarkers and Antioxidant Protocols*
Edited by: D. Armstrong © Humana Press Inc., Totowa, NJ

Table 1
Lipoyl-Containing Proteins in Mitochondria from Mammalian Cells

Multienzyme complex	Component	Lipoyl moiety/ subunit
Pyruvate dehydrogenase complex	Dihydrolipoamide acetyltransferase;	2
	Dihydrolipoamide dehydrogenase-binding protein	1
α-Ketoglutarate dehydrogenase complex	Dihydrolipoamide succinyltransferase	1
Branched-chain α-keto acid dehydrogenase complex	Dihydrolipoamide acyltransferase	1
Glycine cleavage system	H protein	1

E1-phosphatase in the cell and hence activity of PDC. The long-term regulation of PDC results from changes in the levels of the PDC components by transcriptional regulation *(5,6)*. Thus, PDC is regulated at multiple levels and its activity reflects the metabolic state of the mitochondria.

4-Hydroxy-2-nonenal (HNE) is the major product of peroxidation of membrane lipids formed under different pathophysiological conditions *(7)*. It is toxic to the cell and was shown to affect the respiratory process in mitochondria. The inhibition of NADH-linked mitochondrial respiration is achieved by specific modification of α-ketoglutarate dehydrogenase complex and PDC *(8,9)*, specifically by interacting with lipoyl moieties of the component proteins *(see* **Table 1**). Monitoring PDC activity can provide an assay to investigate the effect of oxidative stress generated by peroxidation of membrane lipids on the cell and its possible prevention by specific antioxidants.

2. Materials
2.1. Equipment
1. UV-VIS spectrophotometer with thermo-control unit.
2. Incubator shaker.
3. Microcentrifuge.
4. CO_2 incubator for tissue culture.
5. Hood for tissue culture.

2.2. Reagents
1. Dichloroacetic acid (DCA) (Sigma, St. Louis, MO).
2. 3-[N-Morpholino]propanesulfonic acid (MOPS) (Sigma).
3. Leupeptin (Sigma).

4. Phenylmethylsulfonylfluoride (PMSF) (Sigma).
5. Thiamin pyrophosphate (TPP) (Sigma).
6. Coenzyme A (CoA) (Sigma).
7. Phosphotransacetylase (Sigma).
8. DL-dithiothreitol (DTT) (Sigma).
9. Benzetonium hydroxide (Sigma).
10. Ethylenediaminetetraacetic acid (EDTA) (Fisher Scientific, Fairlawn, NJ).
11. β-nicotinamide adenine dinucleotide (NAD^+) (Boehringer Mannheim, Indianapolis, IN).
12. 4-hydroxy-2-nonenal (HNE) (Calbiochem, La Jolla, CA).
13. Lambda protein phosphatase (New England Biolabs, Beverly, MA).
14. [1-^{14}C] Pyruvic acid (DuPont NEW Life Technology Products, Boston, MA).
15. [1-^{14}C] Ketoglutaric acid (DuPont NEW Life Technology Products).
16. Center wells and stoppers (Kontes, Vineland, NJ).
17. Grade 238 blotting paper (UWR, Boston, MA).
18. HepG2 cells (Cell Lines and Hybridomes, Rockville, MD).

2.3. Buffers and Solutions

1. Phosphate-buffered saline (PBS), 10 X solution: dissolve 0.4 g KH_2PO_4, 16 g NaCl, 4.32 g Na_2HPO_4, 0.4 g KCl in 140 mL of deionized water, adjust pH to 7.4, adjust the final volume to 200 mL, dilute 10 times for working solution.
2. Resuspension buffer: dissolve 595 mg of KCl, 1.045 g of MOPS, 40.6 mg of $MgCl_2$, 18.6 mg of ethylenediaminetetraacetic acid (EDTA) in about 80 mL of deionized water, adjust pH to 7.4 with 1 N NaOH, adjust the final volume to 100 mL. Add protease inhibitors just before use.
3. Protease inhibitors: leupeptin, 0.1 mg/mL; PMSF, 0.1 M or 17.4 mg/mL in ethanol.
4. Lambda protein phosphatase (400,000 U/mL) (shipped with 10X phosphatase buffer, 10X manganese chloride, 100X bovine serum albumin [BSA]).
5. 125 mM dichloroacetatic acid: dissolve 16.7 mg/mL, adjust pH to 6.0–7.0 with NaOH.
6. Supplement A: dissolve 95 mg of NAD^+, 7.1 mg of TPP, 1.42 mL of 0.1 M $MgCl_2$, 3.6 mL of 1 M potassium-phosphate buffer, pH 7.5 in 50 mL of deionized water.
7. Supplement B: dissolve 6.0 mg of coenzyme A, 25 mg of L-cysteine in 5 mL of deionized water.
8. Pyruvate dehydrogenase complex (PDC) reaction mixture: combine 200 µL of 0.1 M $MgCl_2$, 14 mg of NAD^+, 10 µL of phosphotransacetylase (1 U/mL), 2.5 mL 200 mM potassium-phosphate buffer, pH 7.5, 2.5 mL 60 mM potassium oxalate (1.1 g/100 mL), 1.69 mL of deionized water.
9. α-Ketoglutarate dehydrogenase complex (KGDC) reaction mixture: combine 200 µL of 0.1 M $MgCl_2$, 14 mg of NAD^+, 10 µL of phosphotransacetylase (1 U/mL), 2.5 mL 200 mM potassium-phosphate buffer, pH 7.5, 4.19 mL of deionized water.
10. Supplement C: combine 100 µL of 10 mM TPP, 400 µL of 20 mM DTT, 5 mg of coenzyme A, 500 µL of deionized water.

11. 4-Hydroxy-2-nonenal (HNE) should be stored at –80°C and diluted just before use.

3. Methods

3.1. Sample Preparation

1. Purified PDC is reconstituted from individually expressed and purified E1, E2-E3BP and E3 components of human PDC. E1 and E3 are overexpressed in *Escherichia coli* and purified using vectors constructed previously in this laboratory *(10,11)*. E2-E3BP is overexpressed in BL21 cells with pPDHE2/E3BP (a generous gift from Dr. Robert A. Harris of Indiana University School of Medicine) and purified *(12)*. PDC from porcine heart (P7032, P5194) and lipoamide dehydrogenase (L2002), which are commercially available from Sigma or purified from mammalian tissues *(13,14)* can be used to study PDC in the absence of overexpressed components. Inactivation of PDC and its components is performed in the following way: enzymes are pre-incubated with varying amounts of HNE in the absence or presence of the other compounds at 25°C. Aliquots (3–5 µL) are taken from preincubation mixture and added to the cuvet used for assay of the PDC activity.

2. HepG2 cells are grown in modified Eagle's medium containing 10% (v/v) heat-inactivated fetal bovine serum (FBS). At confluency cells are washed with phosphate-buffered saline (PBS), trypsinized, washed with PBS three times, resuspended in PBS with 5 mM glucose (100 µL of PBS per 100 mm plate) and incubated (50 µL of resuspended cells per incubation reaction) at 37°C in an incubator shaker in the presence and absence of HNE and other compounds as indicated. After incubation the cells are separated (by spinning for 2 min at 12,000 rpm in a microcentrifuge), resuspended in the Resuspension buffer (100–200 µL per reaction), divided into 50–100 µL aliquots, freeze-thawed three times, and stored at –80°C.

3.2. PDC Assay with the Purified Components

$$CH_3COCOOH + CoA + NAD^+ \rightarrow Acetyl\text{-}CoA + CO_2 + NADH + H^+$$

1. PDC catalyzes the oxidative decarboxylation of pyruvic acid by the overall equation shown above. PDC activity is measured spectrophotometrically by monitoring the rate of NADH production at 340 nm for an initial 1–2 min at 37°C. Seven hundred microliters of Supplement A, 170–180 µL of deionized water, 1 µg of E1, 1 µg of E2-E3BP, 1 µg of E3, 100 µL of Supplement B are added to a 1 mL cuvet prewarmed at 37°C for 5 min and incubated for 1 min. The reaction is started by addition of 10 µL of 200 mM pyruvic acid. Specific activity is expressed as U/mg of PDC proteins. One unit of PDC activity corresponds to one µmol of NADH formed per min.

3.3. PDC Assay in Treated Cells

1. PDC activity in HepG2 cells is measured by production of CO_2 (*see* equation above) from [1-^{14}C]-pyruvic acid *(15,16)*. PDC is dephosphorylated with lambda protein phosphatase to fully dephosphorylate and hence activate it. For this purpose, 50–100 μg of disrupted HepG2 cells are incubated with 10 μL of 10X phosphatase buffer, 10 μL of 10X $MnCl_2$, 1 μL of 100X BSA, 3.5 μL of 125 mM dichloroacetatic acid, 1 μL of lambda protein phosphatase in 100 μL of total volume (adjusted with Resuspension buffer) for 30 min at 30°C. The suspension of HepG2 cells is frozen after dephosphorylation and kept at –80°C until assayed for activity.

2. To measure PDC activity, 69 μL of PDC reaction mixture and 10 μL of Supplement C are added to each 15 mL plastic tube. Twenty microliters of the activated HepG2 cells suspension is then added to the tube, and the tube is placed in a shaking waterbath at 37°C. After 2 min, 1 μL of [1-^{14}C]-pyruvic acid is added. Immediately, a stopper with a center well is placed on the tube to seal it and prevent the CO_2 evaporation from the tube. The center well contains a paper filter (grade 238 blotting pad) presoaked with 100 μL of benzetonium hydroxide to absorb the released CO_2. After 10 min the reaction is stopped by adding 100 μL of 20% trichloroacetate with 30 mM pyruvic acid. The tube is incubated in the shaking water bath for 1 h to collect liberated CO_2 in the benzetonium hydroxide. The stopper with the center well is then taken out, cut, and placed into the scintillation vial to count the absorbed radioactive CO_2 formed during the reaction.

3. Activity of the α-ketoglutarate dehydrogenase complex (α-KGDC) is measured in HepG2 cells by the decarboxylation assay as described earlier for PDC assay in treated cells *(16)* except: 1) HepG2 cells are not activated with lambda protein phosphatase (α-KGDC is not regulated by phosphorylation-dephosphorylation); 2) α-KGDC reaction mixture is used for reaction; and 3) [1-^{14}C] α-ketoglutaric acid is used as a substrate.

    Specific activity is expressed as munits/mg cellular protein. One unit of activity (PDC or α-KGDC) corresponds to the decarboxylation of 1 μmole of labeled substrate/min.

3.4. Results

1. HNE is shown to modify the reduced lipoyl moiety of the α-ketoglutarate dehydrogenase complex *(9)*. **Figure 1** shows that the inactivation of PDC by 0.5 mM HNE occurs only in the presence of NADH. In the absence of added NADH, the lipoyl moieties of the E2 and E3BP components of PDC are in oxidized form and do not react with HNE. Addition of NADH results in the reduction of the lipoyl moieties which then react with HNE. Addition of cysteine protects PDC from HNE-induced inactivation by interacting rapidly with available HNE in the incubation medium (**Fig. 1**). The formation of a lipoyl-protein-HNE adduct and a cysteinyl-HNE adduct is shown in **Fig. 2**.

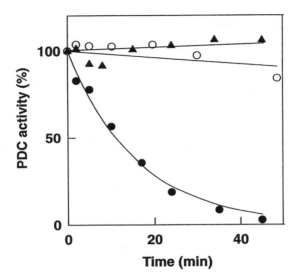

Fig. 1. Inactivation of PDC by HNE. Reconstituted PDC (16 µg of E1, 16 µg of E2-E3BP, and 8 µg of E3) is incubated in 80 µL of potassium-phosphate buffer, pH 7.5, with 0.5 mM HNE at 25° C. ○, no addition; ●, in the presence of 0.66 mM NADH; π, in the presence of 0.66 mM NADH and 3 mM cysteine. Four microliters aliquots are taken from the incubation mixture at indicated times to measure PDC activity.

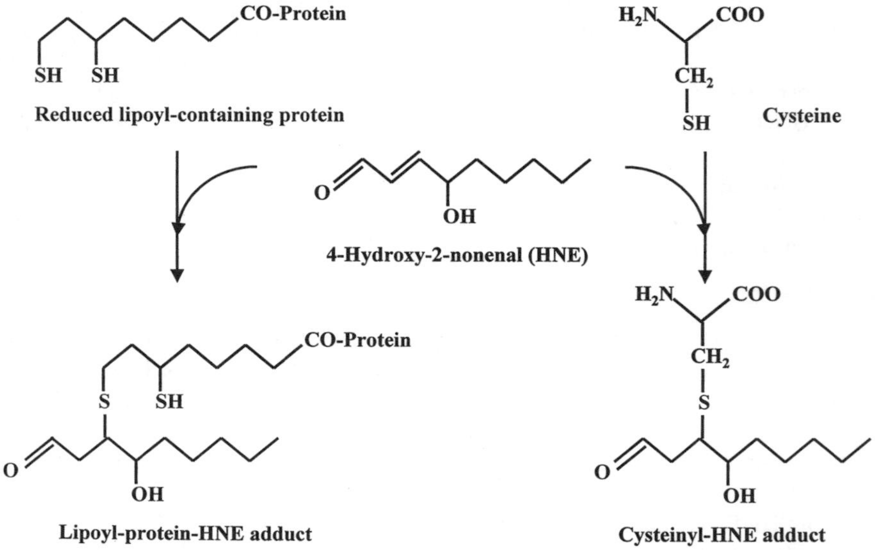

Fig. 2. Reaction of 4-hydroxy-2-nonenal with lipoyl moiety and cysteine.

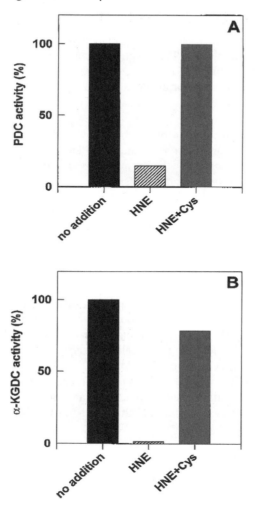

Fig. 3. Effect of HNE on PDC and α-KGDC activities in HepG2 cells. The intact HepG2 cells prewarmed for 10 min at 37° C are incubated for 30 min in the absence or presence of 1 m*M* HNE and 3 m*M* cysteine as indicated. PDC **(A)** and α-KGDC **(B)** activities are determined by measuring decarboxylation of [1-^{14}C] pyruvic acid and [1-^{14}C] ketoglutaric acid, respectively, to $^{14}CO_2$. 100% of PDC activity corresponds to 5.2 mU/mg. 100% of α-KGDC activity corresponds to 2.6 mU/mg.

2. The inhibitory effect of HNE on the activities of PDC and α-KGDC can be demonstrated in intact HepG2 cells. To monitor this effect, HepG2 cells can be incubated in the presence or absence of a known concentration of HNE for 30 min. As shown in **Fig. 3**, the addition of 1 m*M* HNE results in marked reduction in the

activities of PDC and α-KGDC. Again, the simultaneous addition of cysteine to the incubation medium protects HNE-induced inactivation of PDC and α-KGDC activities in HepG2 cells. The results presented here clearly demostrate the usefulness of monitoring of PDC or α-KGDC inactivation by HNE using either purified PDC preparations or intact cells.

4. Notes

1. To avoid thermoinactivation of purified PDC, the experiments are performed at 25°C and a 0.5 mg/mL concentration of PDC is used. Lowering the protein concentration or increasing the temperature would cause a decrease in PDC activity that should be taken into consideration. For example, 40 min incubation of PDC at 0.017 mg/mL results in 20% reduction of its activity.
2. Activity of the other multienzyme α-keto acid dehydrogenase complex, the branched-chain α-keto acid dehydrogenase complex, could be measured in HepG2 cells by the same decarboxylation assay but with a different substrate: [1-^{14}C] α-ketoisovalerate.
3. Concentration of HNE is determined by measuring optical density (OD) at 224 nm and using an extinction coefficient of 13,750 $M^{-1}cm^{-1}$ *(8)*.
4. HepG2 cells without HNE or other compounds used as a positive control should be incubated under the same conditions as the treated cells.
5. Precautions should be taken in handling tissue-culture cells if the experiments are going to be carried out with intact cells. Trypsinization of the cells is preferred over scrubbing the cells from the plate.
6. As we demonstrate, purified PDC and PDC in HepG2 cells respond similarly to HNE-dependent inhibition and cysteine-mediated protection. The same approach can be used to study the effect of other thiols on HNE-induced inactivation of PDC or α-KGDC.

Acknowledgments

The work performed in this laboratory and reported in this article was supported in part by NIH grants DK20478 and DK42885.

References

1. Reed, L. J. (1974) Multienzyme complexes. *Accts. Chem. Res.* **7,** 40–46.
2. Patel, M. S. and Roche, T. E. (1990) Molecular biology and biochemistry of pyruvate dehydrogenase complexes. *FASEB J.* **4,** 3224–3233.
3. Behal, R. H., Buxton, D. B., Robertson, J. G., and Olson, M. S. (1993) Regulation of the pyruvate dehydrogenase multienzyme complex. *Ann. Rev. Nutr.* **13,** 497–520.
4. Ravindran, S., Radke, G. A., Guest, J. R., and Roche, T. E. (1996) Lipoyl domain-based mechanism for the integrated feedback control of the pyruvate dehydrogenase complex by enhancement of pyruvate dehydrogenase kinase activity. *J. Biol. Chem.* **271,** 653–662.

5. Patel, M. S., Naik, S., Johnson, M., and Dey, R. (1996) Long-term regulation and promoter analysis of mammalian pyruvate dehydrogenase complex, in *Alpha-Keto Acid Dehydrogenase Complexes* (Patel, M. S., Roche, T. E., and Harris, R. A., eds.), Birkhauser Verlag, Basel, Switzerland, pp.197–211.

6. Tan, J., Yang, H-S., and Patel, M. S. (1998) Regulation of mammalian pyruvate dehydrogenase α subunit gene expression by glucose in HepG2 cells. *Biochem. J.* **336,** 49–56.

7. Uchida, K., Shiraishi, M., Naito, Y., Nakamura, Y., and Osawa, T. (1999) Activation of stress signaling pathways by the end product of lipid peroxidation. 4-Hydroxy-2-nonenal is a potential inducer of intracellular peroxide production. *J. Biol. Chem.* **274,** 2234–2242.

8. Humphries, K. M., Yoo, Y., and Szweda, L. I. (1998) Inhibition of NADH-linked mitochondrial respiration by 4-hydroxy-2-nonenal. *Biochemistry* **37,** 552–557.

9. Humphries, K. M. and Szweda, L. I. (1998) Selective inactivation of α-ketoglutarate dehydrogenase and pyruvate dehydrogenase: reaction of lipoic acid with 4-hydroxy-2-nonenal. *Biochemistry* **37,** 15,835–15,841.

10. Korotchkina, L. G., Tucker, M. M., Thekkumkara, T. J., Madhusudhan, K. P., Pons, G., Kim, H., and Patel, M. S. (1995) Overexpression and characterization of human tetrameric pyruvate dehydrogenase and its subunits. *Prot. Express. Purif.* **6,** 79–90.

11. Liu, T.-C., Korotchkina, L. G., Hyatt, S., Vettakkorumakankav, N. N., and Patel, M. S. (1995) Spectroscopic studies of the characterization of recombinant human dihydrolipoamide dehydrogenase and its site-directed mutants. *J. Biol. Chem.* **270,** 15,545–15,550.

12. Yang, D., Song, J., Wagenknecht, T., and Roche, T. E. (1997) Assembly and full functionality of recombinantly expressed dihydrolipoyl acetyltransferase component of the human pyruvate dehydrogenase complex. *J. Biol. Chem.* **272,** 6361–6369.

13. Stanley, C. J. and Perham, R. N. (1980) Purification of 2-oxo acid dehydrogenase multienzyme complexes from ox heart by a new method. *Biochem. J.* **191,** 147–154.

14. Linn, T. C., Pelley, J. W., Pettit, F. H., Hucho, F., Randall, D. D., and Reed, L. J. (1972) α-Keto acid dehydrogenase complexes. XV. Purification and properties of the component enzymes of the pyruvate dehydrogenase complex from bovine kidney and heart. *Arch. Biochem. Biophys.* **148,** 327–342.

15. Sheu, K.-F. R., Hu, C.-W. C., and Utter, M. F. (1981) Pyruvate dehydrogenase complex activity in normal and deficient fibroblasts. *J. Clin. Invest.* **67,** 1463–1471.

16. Kerr, D. S., Ho, L., Berlin, C. M., Lanoue, K. F., Towfighi, J., Hoppel, C. L., et al. (1987) Systematic deficiency of the first component of the pyruvate dehydrogenase complex. *Pediatr. Res.* **22,** 312–318.

28

Ceruloplasmin Detection by SDS-PAGE, Immunoblotting, and *In Situ* Oxidase Activity

Leonard A. Levin

1. Introduction

Ceruloplasmin is an acute-phase reactant that functions as a ferroxidase, oxidizing Fe^{++} to Fe^{+++}, the latter being able to combine with apotransferrin to form transferrin, the primary iron transport protein in the plasma. Ferroxidases indirectly reduce formation of hydroxyl radical by preventing the participation of Fe^{++} in the Fenton reaction *(1)*. Ceruloplasmin is primarily produced in the liver and transported in the plasma, but has also been shown to be produced in other tissues *(2–5)*. We describe here methods for detecting ceruloplasmin using sodium dodecyl sulfate polyacrylamide gel electrophoresis (SDS-PAGE) followed by either immunoblotting or assay of *in situ* oxidase activity.

2. Materials

2.1. Equipment

1. Vertical gel electrophoresis apparatus.
2. Blotting apparatus.
3. Gel dryer.

2.2. Reagents and Buffers

1. 1.0 *M* Tris-HCl, pH 6.8: 12.1 g Tris base, 80 mL deionized H_2O. Adjust to pH 6.8 with concentrated HCl. Bring to 100 mL with deionized H_2O.
2. 1.0 *M* Tris-HCl, pH 7.5: 12.1 g Tris base, 80 mL deionized H_2O. Adjust to pH 7.5 with concentrated HCl. Bring to 100 mL with deionized H_2O.
3. 1.5 *M* Tris-HCl, pH 8.8: 18.2 g Tris base, 80 ml deionized H_2O. Adjust to pH 8.8 with concentrated HCl. Bring to 100 mL with deionized H_2O.

From: *Methods in Molecular Biology, vol. 186: Oxidative Stress Biomarkers and Antioxidant Protocols*
Edited by: D. Armstrong © Humana Press Inc., Totowa, NJ

4. 50 m*M* Tris-HCl: 1.25 mL 1 *M* Tris-HCl, pH 7.5. Bring to 25 mL with deionized H$_2$O to 25 mL.
5. Tris-buffered saline (TBS): 7.5 mL 5 *M* NaCl, 12.5 mL 1 *M* Tris-HCl pH 7.5. Bring to 250 mL with deionized H$_2$0.
6. 10X Laemmli running buffer: 30.3 g Tris base, 144.2 g glycine, 10 g SDS, 800 mL deionized H$_2$O. Adjust to pH 8.3 with concentrated HCl. Bring to 1 L with deionized H$_2$O.
7. Sample buffer: 2 mL 10% SDS, 1 mL glycerol, 0.6 mL 1 *M* Tris-HCl, pH 6.8, 1 mL 1 *M* dithiothreitol (DTT), 5.4 mL deionized H$_2$O, 1 mg bromphenol blue.
8. Phosphate-buffered saline (PBS): 16 g NaCl, 0.4 g KCl, 2.88 g Na$_2$HPO$_4$, 0.48 g KH$_2$PO$_4$. Adjust pH to 7.4 with concentrated NaOH or HCl. Bring to 2 L with deionized H$_2$O.
9. 30% acrylamide mix: 30 g acrylamide, 0.8 g N,N'-methylene-bisacrylamide. Bring to 100 mL with deionized H$_2$O. Filter through Whatman #1 paper and store at 4°C in the dark. Acrylamide monomer is a dangerous neurotoxin; mask and gloves should be warn when handling.
10. 10% SDS: 10 g electrophoresis grade SDS in 100 mL deionized H$_2$O. Store at room temperature.
11. 10% ammonium persulfate (APS): 100 mg ammonium persulfate in 1 mL deionized H$_2$O. Store at 4°C for up to 1 wk.
12. N,N,N',N'-tetramethylethylenediamine (TEMED).
13. Gel stain: 2.5 g Coomassie blue (R-250), 100 mL glacial acetic acid, 500 mL methanol, H$_2$O to 1 L.
14. Gel destain: 300 mL glacial acetic acid, 1 L methanol, H$_2$0 to 4 L.
15. Polyvinylidene difluoride (PVDF) blotting material (0.45 µm pore size).
16. Transfer Buffer: 5.8 g Tris base, 29 g glycine, 1 g SDS, 200 mL methanol. Bring to 1 L with deionized H$_2$0. Store at 4°C.
17. Blocking reagent: 1 g nonfat dry milk, 40 µL Tween-20, PBS to 20 mL.
18. First antibody: rabbit anti-rat ceruloplasmin antibody (gift of Robert J. Cousins), diluted 1 : 1000 in 3% bovine serum albumin (BSA) in PBS.
19. Alkaline phosphatase-conjugated goat anti-rabbit antibody (Jackson Immunochemicals, West Grove, PA), diluted 1 : 10000 in 3% BSA in TBS.
20. PBS wash: PBS with 0.2% Tween-20.
21. TBS wash: TBS with 0.2% Tween-20.
22. Nitro blue tetrazolium (NBT) stock: 50 mg in 1 mL 70% dimethylformamide. Aliquot and store at 4°C.
23. 5-bromo-4-chloro-3-indolylphosphate (BCIP) stock: 50 mg in 1 mL 100% dimethylformamide. Aliquot and store at 4°C.
24. Alkaline phosphatase (AP) buffer: 100 m*M* Tris-HCl, pH 9.5, 50 m*M* MgCl$_2$, 100 m*M* NaCl, 0.1% Tween-20.
25. AP substrate solution: 5 mL AP buffer, 33 µL NBT stock, 17 µL BCIP stock. Use within 30 min of preparation.
26. AP stop solution: 20 m*M* EDTA in PBS.

27. Acetate buffer: 63 g Sodium acetate (anhydrous), 13.3 mL glacial acetic acid, 800 mL deionized H_2O. Adjust pH to 5.2, then bring to 1 L with deionized H_2O.
28. PPD solution: 50 mg paraphenylenediamine (Sigma), 100 mL acetate buffer. Prepare fresh each use.

3. Methods

3.1. Sample Preparation and Solubilization

1. Freeze freshly dissected tissue (e.g., liver, retina) on dry ice. Weigh.
2. Add 10 volumes of sample solubilization buffer.
3. Sonicate for 10–20 s, being careful to avoid foaming. Keep tube on ice to avoid overheating.
4. Centrifuge at 10,000g for 10 min, and transfer supernatant to a fresh tube.
5. Calculate the protein concentration using standard methods.
6. Boil sample for 5 min, then either store at –20°C or load onto gel.

3.2. Gel Electrophoresis

1. Several considerations enter into the choice of equipment for PAGE (*see* **Note 1**). The following steps are given for small format gels (mini-gels). The equipment is assembled according to the manufacturer's directions, using freshly cleaned glass plates and seals. The integrity of the seals is paramount, as otherwise neurotoxic acrylamide monomer can leak out.
2. The gel consists of two different concentrations of acrylamide. The lower gel is the separating gel, and contains a concentration of acrylamide, which yields maximal separation of proteins according to protein size (*see* **Note 2**). For ceruloplasmin, a protein of approx 132 kD, a 7.5% acrylamide gel is optimal. The upper gel is the stacking gel, and serves to concentrate into a single band the mixture of proteins placed within the sample well. The use of a stacking gel increases the resolution of bands within the separating gel. Because the separating gel is lower, it is poured first.
3. Mix the following separating gel recipe into an Erlenmeyer flask: 2.6 mL deionized H_2O, 1.25 mL 1.5 M Tris, pH 8.8, 50 μL 10% SDS, 1.25 mL 30% acrylamide mix. De-gas by attaching to a vacuum line for 10 min.
4. In rapid sequence add: 25 μL 10% APS, 2.5 μL TEMED. Carefully mix by swirling, then pipet the solution down one side of the gel until approx 1 cm below where the bottom of the comb would be (i.e., leave 1 cm for the stacking gel). Gently overlay with deionized H_2O, and let polymerize for 15–30 min.
5. Pour off the water, and use a folded piece of stiff filter paper to gently remove residual water from the top of the gel, without touching the gel itself. Place the sample comb at the top of the gel, leaving one side up so that there is room to pour the gel solution. Mix the following stacking gel recipe into a flask or tube: 2.1 mL deionized H_2O , 0.5 mL 30% acrylamide mix, 0.38 mL 1 M Tris, pH 6.8, 0 μL 10% SDS. After de-gassing, rapidly add 30 μL 10% APS and 3 μL TEMED

and pour down the side of the gel, so that it overlays the separating gel. Place the comb in position, making sure that no bubbles form under the wells. Let polymerize for 15–30 min at room temperature.

6. Remove the comb and rinse the wells with 1X Laemmli running buffer (diluted from 10X stock). Assemble the gel apparatus and put running buffer in the upper and lower chambers. Remove air bubbles from below the gel by squirting with a J-shaped pipet (use a Bunsen burner to make this).

7. Load the samples using thin pipet tips or a Hamilton syringe. If the same gel is going to be used for more than one purpose (e.g., immunoblotting or *in situ* oxidase activity detection, then samples should be loaded in duplicate sets). Molecular-weight markers should be loaded at one or both ends of the gel. For ceruloplasmin and other high molecular-weight proteins, high molecular-weight markers should be used (e.g., Bio-Rad high-range SDS-PAGE standards, 161-0303).

8. Attach the electrodes and check for leaks. Run the gel at 100 volts (constant voltage) until the samples have entered the stacking gel, then run at 200 volts until the dye reaches the bottom of the gel. If immunoblotting is to be performed, this is a good time to prepare the blotting apparatus and PVDF membrane.

9. Carefully disassemble the gel apparatus, and gently use a spatula to separate the plates. Use a razor blade to cut off the bottom right corner (closest to last lane) of the gel, so that orientation is maintained during subsequent steps.

10. At this point the gel can be divided into vertical sections with a razor blade, with sections used for staining (below), immunoblotting (**Subheading 3.3.**), or *in situ* detection of oxidase activity (**Subheading 3.4.**). Make sure to cut off the bottom right corner of each section, as before.

11. To stain the gel, gently transfer a section to a plastic or glass container containing gel stain. Cover and place on a slowly rotating or tilting shaker for several hours. Pour off the gel stain (it can be reused) and replace with gel destain. Continue to shake, replacing the destain with fresh solution every few hours, until the gel is virtually clear.

12. Stained gels can be dried in a heated or evaporative gel dryer and mounted in notebooks.

3.3. Immunoblotting

1. The following can be performed while electrophoresis is taking place. Using talc-free gloves, cut a piece of PVDF the same size as the gel section to be blotted. Wet for several seconds in 100% methanol, then 5 min in deionized H_2O, then incubate in transfer buffer for at least 10 min. Cut two pieces of filter paper and place in transfer buffer. If the blotting apparatus has sponges, pads, or other porous supports, place them in transfer buffer as well.

2. After electrophoresis is complete and the gel has been removed from the equipment, place the gel section to be blotted on one of the pieces of wet filter paper, keeping the gel and the PVDF membrane wet at all times. Overlay the PVDF membrane by rolling it onto the gel, taking care not to let bubbles form between

the gel and the membrane. Do not shift the membrane once it is attached to the gel. A wet glove can be used to squeeze bubbles out between the gel and the membrane. Overlay the other piece of filter paper and place the entire sandwich between two soaked support pads. Put the entire package into the blotting apparatus, making sure that the membrane is closest to the positive (red) electrode and the gel is closest to the negative (black) electrode. Fill the blotting apparatus with cold transfer buffer, and put a magnetic stir bar at the bottom of the tank. Move to a cold room and place the entire apparatus on a magnetic stirrer.

3. Transfer according to the manufacturer's directions for the size of the apparatus. For a mini-gel and small blotting apparatus, we typically transfer at 30 mA (constant current) overnight.
4. After disassembling the apparatus, the membrane is cut into strips corresponding to one or more lanes to be stained with each antibody. Orientation can be maintained by cutting off the bottom right corner of each strip. Remember that the blotted proteins are on the gel side of the protein, and so the order of the lanes on the blot will be the reverse of what they were on the gel. Use a 25-gauge needle to poke a code in each blot, to avoid confusion later on.
5. Wash the blots in PBS for 5 min by gently agitating for 5–10 min at room temperature.
6. Incubate blots in blocking reagent for 1 h at room temperature or overnight at 4°C.
7. Incubate blots in first antibody solutions (*see* **Note 3**). To preserve reagents, remove each strip and place in a fresh petri dish. Gently pipet the diluted antibody onto the strip; surface tension should keep the liquid from spilling to the adjacent plastic. Incubate overnight at 4°C.
8. Wash the strips 4 times, using TBS for the last wash (*see* **Note 4**).
9. Remove each strip and place in a fresh petri dish. Incubate blots in second antibody solution for 2 h at room temperature.
10. Wash the strips in 4 changes of TBS. Do not use PBS.
11. Cover blots with AP substrate solution and incubate at room temperature until bands are visible.
12. Stop the reaction with AP stop solution. Blots can be dried and stored in a notebook.

3.4. In Situ Detection of Protein Activity

In some cases the biological activity of a protein is maintained even when subject to SDS-PAGE. Although SDS will considerably change the conformation of a protein, renaturation can occur if the SDS is diluted out, allowing the protein to become active. The following technique relies on the oxidation of paraphenylenediamine to an insoluble black product.

1. After electrophoresis is complete and the gel has been removed from the equipment, use a razor blade to cut the section of the gel to be assayed for oxidase activity from the rest of the gel. Cut off the bottom right corner as a marker.

Ordinarily, the other sections of the gel will be either stained with Coomassie blue or immunoblotted.

2. Equilibrate the gel by incubating for 1 h in acetate buffer at room temperature on a rotating shaker.

3. Pour off the acetate buffer and add freshly prepared paraphenylenediamine (500 µg/mL) in acetate buffer. Incubate overnight at room temperature on a rotating shaker.

4. Bands corresponding to oxidase activity should appear at a position corresponding to the size of ceruloplasmin. At this point the gel can either be dried down or further incubate. Longer incubation times will increase the density of the band, but will increase background.

4. Notes

1. A variety of gel apparatuses are available for pouring and running polyacrylamide gels. Considerations for selecting the specific gel equipment include size of the gel, ease of use, and ability to run commercially available prepoured gels. The advantages of a large gel are the ability to separate molecules that have similar molecular weights (resolution), the ability to load more sample and still see individual bands, and the ability to run more samples on a single gel. The advantages of a small gel are the ease and speed of preparation and conservation of reagents. The choice of specific manufacturer is less important than other factors, and we have successfully run gels on a variety of different rigs. The use of prepoured gels makes for a efficient and rapid preparation time before electrophoresis, but is more expensive.

2. The acrylamide concentration in the gel will determine the resolution of similar-sized proteins, as well as the efficiency of transfer during blotting. For ceruloplasmin, a 132 kD protein, a 6 or 7.5% acrylamide concentration is optimal. Using a higher percentage acrylamide gel will inhibit the ability of the ceruloplasmin to go into the gel, and impede transfer through the gel to the PVDF membrane.

3. Dilutions of first antibodies and second depend on a variety of factors, and usually need to be optimized by trial and error. We typically use 1:1000 for antigen-specific antisera. Control solutions can be no antibody or normal serum from the animal in question (preferred).

4. PBS should not be used with alkaline phosphatase-coupled antibodies, nor sodium azide (to prevent bacterial growth) with horseradish peroxidase-coupled antibodies.

Acknowledgments

Supported by the Retina Research Foundation, the Glaucoma Foundation, NIH EY00340, and an unrestricted departmental grant from Research to Prevent Blindness, Inc. Dr. Levin is a Research to Prevent Blindness Dolly Green scholar.

References

1. Gutteridge, J. M. (1985) Inhibition of the Fenton reaction by the protein caeruloplasmin and other copper complexes. Assessment of ferroxidase and radical scavenging activities. *Chem. Biol. Interact* **56,** 113–120.
2. Fleming, R. E., Whitman, I. P., and Gitlin, J. D. (1991) Induction of ceruloplasmin gene expression in rat lung during inflammation and hyperoxia. *Am. J. Physiol.* **260,** L68–L74.
3. Jaeger, J. L., Shimizu, N., and Gitlin, J. D. (1991) Tissue-specific ceruloplasmin gene expression in the mammary gland. *Biochem. J.* **280,** 671–677.
4. Klomp, L. W. J., Farhangrazi, Z. S., Dugan, L. L., and Gitlin, J. D. (1996) Ceruloplasmin gene expression in the murine central nervous system. *J. Clin. Invest.* **98,** 207–215.
5. Levin, L. A. and Geszvain, K. M. (1998) Expression of ceruloplasmin in the retina: induction after retinal ganglion cell axotomy. *Invest. Ophthalmol. Vis. Sci.* **39,** 157–163.

29

Metallothionein Determination by Isocratic HPLC with Fluorescence Derivatization

Shinichi Miyairi and Akira Naganuma

1. Introduction

Metallothionein (MT) is reported as having an intensive affinity to a number of heavy metal ions such as Zn^{2+}, Cd^{2+}, Cu^{2+}, and Hg^{2+}, which suggests a relation with zinc and copper homeostasis (1). The structural uniqueness of MT is expressed by its high cysteine content without intramolecular cystine bonding; 20 cysteins in 61 amino acids, and all the sulfuhydryl groups of cysteine residues coordinate to metal ions (**Fig. 1**). The importance of MT in protection against oxidative stress (2) in addition to detoxification of heavy metals (1) is reviewed. In oxidative stress and heavy-metal toxicology studies, MT has been considered to be one of the important factors that should be monitored.

In this chapter a convenient method for direct determination of MT using high-performance liquid chromatography (HPLC) with fluorescence detection using an isocratic solvent system is described. Fluorescence detection is one of the most convenient techniques with high sensitivity in HPLC. In addition, HPLC with an isocratic solvent system is generally useful to determine a specific component in numerous specimen rather than using a gradient solvent system.

2. Materials

2.1. Equipment

1. Equipment for organic synthesis.
   a. Magnetic stirrer.
   b. Water bath.
   c. Rotary vacuum evaporator system.
   d. Aspiration system.
   e. Separation funnel.

From: *Methods in Molecular Biology, vol. 186: Oxidative Stress Biomarkers and Antioxidant Protocols*
Edited by: D. Armstrong © Humana Press Inc., Totowa, NJ

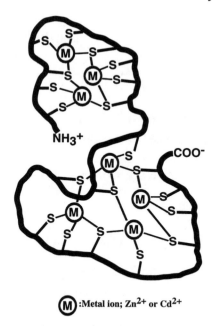

(M) :Metal ion; Zn^{2+} or Cd^{2+}

Fig. 1. Structure of MT.

f. Sintered glass funnel.
g. Filtration unit for sintered glass funnel.
2. Cell-culture unit.
a. Clean bench.
b. CO_2 incubator.
3. Polytron-type homogenizer.
4. Ultrasonic homogenizer.
5. Centrifuge with cooling system.
6. Microplate reader.
7. pH meter.
8. Nitrocellulose membrane, 0.45 μm.
9. Filtration unit for nitrocellulose membrane.
10. Ultrasonic generator with water bath.
11. Analytical HPLC instrumentation.
a. Shimadzu (Kyoto, Japan) LC-10AD HPLC pump.
b. Shimadzu RF-10AXL fluorescence detector.
c. Shimadzu SCL-10A system controller.
d. Shimadzu SIL-10AXL auto injector.
e. Shimadzu C-R6A Chromatopac data processor.
f. Puresil C18 column (250 × 4.6 mm *i.d.*; Millipore-Waters, Milford, MA).
g. Shodex RSpak RP-18-413 column (150 × 4.6 mm *i.d.*; Showa Denko, Tokyo, Japan).

2.2. Reagents

1. Cystamine hydrochloride.
2. Propionic anhydride.
3. Pyridine.
4. Ammonium 7-fluorobenz-2-oxa-1,3-diazole-4-sulfonate (Dojindo Laboratories, Kumamoto, Japan).
5. Tri-*n*-butylphosphine.
6. Tris(2-carboxyethyl)phosphine hydrochloride (Tokyo-Kasei Kogyo Ltd., Japan).
7. Metallothionein II from rabbit liver (Sigma Chemical Co., St. Louis, MO).
8. Kanamycin sulfate.
9. Fetal calf serum.
10. Dulbecco's modified Eagle's medium (DMEM).
11. Dye-binding assay reagent (Bio-Rad Laboratories, Hercules, CA).
12. Acetonitrile, HPLC grade.

3. Methods

3.1. Preparation of N,N'-Dipropionylcystamine (NPC-dimer)

1. Stir a mixture of 20 g of cystamine hydrochloride and 40 mL of propionic anhydride in 100 mL of pyridine at room temperature for 3 h using a magnetic stirrer with occasional heating at 70°C in a water bath.
2. Add 30 mL of water to the reaction mixture and stir at room temperature for 20 min.
3. Evaporate the solvent using a rotary vacuum evaporation system.
4. Extract the product with ethyl acetate (EtOAc) using a separation funnel.
5. Wash the organic layer sequentially with 5% (w/v) sodium hydroxide, 5% (w/v) hydrochloric acid (HCl), 5% (w/v) sodium bicarbonate ($NaHCO_3$) and water in the separation funnel.
6. Dry the organic phase over anhydrous sodium sulfate.
7. Evaporate the organic solvent using a rotary vacuum evaporation system after filtrate.
8. Crystallize the residue from EtOAc.
9. Collect the crystals on a sintered glass funnel using a filtration unit. The crystals, colorless needles, are the dimer of N-propionylcysteamine (NPC-dimer) (melting point; 116°C *[3]*).
10. Dissolve approx 180 mg of the NPC-dimer in 10 mL of 10% (v/v) aqueous ethanol as the NPC-dimer stock solution.

3.2. Sample Preparation

3.2.1. Tissues

1. Inject zinc sulfate ($ZnSO_4$) solution in saline (200 μmol/kg of weight), cadmium sulfate ($CdSO_4$) solution in saline (200 μmol/kg of weight) or vehicle into ICR/Slc mice (10–13 wk old, male) intraperitoneally once a day for 2 or 5 d.

2. Collect the liver and kidney at 16–24 h after the last administration.
3. Homogenize the tissue (50 mg of wet weight) using a polytron type homogenizer at 0°C in 1 mL of 150 mM potassium chloride (KCl) containing 10 μL of the NPC-dimer stock solution in 100 mL of the KCl solution.
4. Heat the homogenate in boiling water for 5 min.
5. Centrifuge the denatured homogenate at 1000g for 10 min at 4°C.
6. Collect the supernatant as the sample for MT determination.

3.2.2. Cells

1. Culture HepG2 cells (8.5×10^5 cells) in a 6-well cluster dish with 3 mL of DMEM in the presence of fetal calf serum (FCS) (10% v/v), kanamycin sulfate (60 μg/mL), L-glutamine (316 μg/mL) and $NaHCO_3$ (1 mg/mL) at 37°C for 24 h under a 5 % CO_2 atmosphere.
2. Remove the culture medium by aspiration.
3. Incubate the cells at 37°C for 24 h under a 5 % CO_2 atmosphere after addition of 3 mL of the culture medium described in **steps 1** and **2** with or without additional zinc chloride ($ZnCl_2$).
4. Recover the cells by centrifugation at 250g for 5 min at 4°C after scraping from the dish in phosphate-buffered saline (PBS).
5. Homogenize the cells using an ultrasonic homogenizer at 6 watts for 10 s for 3 times at 0°C in 0.5–1.0 mL of 150 mM KCl containing 5 μL of the NPC-dimer stock solution in 100 mL of the KCl solution.
6. Determine protein concentration of the cell lysate. Mix 10 μL of the cell lysate with 100 μL of fivefold diluted the dye-binding assay reagent (**4**) in a 96-well cluster dish. Use bovine serum albumin (BSA) solutions at 0, 0.1, 0.3, 0.5, 0.75, and 1.0 mg/mL of concentrations as the standard protein solutions. Measure light absorption at 620 nm (595 nm is recommended) using a microplate reader.
7. Heat the residual cell lysate in boiling water for 5 min.
8. Centrifuge the denatured cell lysate at 1000g for 10 min at 4°C.
9. Collect the supernatant as the sample for MT determination.

3.3. General Procedure for Fluorescence Derivatization of MT

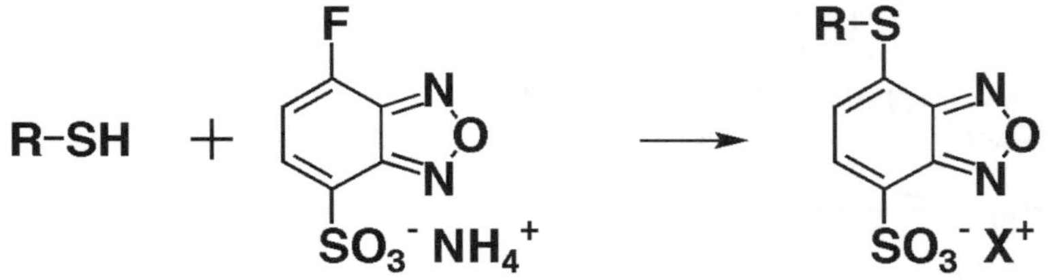

3.3 Reaction

Fig. 2. Derivatization of thiol compound with SBD-F.

3.3.1. Preparation of Reaction Buffer (1 M borate buffer at pH 10.5 containing 5 mM ethylenediaminetetraacetic acid)

1. Dissolve 6.183 g of boric acid, 146 mg of ethylenediaminetetraacetic acid (EDTA) and 4.5 g of potassium hydroxide (KOH) in 90 mL of water.
2. Adjust the pH of the solution at 10.5 with 1–5 M KOH solution.
3. Adjust the volume of the solution to 100 mL with water.

3.3.2. Derivatization

1. Transfer 150 µL of the supernatant of the denatured homogenate or the cell lysate into a glass test tube and mix it with 350 µL of the reaction buffer described in **Subheading 3.3.1.** (*see* **Note 1**), 10 µL of 20% (v/v) tri-n-butylphosphine (TBP) in 2-propanol (*see* **Note 2**) and 40 µL of 0.5% (w/v) ammonium 7-fluorobenz-2-oxa-1,3-diazole-4-sulfonate (SBD-F) *(5)* in water using a vortex-mixer.
2. Heat the mixture for 30 min at 50°C in a water bath (*see* **Note 3**).
3. Add 50 µL of 4 M HCl to the mixture to terminate the reaction.
4. Centrifuge the reaction mixture at 1000g for 10 min at 4°C.
5. Subject the supernatant of the resultant mixture to HPLC after appropriate dilution with water (**Note 4**).

3.4. HPLC Procedure

3.4.1. Mobile Phase

A mixture of 20 mM potassium phosphate buffer at pH 7.5, acetonitrile and methanol (80:18:2, v/v/v) is shaken well and the air in the mixture evacuates using sonication. The potassium phosphate buffer should be filtered through 0.45-µm nitrocellulose membrane using a suitable filtration unit prior to use.

3.4.2. HPLC Analysis

4-sulfonylbenz-2-oxa-1,3-diazol-7-yl (SBD)-labeled MT is separated from the interfering substances using a tandem column system with a Shodex RSpak RP18-413 column and a Puresil C18 column, in that order. Forty µL of the each derivatized sample described in **Subheading 3.3.2.** is injected by SIL-10AXL auto injector under the control of the SCL-10 system controller, at 15 min interval. The samples are eluted isocratically using the mobile phase described in **Subheading 3.4.1.**, at a flow rate of 0.7 mL/min. Fluorescence intensity of SBD derivatives of MT and NPC are monitored with an excitation wavelength of 384 nm and emission wavelength of 510 nm using the RF-10AXL fluorescence detector.

3.5. Results

3.5.1. Chromatographic Separation

Chromatograms of SBD-labeled components prepared from liver and kidney of a mouse treated with $ZnSO_4$ given intraperitoneally for 2 d and those of

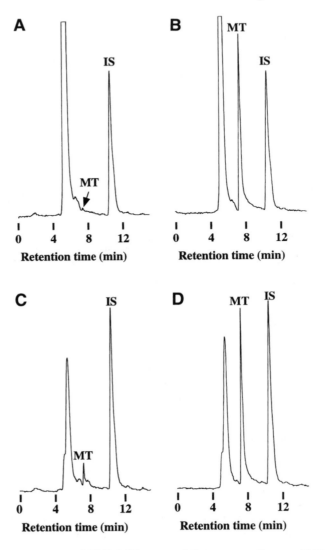

Fig. 3. Chromatograms of SBD-MT prepared from mouse liver and kidney. Samples
are prepared from (A) liver of nontreated mouse, (B) liver of ZnSO₄-treated mouse, (C)
kidney of nontreated mouse, (D) and kidney of ZnSO₄-treated mouse.

a nontreated mouse are shown in **Fig. 3**. A chromatographic profile of SBD
derivatives in the sample prepared from human kidney is depicted in **Fig. 4**.
The elution of SBD-labeled MT in the samples prepared from HepG2 cells
cultured with or without ZnCl₂ are shown in **Fig. 5**. In all cases examined the
peaks corresponding to SBD-labeled MT and SBD-labeled NPC used as an

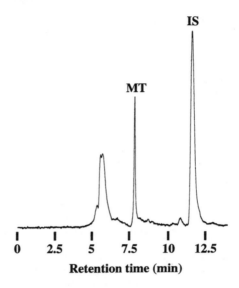

Fig. 4. Chromatogram of SBD-MT prepared from human kidney. The supernatant of heat-denatured human kidney homogenate is subjected to the derivatization reaction with SBD-F as described in **Subheading 3.3.2.**

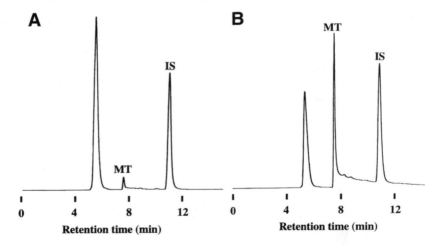

Fig. 5. Chromatograms of SBD-MT prepared from HepG2 cells. Samples are prepared from HepG2 cells cultured without $ZnCl_2$ (**A**) or with 400 μM of $ZnCl_2$ (**B**).

internal standard appeared at approx 7.2 min and 11.5 min of retention time, respectively. The peaks preceding MT appeared at 5.4–5.7 min of retention time might be SBD derivatives of thiol containing small molecules such as glutathione.

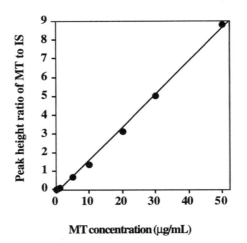

Fig. 6. Relationship of amount of MT and the ratio of peak height of MT to NPC (*IS*). Rabbit MT-2 was derivatized using SBD-F in the presence of NPC-dimer as an internal standard (IS) and subjected to HPLC (r = 0.999).

3.5.2. Calibration Curve for MT Determination

Commercially available rabbit metallothionein II (rabbit MT-2) can be used for preparation of the MT calibration curve (*see* **Note 5**). The relation of the amount of MT and the ratio of peak height of MT to NPC is depicted in **Fig. 6**. The detection limit of SBD-labeled MT is a few hundred picograms per injection. The calibration curve shows good linearity in a range up to 50 μg/mL for rabbit MT-2 (*see* **Note 6**). Recovery (>93%) of rabbit MT-2 from mouse liver homogenate is satisfactory in the range of 1–30 μg/mL. The accuracy of the obtained values is clarified by the intra- and interassay coefficients of variation that are 5.1 % (*n* = 7) and 3.9 % (*n* = 3), respectively. The peak area ratio of SBD-labeled MT to SBD-labeled NPC (*IS*) is constant and can be maintained for at least for 5 d when the mixture is stored at 4°C.

3.5.3. Determination of MT in Mouse Tissues and HepG2 Cells

1. MT concentrations in liver and kidney of mice given 0.1 mL saline solution of $ZnSO_4$, $CdSO_4$ or vehicle intraperitoneally once a day for 5 d are determined using this HPLC method. The basal MT levels of liver and kidney are 15 and 38 μg/g of wet tissue, respectively. The MT levels of these tissues are elevated to 655 and 261 μg/g after treatment with $ZnSO_4$ and 1163 and 298 μg/g after treatment with $CdSO_4$.
2. MT concentration in HepG2 cells cultured with or without $ZnCl_2$ is also determined using this HPLC method (**Fig. 7**). The basal level of MT in HepG2 cells is 1.4 μg/mg of protein. The cell viability is not affected by addition of

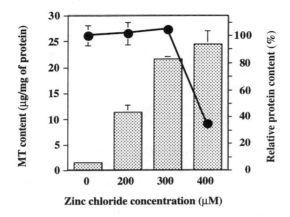

Fig. 7. Determination of cellular MT level in HepG2 cells. Sample is prepared from HepG2 cells after 24 h incubation with various concentrations of ZnCl$_2$. Data are presented as mean ± SD (n = 3). Symbols: (▨), MT content in HepG2 cells; ●, relative protein content.

ZnCl$_2$ up to 300 μM as determined from the protein content of the adherent cells on the culture dish, which then decreases to approx 35% after addition of 400 μM of ZnCl$_2$. The MT contents in HepG2 cells are increased dose-dependently after treatment with ZnCl$_2$ to 11.2, 21.4, and 24.4 μg/mg of protein at 200, 300, and 400 μM of ZnCl$_2$, respectively.

3.5.4. Application of The HPLC Method to Determination of Cu-MT

MT is presumed to be a physiological reservoir for zinc and copper, which are essential metals (**6**). Furthermore, abnormal accumulation of Cu^{2+} in liver and kidney is observed in several animal models (*see* **Note 7**). Accordingly, in some cases, tissue containing MT protein complexed with Cu^{2+} in addition to Zn^{2+} and Cd^{2+} would be a sample for this HPLC method. Removal of metal ions from the MT molecule using chelating agents is essential for derivatization of MT with SBD-F (*see* **Note 8**). Although MT protein complexed with Zn^{2+} and Cd^{2+} are derivatized quantitatively with SBD-F in the derivatization condition described in **Subheading 3.3.2.**, MT protein complexed with Cu^{2+} results in only 50–60% of the actual amount. When the derivatization reaction is carried out in a borate buffer at pH 10.5 containing 1 mM bathophenanthroline disulfonic acid disodium salt (BPS) in addition to 5 mM EDTA, MT protein complexed with Cu^{2+} is converted to an SBD derivative quantitatively as well as MT protein complexed with Zn^{2+} and Cd^{2+}. This indicates that utilization of BPS in combination with EDTA as a chelating agent is effective for the derivatization of Cu-MT with SBD-F in tissues or cultured cells from the model animals described previously for the present HPLC method.

However, the general derivatization method is sufficient to determine MT in human kidney as well as ordinary animal tissue and cells, because the apparent MT concentrations (350 ± 190 μg/g of wet tissue; mean ± SD in the range of 147–768 μg/g of wet tissue) in the supernatants of heated homogenates prepared from 10 human kidneys are not influenced by addition of BPS to the reaction buffer.

4. Notes

1. The height of the peak corresponding to SBD-labeled MT is increased by derivatization of MT with SBD-F in the presence of 1 mM EDTA in the borate buffer and is constant at the concentration up to 20 mM. EDTA may be effective for removing coordinated metal ions such as Zn^{2+} and Cd^{2+}.
2. The height of the peak corresponding to SBD-labeled MT is increased by addition of 10 % (v/v) TBP solution (10 μL) and is constant in the range of 10–40% (v/v) of TBP in the solution.
3. The height of the peak corresponding to SBD-labeled MT is increased with a reaction time up to 20 min and is constant up to 40 min.
4. The dilution should not be less than 10-fold, because 2-propanol may affect to the resolution in concentrations higher than 0.17 % (v/v). When a more concentrated solution is subjected to HPLC, we recommend using 20% (w/v) tris (2-carboxyethyl)phosphine hydrochloride in water instead of the TBP solution.
5. Retention times of SBD derivatized isoforms of MT (MT-1 and MT-2) from rabbit, rat, and mouse (and human; *see* **Fig. 4**) are identical when using the tandem column method. Fluorescence intensities of these MT isoforms after derivatization with SBD-F are almost equivalent. Therefore, we suggest that the total amount of MT can be determined by this HPLC system and commercially available rabbit MT can be used as a standard sample for determination of MT in several animal species.
6. Good linearity of the calibration curve in the range up to 100 μg/mL has been confirmed in other experiments.
7. The Long-Evans Cinnamon rat is an animal model of liver cancer that accumulates Cu^{2+} in liver as a complex of MT, which spontaneously develops hepatoma *(7)*. Animal models of Menkes disease *(8)* also accumulate excessive concentrations of copper in their kidneys.
8. The efficiency of removal of metal ions from MT may depend on the affinity of the chelating agent to the metal ion and its concentration. Metal ions such as Zn^{2+}, Cu^{2+}, Cd^{2+}, and Hg^{2+} can bind to MT in vivo, and the relative magnitude of affinity of the metals to MT is in the order of $Zn^{2+} < Cd^{2+} < Cu^{2+} < Hg^{2+}$ *(9)*.

References

1. Kägi, J. H. R. (1993) Evolution, structure and chemical activity of class I metallothioneins: an overview, in *Metallothionein III* (Suzuki, K. T., Imura, N., and Kimura, M., eds.), Birkhäuser Verlag, Basel, Switzerland, pp. 29–55.

2. Sato, M. and Bremner, I. (1993) Oxygen free radicals and metallothionein. *Free Rad. Biol.* **14,** 325–337.
3. Miyairi, S., Shibata, S., and Naganuma A. (1998) Determination of metallothionein by high-performance liquid chromatography with fluorescence detection using an isocratic solvent system. *Anal. Biochem.* **258,** 168–175.
4. Bradford, M. M. (1976) A rapid and sensitive method for the quantitation of microgram quantities of protein utilizing the principle of protein-dye binding. *Anal. Biochem.* **72,** 248–254.
5. Imai, K., Toyo'oka, T., and Watanabe, Y. (1983) A novel fluorogenic reagent: ammonium 7-fluorobenzo-2-oxa-1,3-diazole-4-sulfonate. *Anal. Biochem.* **128,** 471–473.
6. Cherian, M. G. and Chan, H. M. (1993) Biological functions of metallothionein, in *Metallothionein III* (Suzuki, K. T., Imura, N., and Kimura, M., eds.), Birkhäuser Verlag, Basel, Switzerland, pp. 87–109.
7. Mori, M., Hattori, A., Sawaki, M., Tsuzuki, N., Sawada, N., Oyamada, M., et al. (1993) The LEC rat: a model for human hepatitis, liver cancer, and much more. *Am. J. Pathol.* **144,** 200–204.
8. Hunt, D. M. (1974) Primary defect in copper transport underlies mottled mutants in the mouse. <ctItalic>Nature **249,** 852–854.
9. Webb, M. (1979) Metallothionein, in *The Chemistry, Biochemistry and Biology of Cadmium* (Webb, M., ed.), Elsevier/North Holland, Amsterdam, pp. 195–266.

30

Quantification of Oxidized Metallothionein by a Cd-Saturation Method

Dominik Klein, Uma Arora, Shin Sato, and Karl H. Summer

1. Introduction

One of the most prominent characteristics of metallothionein (MT) is its high cysteine content of 30%. The sulfhydryl groups, normally involved in metal binding, render the protein particularly sensitive towards oxidation. Oxidation of sulfhydryl groups in MT may occur in vivo and in vitro, either intra- or intermolecularly, the latter leading to dimeric or polymeric forms of MT. Thiol oxidation and oligomerization of MT may be induced by various radicals *(1–3)* and free metal ions *(4)*. The sensitivity to oxidation of thiol groups in MT to a great extent depends on the metals that are bound to the protein. Particularly the Cu-containing MT easily polymerizes *(3,5)*. Accordingly, Cu-rich polymerized MT has been isolated from the livers of newborn animals *(6)* and human fetuses *(7,8)*. Insoluble polymerized forms of MT were shown to be mainly located in the heavy lysosomal fraction of the liver *(6)*. In agreement, elevated levels of Cu associated with oxidized forms of MT have been described in lysosomes of patients with Wilson disease and dogs with inherited Cu-toxicosis *(9–12)*. More recently, polymeric MT has been also detected in lysosomes of hepatocytes and Kupffer cells from LEC rats, the rodent model for Wilson disease *(13)*.

The present test protocol describes a fast and easy to perform method for quantifying oxidized MT. In principle, oxidized MT is converted into native MT with 2-mercaptoethanol as reducing agent and $ZnSO_4$ as metal-donor, and MT is subsequently determined via Cd-saturation as described previously *(14)*.

From: *Methods in Molecular Biology, vol. 186: Oxidative Stress Biomarkers and Antioxidant Protocols*
Edited by: D. Armstrong © Humana Press Inc., Totowa, NJ

2. Materials

2.1. Equipment

1. Potter-Elvehjem homogenizer (Braun, Melsungen, Germany).
2. Ultracentrifuge Optima L-70, 50 Ti rotor, HPLC Spherogel TSK 2000 SW column (7.5 × 300 mm, 10 μm) and centrifuge Microfuge E (Beckman, Munich, Germany).
3. Minaxi Autogamma 5000 Counter with 2" NaJ detector (Canberra Packard, Frankfurt, Germany).
4. Dialyzing tube Thomapor®, exclusion limit 2,000 Da (Reichelt-Chemie-Technik, Heidelberg, Germany).
5. Mixer 5432 (Eppendorf, Hamburg, Germany).

2.2. Chemicals

1. ^{109}Cd (37 MBq/mg Cd) in 0.1 M HCl (Amersham Buchler, Braunschweig, Germany).
2. Ammonium tetrathiomolybdate (Ventron-Alfa, Karlsruhe, Germany).
3. 5,5'-Dithio-bis(-2-nitrobenzoic acid) (DTNB) (Roche, Mannheim, Germany).
4. Cd_5,Zn_2-thionein from rabbit liver, 2-mercaptoethanol (2-ME), $CuSO_4 \cdot 5H_2O$, $ZnSO_4 \cdot 7H_2O$, Carboxymethyl (CM)-Sephadex (40–200 μm), Diethylaminoethyl (DEAE)-Sephacel (40–150 μm), acetonitrile (high-performance liquid chromatography [HPLC]-grade), $CuCl_2 \cdot 2H_2O$, bovine serum albumin (BSA) (radioimmunoassay grade) (Sigma Chemie, Deisenhofen, Germany).
5. $CdCl_2$ (standard solution for atomic absorption spectroscopy) (Aldrich, Steinheim, Germany).
6. Chelex-100 (100–200 mesh) (Bio-Rad, Munich, Germany).
7. Assay buffer: 10 mM Tris-HCl, 85 mM NaCl, pH 7.4.
8. CM-Sephadex, DEAE-Sephacel, and Chelex-100 were washed with 30 volumes of 10 mM Tris-HCl, 1 M NaCl, pH 7.4, subsequently equilibrated with 30 volumes of assay buffer and were kept as 66% (v/v) suspension in the assay buffer.
9. $Cu(CH_3CN)_4ClO_4$ was synthesized according to Hemmerich and Sigwart *(15)*, recrystallized twice from acetonitrile, dried in a desiccator over silicagel, and stored under argon at 4°C.
10. All solutions were degassed by sonification and saturated with nitrogen at 4°C.

3. Methods

3.1. Preparation of Rat and Human Liver Cytosol

1. A male Wistar rat weighing 200 g was injected i.p. daily on two consecutive days with 10 mg Zn^{2+} per kg b.w. in 0.9% NaCl and killed 18 h after the last treatment. A sample of human liver was obtained from a 41-yr-old male after suicidal death.
2. Liver samples were homogenized with 4 volumes of assay buffer, at 4°C with a Potter-Elvehjem homogenizer.

3. The homogenates were centrifuged at 100,000g (ultracentrifuge) for 60 min at 4°C and the supernatant fractions (cytosol) were kept at –80°C.

3.2. In Vitro Preparation of Oxidized MT

3.2.1. Oxidation with Cu^{2+}

Cd$_5$,Zn$_2$-thionein from rabbit liver, rat liver cytosol of the Zn-treated rat, and human liver cytosol, respectively, were incubated for 10 min with CuCl$_2$•2H$_2$O in molar ratios of Cu/MT of 6 for Cd$_5$,Zn$_2$-thionein and 30 for the rat and human liver cytosol.

3.2.2. Oxidation of Apothionein

Cd$_5$,Zn$_2$-thionein from rabbit liver, human liver cytosol and cytosol from the Zn-treated rat were dialyzed against 15 mM HCl, 150 mM NaCl at 4°C for 12 h to obtain apothionein. Subsequently the solution was neutralized with 20 mM Tris-HCl, pH 8.3.

3.3. Characterization of Oxidized MT

1. Cd$_5$,Zn$_2$-thionein (10 μg) from rabbit liver was oxidized by both Cu^{2+} and neutralization of the apothionein (*see* **Subheading 3.2.2.**). In addition, Cd$_5$,Zn$_2$-thionein was incubated with Cu^{1+} (Cu(CH$_3$CN)$_4$ClO$_4$ molar ratio Cu/MT = 6) for 10 min. After incubation with either Cu^{2+} or Cu^{1+}, MT-bound Cu was removed with ammonium tetrathiomolybdate (20 nmol) and DEAE-Sephacel suspension (0.1 mL, shaking time 3 min).
2. The degree of oxidation in the different samples was verified by determining the thiol concentration according to the method of Ellman (*16*) with a final DTNB-concentration of 80 μM.

3.4. MT Determination

1. Mix 0.05 mL of sample in a 1.5-mL vial with 0.01 mL of 300 mM ZnSO$_4$ • 7 H$_2$O and subsequently incubate at room temperature for 30 min with 0.01 mL of 140 mM 2-ME or assay buffer.
2. Incubate the solution with 0.07 mL of acetonitrile for 3 min, then add 0.5 mL of assay buffer and 0.1 mL of the CM-Sephadex suspension.
3. Shake the mixture for 3 min and incubate first with 0.05 mL of BSA solution (30 mg/mL in assay buffer, freshly prepared) for 2 min and then with 0.02 mL of ammonium tetrathiomolybdate (500 μM in assay buffer, freshly prepared) for another 2 min.
4. Add 0.1 mL of DEAE-Sephacel suspension and shake for 3 min.
5. Centrifuge at 8,000g (Microfuge E) for 5 min, and then incubate 0.6 mL of the supernatant for 5 min with 0.01 mL of ^{109}Cd-labeled CdCl$_2$ solution (50 ppm, 1 μCi/μg Cd).

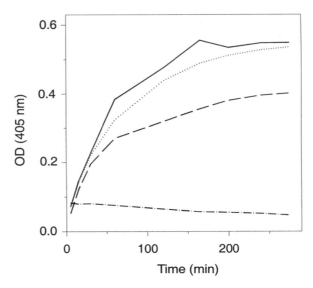

Fig. 1. Thiol content of Cd_5,Zn_2-thionein after oxidation in vitro. Thiol groups were detected at 405 nm using 5,5'-dithiobis(2-nitrobenzoic acid) *(16)* with a molar ratio of DTNB/MT of 50. (—) Native MT; (·····) MT after incubating with Cu^{1+} ($Cu[CH_3CN]_4ClO_4$); (- - - - -) MT after oxidation with Cu^{2+} ($CuCl_2 \cdot 2H_2O$); (- · - · -). MT oxidized by neutralizing apothionein. Adapted with permission from ref. *19*.

6. Add 0.1 mL of Chelex-100 suspension, and shake the mixture for 15 min.
7. Centrifuge at 8,000g (Microfuge E) for 5 min, and then incubate 0.5 mL of the supernatant fraction with 0.5 mL of acetonitrile for 3 min.
8. After centrifugation at 8,000g (Microfuge E) for 5 min, analyze 0.9 mL of the supernatant fraction for ^{109}Cd.
9. The concentration of total MT and the metal containing MT is calculated from the results in the presence and absence of 2-ME, respectively. The concentration of oxidized MT is then obtained by subtracting the concentration of metal containing MT from the total MT.
10. All concentrations of MT are calculated assuming a molar ratio of Cd/MT of 7 *(17)*.

3.5. Results

1. The addition of Cu^{2+} to purified Cd_5,Zn_2-thionein resulted in a decrease of the thiol content of MT by almost 40%. After incubating the rat and human liver cytosol with Cu^{2+}, oxidation was up to 25 and 45%, respectively.

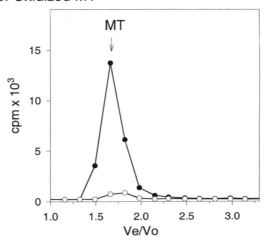

Fig. 2. Specificity of the MT-determination. MT in rat liver cytosol was oxidized in vitro by neutralizing apothionein and subsequently analyzed following the standard protocol in the absence (○—○) and presence (●—●) of 2-ME. The chromatographic profile shows the distribution of ^{109}Cd in the final supernatant fractions of the test procedures. Gelchromatography was performed using a Spherogel TSK 2000 SW column and 50 mM Tris-HCl, 150 mM NaCl, pH 7.0, as mobile phase (flow rate 1.0 mL/min). The column was calibrated with Cd_5,Zn_2-thionein from rabbit liver (arrow). Adapted with permission from ref. *19*.

2. After neutralization of the corresponding apothionein, no thiol groups could be detected, indicating a complete oxidation of MT (**Fig. 1**).
3. As shown by the gel-filtration HPLC, the method is highly specific for MT. Accordingly, in the absence of 2-ME, Cd was virtually absent in the HPLC fractions, whereas in presence of 2-ME, Cd was almost exclusively recovered in the MT-specific fractions (**Fig. 2**).
4. The accuracy and linearity of the assay were investigated through the recovery of the oxidized MT. In presence of 2-ME, recovery of MT was complete and independent of the oxidation procedure, the degree of oxidation, origin, and the amount (0.118–1.465 µg) of MT (**Table 1**).
5. The method offers the possibility of evaluating the portion of MT oxidized simply by calculating the difference in the results of the tests in presence and absence of 2-ME. In combination with the Cd-Chelex assay *(18)*, information on the portion of MT that binds Cu is also accessible. Due to the similar test protocols, the various pieces of information may be obtained simultaneously.

Table 1
Recovery of Oxidized MT

Oxidation method	Sample	Initial MT[a] (μg)	Oxidized MT (%)	Recovery[b] (%)	Mean ± SD (%)
Cu^{2+c}	Rat liver cytosol	1.19	25	92	
		0.54	18	100	97 ± 4
		0.12	9	100	
	Human liver cytosol	1.47	36	104	
		0.75	45	101	104 ± 3
		0.19	42	108	
Apo-MT[d]	Rat liver cytosol	1.14	99	86	
		0.58	88	98	91 ± 5
		0.12	100	89	
	Human liver cytosol	0.81	85	100	
		0.54	91	97	98 ± 1
		0.21	97	98	

[a]Before oxidation.
[b]After oxidation and subsequent reduction with 2-ME.
[c]Oxidation with Cu^{2+} (molar ratio Cu/MT = 30).
[d]Oxidation of Apothionein.
Adapted with permission from ref. *19*.

References

1. Suntres, Z. E. and Lui, M. K. (1990) Biochemical mechanism of metallothionein-carbon tetrachloride interaction in vitro. *Biochem. Pharmacol.* **39**, 833–840.
2. Thornalley, P. J. and Vasak, M. (1985) Possible role for metallothionein in protection against radiation-induced oxidative stress. Kinetics and mechanism of its reaction with superoxide and hydroxyl radicals. *Biochim. Biophys. Acta* **827**, 36–44.
3. Hartmann, H. J. and Weser, U. (1977) Copper-thionein from fetal bovine liver. *Biochim. Biophys. Acta* **491**, 211–222.
4. Otvos, J. D., Engeseth, H. R., and Wehrli, S. (1985) Preparation and 113Cd NMR studies of homogeneous reconstituted metallothionein: reaffirmation of the two-cluster arrangement of metals. *Biochemistry* **24**, 6735–6740.
5. Bremner, I. and Young, B. W. (1977) Copper thionein in the kidneys of copper-poisoned sheep. *Chem. Biol. Interact.* **19**, 13–23.
6. Porter, H. (1974) The particulate half-cystine-rich copper protein of newborn liver. Relationship to metallothionein and subcellular localization in non-mitochondrial particles possibly representing heavy lysosomes. *Biochem. Biopys. Res. Commun.* **56**, 661–668.

7. Riordan, J. R. and Richards, V. (1980) Human fetal liver contains both zinc- and copper-rich forms of metallothionein. *J. Biol. Chem.* **255**, 5380–5383.

8. Ryden, L. and Deutsch, H. F. (1978) Preparation and properties of the major copper-binding component in human fetal liver. Its identification as metallothionein. *J. Biol. Chem.* **253**, 519–524.

9. Hanaichi, T., Kidokoro, R., Hayashi, H., and Sakamoto, N. (1984) Electron probe X-ray analysis on human hepatocellular lysosomes with copper deposits: copper binding to a thiol-protein in lysosomes. *Lab. Invest.* **51**, 592–597.

10. Lerch, K., Johnson, G. F., Grushoff, P. S., and Sternlieb, I. (1985) Canine hepatic lysosomal copper protein: identification as metallothionein. *Arch. Biochem. Biophys.* **243**, 108–114.

11. Johnson, G. F., Morell, A. G., Stockert, R. J., and Sternlieb, I. (1981) Hepatic lysosomal copper protein in dogs with an inherited copper toxicosis. *Hepatology* **1**, 243–248.

12. Sternlieb, I. (1987) Hepatic lysosomal copper-thionein in *Metallothionein II* (Kägi, J. H. R., and Kojima, Y., eds.), Birkhäuser, Basel, pp. 647–653.

13. Klein, D., Lichtmannegger, J., Heinzmann, U., Müller-Höcker, J., Michaelsen, S., and Summer, K. H. (1998) Association of copper to metallothionein in hepatic lysosomes of Long-Evans cinnamon (LEC) rats during the development of hepatitis. *Eur. J. Clin. Invest.* **28**, 302–310.

14. Klein, D., Bartsch, R., and Summer, K. H. (1990) Quantitation of Cu-containing metallothionein by a Cd-saturation method. *Anal. Biochem.* **189**, 35–39.

15. Hemmerich, P. and Sigwart, C. (1963) Cu(CH$_3$CN)$_2$$^+$, ein Mittel zum Studium homogener Reaktionen des einwertigen Kupfers in wässriger Lösung. *Experientia* **19**, 488–489.

16. Ellman, G. L. (1959) Tissue sulfhydryl groups. *Arch. Biochem. Biophys.* **82**, 70–77.

17. Winge, D. R. and Miklossy, K. A. (1982) Domain nature of metallothionein. *J. Biol. Chem.* **257**, 3471–3476.

18. Bartsch, R., Klein, D., and Summer, K. H. (1990) The Cd-Chelex assay: a new sensitive method to determine metallothionein containing zinc and cadmium. *Arch. Toxicol.* **64**, 177–180.

19. Klein D., Sato, S., and Summer, K. H. (1994) Quantification of oxidized metallothionein in biological material by a Cd saturation method. *Anal. Biochem.* **221**, 405–409.

31

Fractionation of Herbal Medicine for Identifying Antioxidant Activity

Mohammed Afzal and Donald Armstrong

1. Introduction
1.1. Modern Medicine Owes Grandma Recipes

Herbal medicine is gaining rapid acceptance all over the world. Our earliest knowledge of herbal medicine, also termed as "naturomedicine," dates back to about 3000 BC when Chinese, Islamic-tibb, and Ayurvedic medicine in India were widely practiced. It is possible to trace back written records as well as unwritten knowledge of the rural people of herbal medicine and its use in the treatment of many physiological disorders. Thus in Mesopotamia, the cradle of Western culture, Sumerian tablets record 1000 medicinal plants. Egyptian, Greek, Roman, and Persian civilizations have been particularly rich in herbal medicine.

Conspicuously, man has used herbal medicine since the dawn of history and we have possessed recipes from the medieval times. Thus, common plants such as concoction of chamomile, maillot flowers, lavender, rosemary, sage, thyme, and extracts from red roses have been used for many ailments *(1)*. Chronic disorders such malaria have been treated by Cinchona bark, which contains quinine. Childhood leukemia has been treated with Periwinkle containing two antitumor drugs, vincristine and vinblastine. Many other important drugs such as strychnine, sennosides A & B, morphine, canabinoids, aloes, rhubarb, and vitamins such as biotin, cobalamin, vitamin B series all owe their existence to natural sources. Recently *Thunningia Sanguinea*, an herbal medicine commonly used for the management of asthma, has exhibited a scavenging action for free radicals *(2)*. This chapter is intended to provide guidance for the extraction and isolation of antioxidants from plant material, considering their solubility and stability.

From: *Methods in Molecular Biology, vol. 186: Oxidative Stress Biomarkers and Antioxidant Protocols*
Edited by: D. Armstrong © Humana Press Inc., Totowa, NJ

1.2. The Oxygen Paradox

Aerobic life has oxygen paradox, which can support as well as damage life, since it is implicated in many diseases and degenerative conditions *(3)*. Oxygen-derived free radicals appear to be involved in cancer initiation, aging, arthritis, myocardial infarction, atherosclerosis, diabetes, neurological disorders, and many other chronic diseases. Most of these radical-generated reactions take place in mitochondria, which is considered to be the energy storehouse of the living cell. These oxygen-mediated radicals can be encountered by the body's natural defenses using various types of radical scavengers, here called antioxidants, such as glutathione peroxidase, superoxide dismutase (SOD), catalase, and melatonin, produced mainly by the pineal gland in human brain *(4,5)*. Melatonin enters our brain cells more readily than other antioxidants and protects many different molecules. Dietary antioxidants such as ascorbic acid, carotenoids, and flavonoids also protect us from damaging effects of free radicals.

1.3. What are Free Radicals?

A free radical in any molecule is defined as an unpaired electron that occupies an atomic or molecular orbital on its own. This confers considerable reactivity on the molecule. Since this reactive molecule seeks another electron to pair, this initiates an uncontrolled chain reaction that can damage the natural function of the living cell, resulting in various diseases. However one must not be frightened of the consequences of free-radical generation, because these are produced as normal intermediates in metabolic processes and are as old entities as life itself. However unchecked overproduction of free radicals can lead to undesirable consequences.

1.4. How to Combat Free Radicals

Combating free radicals through diet is the best way to keep these dangerous species under control. Most herbal medicines consist of different parts of fruits and vegetables that contain natural antioxidants. These can abstract the lone electron from free-radical molecules and help humans to keep control on these injurious species. Most of these antioxidants in plants are colored materials and are produced for their own defenses and energy production and can combat free radicals. For example, highly colored anthocyanines, proanthocyaninidins, flavans, flavonoids, and their glycosides, carotenoids, like β-carotene and lycopene, are abundantly present in numerous plant materials. In addition, ascorbic acid is a common component in citrus plants, which can help in resisting damage by free radicals. Natural antioxidants can be broadly divided into three main classes:

1. Water-soluble antioxidants: these include ascorbic acid, anthocyanidins, catechins, epicatechins, flavonoids and other phenolic glycosides. Water-soluble antioxidants, i.e., Ascorbic acid (vitamin C), are the most common antioxidants present in citrus fruits. In plants, this vitamin is synthesized from glucose, however, for lack of an enzyme, it cannot be synthesized by human. In addition to its antioxidant activity, it is essential for prevention of gingivitis and scurvy. Due to its free radical-scavenging potential, mega doses of ascorbic acid have been solicited for prevention and treatment of cancer. Glycosides of phenolic compounds such as flavonoids have many therapeutic activities including augmentation of immune system, cytotoxic, and as powerful antioxidants (**6**). Beet root, pomegranate, and even onions are all rich in anthocyanins and can be powerful antioxidants. Thus silybin from traditional European medicine *Silybum marianum* (Compositea) is known to protect liver against the damaging effect of carbon tetrachloride, which is recognized to impair liver function through free radicals. Additionally, glycosides of phenolic quinonoid compounds such as phylloquinone, present in many common plants, can be invaluable for scavenging free radicals.

2. Fat-soluble antioxidants: Fat-soluble antioxidants include vitamins A and E; carotenoids, including β-carotene; powerful antioxidant lycopene; and many quinonoid compounds. Many of these compounds contain long-chain hydrocarbons with conjugated double bonds and β-ionone rings. To date, hundreds of different carotenoids have been identified from plants. Lycopene, from tomato, is recognized as the most potent antioxidant; β-carotene from carrots ranks second to lycopene. Other isomers of β-carotene are also present in plants but are less effective antioxidants. Vitamin A is present in two forms: retinol and retinal, and both are considered good antioxidants. Vitamin A, in addition to its antioxidant activity, is also essential for growth and development, especially in children. Vitamin E is present in four isomeric forms called α, β, γ, and δ-tocopherols. α-Tocopherol is the most potent antioxidant of all the other tocopherols present in plants.

3. Antioxidant metals such as selenium is also found in many plants like onion and garlic.

All these antioxidants can be extracted and evaluated using different methodologies. Here we have attempted to describe some of these methods in brief to give an insight to the reader about how to adopt these techniques to extract and evaluate antioxidants from natural sources.

2. Material and Methods

2.1. Equipment

1. Ordinary home blender or plant-powdering machine.
2. Glass bottle of 1–5 L capacity with a tap fixed at the side near the bottom.
3. A low-speed refrigerated centrifuge.
4. Rotary evaporator, for organic solvent evaporation at low temperature and reduced pressure.

5. Glass columns of various lengths and diameters.
6. UV cabinet.
7. Aluminium-coated thin-layer chromatography (TLC) plates (Merck).
8. Inductively coupled plasma chromatography (ICP) for trace-metal analysis.
9. Analytical HPLC instrument with chromatography data station.
   a. Waters 560 pump.
   b. High-performance liquid chromatography (HPLC) solvent de-gassing system.
   c. Photodiode array detector.
   d. Manual Rehodine injector with 200 µL loop.
   e. Reverse-phase (RP) C18 column (Chrompack).

2.2. Reagents

1. HPLC-grade solvents (Merck).
2. Chromatographic silica gel 200–400 mesh (Merck).
3. LH-20 Sephadex (*see* **Note 1**).
4. Buffer reagents: Ammonium bicarbonate, Ammonium acetate, Acetic acid, Pyridine, Picoline, Collidine, N-ethylmorpholine.
5. Standard antioxidants.

3. Methods
3.1. Planning the Isolation Method

Isolation of antioxidants from plant material depends on the polarity of these compounds. A number of methods can be used to determine the polarity of the compound. First distribution of antioxidants between a polar and a medium polar solvent (n-butanol : ethylacetate) can be used to determine the distribution coefficient of the compounds between two phases. Alternatively, simple TLC analysis can determine the relative polarities of the compounds.

3.2. Sample Extraction

Since many of the colored materials, such as chloroplasts, are located inside the cell compartments, plant material needs to be homogenized before any type of extraction. This assists the penetration of the solvent to the cellular structure of the plant tissues, thereby increasing the yield of the extract. Plants can be homogenized fresh or dried, under shade in a ventilated place (*see* **Note 2**). A common kitchen blender may be used for this purpose. Different extraction techniques are used for different classes of antioxidants. Care must be taken to avoid chlorophyll and vegetable tannins, which are ubiquitous in plants.

3.2.1. Water-Soluble Antioxidants

These antioxidants can be extracted simply by homogenizing the fresh plant material in water or a suitable buffer (*see* **Note 3**). Water-soluble materials are not often extracted using water as a solvent; they are better extracted with methanol or ethanol. This eliminates the possibility of microbial growth, which is a common problem in water extracts. The use of azide to prevent microbial growth has been recommended, but it has its own problems of reacting with some of the extracted components of the plant. Moreover, water extracts may also react with some of the enzymes that are co-extracted, resulting in modified antioxidants (*see* **Note 4**).

Usually 1:5 (w/v) ratio of powdered plant and solvent is safely used for extraction. Aqueous extracts are generally difficult to filter and therefore a low-speed centrifugation at 300 rpm for 30 min is recommended. The aqueous extract thus obtained can be freeze-dried to get the water-soluble part of the plant, which will, of course, contain the antioxidants present in this plant. In the case of volatile buffers, evaporation under reduced pressure at low temperature is recommended. A general fractionation scheme for water-soluble materials is given in **Subheading 1.4.**

Plants may also contain salts that can be co-extracted with water or alcohol. Salts may also be present if the extraction is carried out in volatile buffers *(7)*. Therefore desalting must be done before any attempt for purification is made (*see* **Note 1**). This can be achieved on ion-exchange resins Sephadex G-10, bio-gel P-2 or preparative HPLC using an RP column, using methanol:acetic acid-1% (1:4 v/v) as a mobile phase. LH-20 Sephadex in combination with high-porosity polystryrene gel CHP-20P can also be used for the purification of water soluble antioxidants. A combination of ion exchange and RP-HPLC purification results in great separations. Flavone glycosides can be monitored at 280 and 330 nm using photodiode array detector. Identification of the purified antioxidants can be made by the help of nuclear magnetic resonance (NMR) with Chelex 100 present in the sample solution *(7)*. Chelex 100 resin, H^+ acetate form, chelates with metal contaminants and improves spectrum quality of all water-soluble antioxidants.

3.2.2. Fat-Soluble Antioxidants

These can be extracted from the plant by dichloromethane. Care must be taken because this solvent will also extract chlorophyll and other undesired photosynthetic pigments present in the plant material. These pigments can be eliminated by solvent partitioning and passing the extract through a small neutral alumina column (*see* **Note 5**). The resulting cholorophyll-free eluate can be separated by HPLC using a standard method described by Browne and Armstrong *(8)*.

Antioxidant trace metals like selenium can be quantitated by digestion of the whole plant material in nitric acid followed by heating it in a Muffle furnace at high temperature. Quantification is done using ICP or atomic absorption spectrometer.

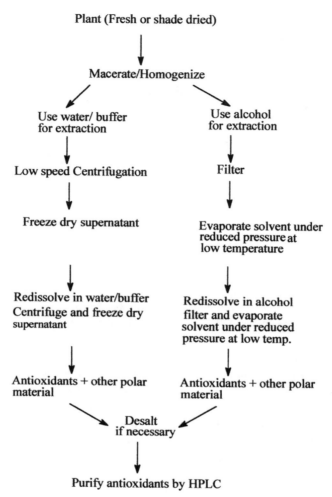

Scheme 1. Summary flowchart for isolation of phyto-antioxidants.

3.3. General Approach

Polar antioxidants can be isolated according to **Scheme 1.**

4. Notes

1. Excellent resolution of eight flavonol glycosides has been achieved in our laboratories using polyamide TLC plates with water-ethanol-butanone-acetylacetone (65:15:15:5 v/v) as a mobile phase. Chromatograms can also be run in a second direction using chloroform-methanol-butanone (60:26:14 v/v) for complex mixtures of flavonoid glycosides.

2. Plant material stored for prolonged periods of time may result in decomposition of its active components.
3. Plant enzymes can be denaturated by soaking plant material in methanol or ethanol.
4. Desalting of extracts is a difficult task. Volatile buffers may also pose problems. Some solvents used such as pyridine may be noxious and may not be safe for use in laboratory. The aqueous buffers may be freeze-dried.
5. Use aqueous methanol or ethanol for elution of the water-soluble antioxidants from the column.

References

1. Farnsworth, N. R. and Soejarto, D. D. (1991) Global importance of medicinal plants, in Conservation of medicinal plants (Akerele, O., Heywood, V., and Synge, H., eds.), Cambridge University Press, NY, pp. 25–51.
2. Gyamfi, M. A., Yonamine, M., and Aniya, Y. (1999) Free-radical scavenging action of medicinal herbs from Ghana *Thonningia sanguinea* on experimentally-induced liver injuries. *Gen. Pharmacol.* **32,** 661–667.
3. Mylonas, C. and Kouretas, D. (1999) Lipid peroxidation and tissue damage. *In Vivo* **13(3),** 295–309.
4. Reiter, R. J., Tan, D. X., Cabrera, J., D'Arpa, D., Sainz, R. M., May, J. C., and Ramos, S. (1999) The oxidant/antioxidant network: role of melatonin. *Biol. Sign. Recept.* **8(1–2),** 56–63.
5. Tan, D. X., Manchester, L. C., Reiter, R. J., and Plummer, B. F. (1999) Cyclic 3-hydroxymelatonin: a melatonin metabolite generated as a result of hydroxyl radical scavenging. *Biol. Sign. Recept.* **8(1–2),** 70–74.
6. Harborne, J. B. (1988) *The flavinoids.* Chapman and Hall, NY, pp. 53–54.
7. Shimizu, Y. (1988) Purification of water-soluble natural products, in *Natural Products Isolation* (Cannell, R. J. P., ed.), Humana Press, NJ, pp. 329–341.
8. Browne, R. W. and Armstrong, D. (1995) EPR measurement of nitric oxide-induced chromanoxyl radicals of vitamin E: Interaction with vitamin C in: *Methods in Molecular Biology, vol. 108: Free Radicals and Antioxidant Protocols* (Armstrong, D., ed.), Humana Press Inc., Totowa, NJ, pp. 269–275.

32

Designing Safer (Soft) Drugs by Avoiding the Formation of Toxic and Oxidative Metabolites

Nicholas Bodor and Peter Buchwald

1. Introduction

Living organisms possess not only fine-tuned metabolic mechanisms for endogenous chemicals but also several defensive mechanisms to detoxify xenobiotics. Most metabolic processes that eliminate invading foreign chemicals by transforming them into more hydrophilic or more easily conjugated compounds are oxidative in nature. Unfortunately, many of these mechanisms are indiscriminate, and detoxifying enzymes, such as cytochrome P450 or N-acetyltransferase, can generate toxic reactive intermediates such as epoxides or radicals from otherwise nontoxic compounds (1,2). Chemicals and xenobiotics, therefore, are not always metabolized only into more hydrophilic and less toxic substances, but also into highly reactive chemical species, which then can react with various macromolecules and cause tissue damage or elicit antigen production.

In general, metabolic conversion of a drug (D) can generate analog metabolites $(D_1...D_m)$ that have structures and activities similar to the original drug but have different pharmacokinetic properties, other metabolites $(M_1...M_k)$ including inactive ones (M_l), and potential reactive intermediates $(I^*_1...I^*_n)$ that are mainly responsible for various kinds of cell damage by forming toxic species (IC_l) (**Fig. 1**). As all these compounds may be present simultaneously and in various time-dependent concentrations, toxicity (T) in a rigorous approach has to be described as a combination of intrinsic toxicity-selectivity, $T_I(D)$, and toxicities due to various metabolic products:

$$T(D) = T_I(D) + T(D_1...D_m) + T(M_1...M_k) + T(I^*_1...I^*_n) \qquad (1)$$

From: *Methods in Molecular Biology, vol. 186: Oxidative Stress Biomarkers and Antioxidant Protocols*
Edited by: D. Armstrong © Humana Press Inc., Totowa, NJ

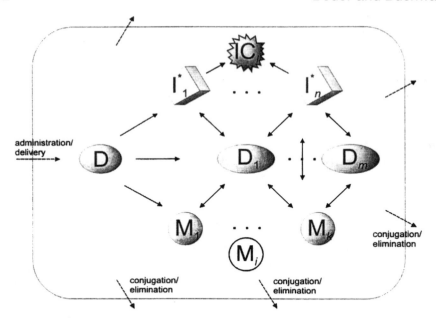

Fig. 1. Schematic representation of possible metabolic pathways for traditional drugs. In general, metabolic conversion of a drug (D) can generate analog metabolites ($D_1...D_m$) that have structures and activities similar to the original drug but have different pharmacokinetic properties, other metabolites ($M_1...M_k$) including inactive ones (M_l), and potential reactive intermediates ($I^*_1...I^*_n$), that are mainly responsible for various kinds of cell damage by forming toxic species (IC_l).

Intrinsic toxicity $T_1(D)$ is a molecular property inherent to the drug candidate and is a combination of: 1) biological mechanism-based, 2) other biological mechanism-based (actions via other receptor/enzyme systems), and 3) nonspecific chemical-based inputs. Soft drug design addresses the other toxicity terms $[T(D_1...D_m), T(M_1...M_k), T(I^*_1...I^*_n)]$ of **Eq. 1**. Inclusion of a metabolically sensitive site into the drug molecule makes possible the design and the prediction of the major metabolic pathway and avoids the formation of undesired toxic, active, or high-energy intermediates. Hence, in soft drug design, the goal is not to avoid metabolism, but rather to control and direct it. If possible, inactivation should take place as the result of a single, low-energy, high-capacity step that yields inactive species subject to rapid elimination.

Most critical metabolic pathways are mediated by oxygenases that exhibit not only interspecies but also interindividual variability and are subject to inhibition and induction (*1*). In different individuals, half-lives of foreign compounds may vary as much as 10–50 fold (*1*). Furthermore, the rates of hepatic monooxygenase reactions are at least two orders of magnitude lower than the slowest of the other enzymatic reactions *(3)*. For example, while the turnover numbers of representative enzymatic reactions such as those catalyzed by carbonic anhydrase *c* or β-galactoside are 36,000,000 and 12,500, respectively (nmol substrate transformed/nmol enzyme/min), the reaction rates of monooxygenase reactions are usually within the 1–10 range (nmol substrate transformed/nmol P450/min). These monooxygenase reactions are slow because they only have very low concentrations of enzyme substrate to react with. The substrate for NADPH-cytochrome P450 reductase is not the exogeneous substrate *per se*, but the ferricytochrome P450-substrate complex, which is present in considerably lower concentrations *(3)*.

We, therefore, advocated for a long time the avoidance of oxidative pathways and slow, easily saturable oxidases and the design of soft drugs that are inactivated by hydrolytic enzymes. Mammalian carboxylesterases (EC 3.1.1.1) comprise a multigene family and have low substrate specificity. Together with other carboxylic ester hydrolases (EC 3.1.1), such as arylesterase (EC 3.1.1.2) or cholinesterase (EC 3.1.1.8), these enzymes efficiently catalyze the hydrolysis of a variety of ester- and amide-containing chemicals to the respective free acids *(4,5)*.

2. Materials

This chapter describes general drug-design principles that require no specific instrumentation or materials.

3. Methods
3.1. Drug Design
3.1.1. Design Principles

Soft drug design aims to design safer drugs with an increased therapeutic index by integrating metabolism considerations into the drug-design process (*see* **Note 1**) *(6–8)*. Soft drugs are active isosteric-isoelectronic analogs of a lead compound that are deactivated in a predictable and controllable way after achieving their therapeutic role. They are designed to be rapidly metabolized into inactive species and, hence, to simplify the transformation-distribution-activity profile of the lead. Consequently, soft drugs are new therapeutic agents obtained by building in the molecule, in addition to the activity, the most desired way in which the molecule is to be deactivated and detoxified subsequent to exerting its biological effects (*see* **Note 2**). The desired activity is generally local, and the soft drug is applied or administered near the site of action. Therefore, in most cases, they produce pharmacological activity locally, but their distribution away from the site results in a prompt metabolic deactivation that prevents any kind of undesired pharmacological activity or toxicity.

The soft drug concept was introduced in 1976 *(6)* and reiterated in 1980 *(9–11)*. Since then, the following four major types have been identified:

1. Soft analogs: close structural analogs of known active drugs that have a specific metabolically sensitive moiety built into their structure to allow a facile, one-step controllable deactivation and detoxication after the desired therapeutic role has been achieved.
2. Inactive metabolite-based soft drugs: active compounds designed starting from a known (or hypothetical; *see* **Note 3**) inactive metabolite of an existing drug by converting this metabolite into an isosteric/isoelectronic analog of the original drug in such a way as to allow a facile, one-step controllable metabolic conversion after the desired therapeutic role has been achieved back to the very inactive metabolite the design started from.
3. Active metabolite-based soft drugs: metabolic products of a drug resulting from oxidative conversions that retain significant activity of the same kind as the parent drug. If activity and pharmacokinetic considerations allow it, the drug of choice should be the metabolite at the highest oxidation state that still retains activity.
4. Pro-soft drugs: inactive prodrugs (chemical delivery forms) of a soft drug of any of the above classes including endogenous soft molecules. They are converted enzymatically into the active soft drug, which is subsequently enzymatically deactivated (*see* **Note 4**).

Examples for each of these classes were provided in the literature *(7,8)*, but the inactive metabolite-based and the soft-analog approaches have been the most useful and successful strategies for designing safe and selective drugs. A number of already marketed drugs, such as esmolol (**2**) *(12,13)*, remifentanil (**5**) *(14,15)*, or loteprednol etabonate (**8**) *(16–18)*, resulted from the successful application of such design principles (**Fig. 2**). Many other promising drug candidates are currently under investigation in a variety of fields, including possible soft antimicrobials, anticholinergics, corticosteroids, β-blockers, analgetics, ACE inhibitors, anti-arrhythmics, and others *(8)*.

3.2. Results

3.2.1. Bufuralol

A particularly suited example for the present article is provided by bufuralol (**10**, **Fig. 3**), for which formation of oxidative metabolites can be avoided through soft drug design in two entirely different ways: by using the inactive metabolite-based and the active metabolite-based approaches, respectively.

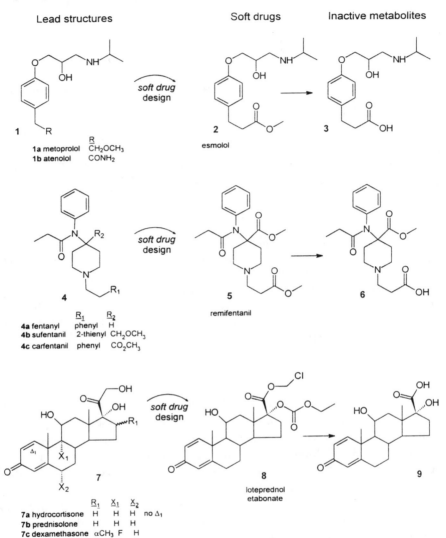

Fig. 2. Structures of esmolol (**2**), an ultra-short-acting β-blocker, remifentanil (**5**), an ultra-short-acting opioid analgetic, and loteprednol etabonate (**8**), a soft corticosteroid. For each of them, some of the corresponding (hard) lead compounds (**1a,b**, **4a–c**, and **7a–c**) and the inactive metabolites resulting from the hydrolytic deactivation (**3**, **6**, and **9**) are also shown.

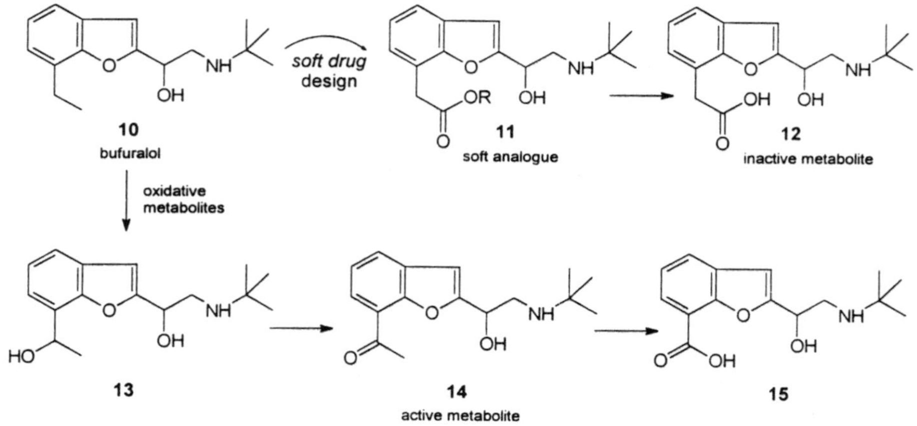

Fig. 3. Bufuralol (**10**), its oxidative, active metabolites (**13, 14**), and the final, inactive acid (**15**). As the oxidative metabolites are still active and even have longer elimination half-lives, the active metabolite-based soft drug of choice should be the ketone (**14**), the highest active oxidative metabolite, to avoid mixtures of active components. On the other hand, starting from a hypothetical inactive metabolite (**12**), a variety of soft drugs (**11**) can be designed using the inactive metabolite-based approach that will undergo one-step hydrolytic deactivation.

3.2.1. Active Metabolite-Based Design

Bufuralol is a potent, nonselective β-adrenoceptor antagonist with β_2 partial agonist properties. Its effectiveness in the treatment of essential hypertension is probably due to a favorable balance of β-blockade and β_2-agonist-mediated vasodilatation. Bufuralol undergoes complex metabolism in humans, including stepwise oxidation to corresponding hydroxy (**13**) and keto (**14**) intermediates, which are still active *(19,20)* and have different (interestingly, longer) elimination half-lives *(21)*. Not only does a differential metabolism of the two enantiomers occur, but in a small percentage of subjects, differences due to genetic polymorphism are also encountered. Applying the principle of active metabolite-based design, the ketone **14**, the highest active oxidative metabolite, should be the drug of choice *(7)* as it retains most of the activity, but is deactivated by one-step oxidation to structure **15**.

3.2.2 Inactive Metabolite-Based Design

On the other hand, starting from a hypothetical inactive metabolite (**12**), a variety of soft drugs (**11**) can be designed using the inactive metabolite-based approach. Recently, a number of these **11** esters have been synthesized and tested for β-antagonist activity by recording electrocardiogram (ECG) and

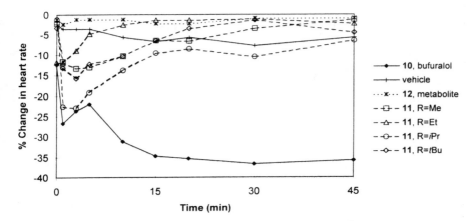

Fig. 4. Effects of bufuralol (**10**) (3.8 µmol/kg), vehicle (10% DMSO in 30% hydroxy-propyl β-cyclodextrin), and some representative soft drugs (**11**) on isoproterenol-induced tachycardia in rats. Data represent the mean of at least three animals; error bars were omitted for better visibility.

intra-arterial blood pressure in rats *(22)*. While in the isoproterenol-induced tachycardia model bufuralol at an i.v. dose of 1 mg/kg diminished heart rate for at least 2 h, the effects of equimolar soft drugs lasted only 10–30 min (**Fig. 4**). The inactive metabolite did not decrease the heart rate significantly. These, together with the other experimental evidences obtained *(22)*, provide additional support for the correctness of the used design principles.

3.2.3. Soft Chemicals

3.2.3.1. CHLOROBENZILATE: FORMAL DESIGN AS SOFT CHEMICAL

The very same concepts used for soft drug design can be extended to the design of less toxic commercial chemical substances, provided that adequate structure-activity relationship (SAR) and metabolism data of analogous substances can be gathered. Such approaches can provide good examples for the change in thinking required to design environmentally safe, nontoxic chemicals (green chemistry). One such example in which these principles have been accidentally used for the design of nonpharmaceutical products is the well-known pesticide dichlorodiphenyltrichloroethane (DDT, chlorophenothane). As recently summarized *(8)*, DDT undergoes in vivo oxidation, oxidative dehaloformation, iterative dehydrohalogenation/reduction cycles, and hydrolysis. The inactive acid metabolite formed by this hydrolytic transformation is of low toxicity, can be excreted as a water-soluble species, and is indeed a major metabolite detected in feces and urine. Therefore, it is an ideal lead compound for a formal inactive metabolite-based approach. It was found that the ester-containing chlorobenzilate is also active as a pesticide, but has much lower carcinogenicity than DDT or dicofol (kelthane). The ethyl ester moiety of this compound apparently functions similarly to that of the trichloromethyl group of DDT in restoring pesticidal activity. However, chlorobenzilate is considerably less toxic than DDT because its labile ethyl ester group enables rapid metabolism in exposed subjects to the free, nontoxic carboxylic acid.

3.2.3.2. MALATHION: EXPLOITING DIFFERENTIAL ENZYME DISTRIBUTIONS

There is one additional important aspect related to the design of soft chemicals that has not yet been sufficiently explored. As described earlier, most soft chemicals are designed to be deactivated by carboxylesterases. In addition to the usual advantages of soft drug design, the differential distribution of these enzymes between vertebrates and insects may also provide selectivity based on metabolism for soft chemicals designed as possible pesticides. Malathion (16), one of the most generally useful loco-systemic compounds, provides an elegant example for this selectivity achieved by exploiting differences in the enzymatic constitution of different organisms (Fig. 5). It is detoxified through a variety of metabolic pathways, one of the most prominent being the ester hydrolysis of one of the two ethyl carboxylester groups. The carboxylesterase that hydrolysis and thereby detoxifies malathion is widely distributed in mammals, but only sporadically in insects, where in some rare cases is responsible for insecticide resistance (*[23,24]* and references therein). In the meantime, insects seem to posses a very active oxidative enzyme system that transforms malathion (16) into malaoxon (17), a much more active cholinesterase inhibitor. Probably, all insects and all vertebrates possess both an esterase and an NADPH-dependent oxidase system; it is the balance of action of these two systems that varies from one organism to another and provides the selectivity of action. Similar mechanism may provide considerable selectivity for the soft chemicals to be designed and may result in safer "soft" insecticides, for example, in the parathion family. These compounds have unacceptably high mammalian toxicities (acute oral LD_{50} of ~5 mg/kg in male rats) and are also activated by a similar oxidation.

3.2.4. Computerized Design

Starting from a lead compound, soft drug design approaches can provide a variety of new, active compounds that already have metabolic considerations built into their structure. By being susceptible to low-energy, high-capacity hydrolytic deactivation after achieving the desired therapeutic effect, soft drugs are safe, nontoxic compounds, and they can provide a viable alternative

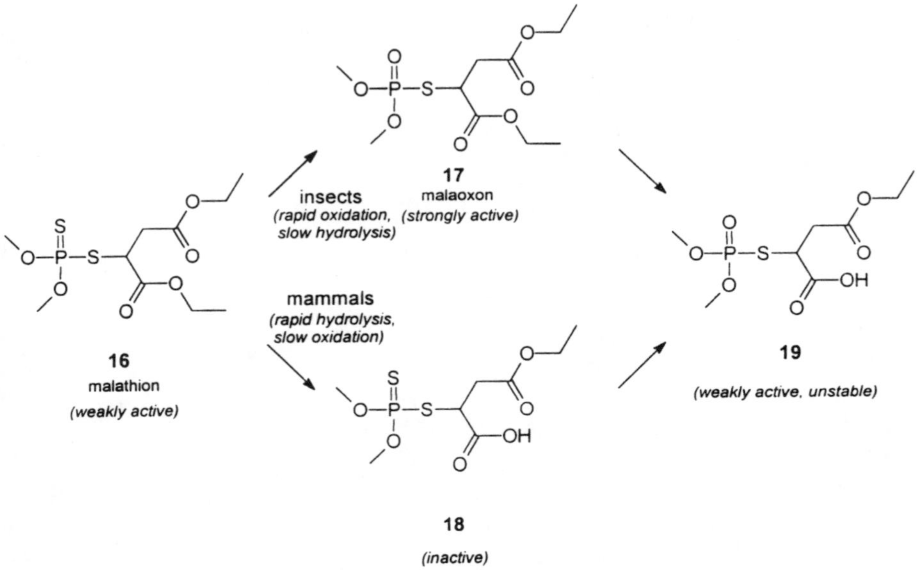

Fig. 5. Malathion (**16**), its oxidative activation to malaoxon (**17**) an its deactivation by carboxylesterases. The carboxylesterase that hydrolyses and thereby detoxifies malathion is widely distributed in mammals, but only sporadically in insects, thus accounting for the selective toxicity to insects and the relative safety to mammals *(23,24)*.

to avoid formation of oxidative metabolites that are often a main source of toxicity. However, owing to the considerable flexibility of soft drug design, for certain lead compounds a large number of possible soft structural analogs can be designed, and finding the best drug candidate among them may prove tedious and difficult. As recently summarized *(8,25)*, we developed a variety of computer methods to calculate various molecular properties and make possible more quantitative design. The capabilities of quantitative design have been further advanced by developing expert systems that combine the various structure-generating rules of soft drug design with the developed predictive softwares to provide a ranking order based on isosteric/isoelectronic analogy. Most recently, by introducing the inaccessible solid angle Ω_h as a novel measure of steric hindrance, we developed a quantitative structure-metabolism relationship (QSMR) equation to predict hydrolytic lability for in vitro human blood data *(26)* (*see* **Note 5**), and this was integrated within the expert system. The interface of a recent version of the corresponding computer program is shown in **Fig. 6**.

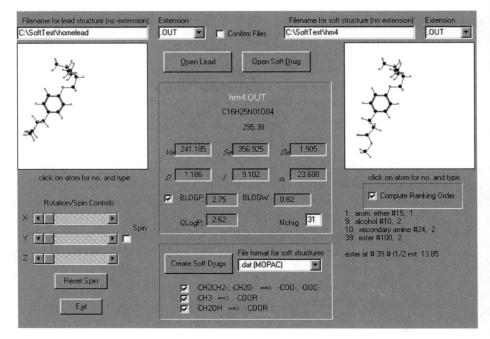

Fig. 6. Interface of the current version of the program for computer-aided soft drug design.

4. Notes

1. In soft drug design, wherein improvement of the therapeutic index (usually defined as the ratio between median toxic dose and median effective dose, TI = TD_{50}/ED_{50}) and not of the intrinsic activity is the main objective, higher doses of less potent, but much less toxic soft compounds are often preferred over lower doses of more potent, but also more toxic compounds.

2. Since they are often confused, it is important to note that, at least theoretically, the prodrug and the soft drug concepts are opposite to each other. Whereas prodrugs are, ideally, inactive by design and are converted by a predictable mechanism to the active drug, soft drugs are active *per se* and are designed to undergo a predictable and controllable metabolic deactivation.

3. The design can start with an inactive metabolite that was not isolated and identified in previous studies. By using the general rules of metabolic transformations, one can design a useful inactive metabolite (hypothetical inactive metabolite). Obviously, in a later stage of the development, the inactivity of this metabolite has to be proven.

4. As results from their definitions, one can design a "pro-soft drug" (soft drugs, as any other drugs, can be the subject of prodrug design), but one can never design a "soft prodrug."

5. Since most likely a number of different enzymes are involved in the hydrolysis of these compounds, one cannot predict accurate hydrolytic half-lives for arbitrary structures, and calculated values have to be treated with caution. Nevertheless, the present method may become useful in distinguishing among compounds whose hydrolysis is fast, medium, or slow based on chemical structure alone.

References

1. Gillette, J. R. (1979) Effects of induction of cytochrome P-450 enzymes on the concentration of foreign compounds and their metabolites and on the toxicological effects of these compounds. *Drug Metab. Rev.* **10,** 59–87.
2. Picot, A. and Macherey, A.-C. (1996) Chemical aspects of biotransformations leading to toxic metabolites, in *The Practice of Medicinal Chemistry* (Wermuth, C. G., ed.), Academic Press, London, pp. 643–670.
3. Mannering, G. J. (1981) Hepatic cytochrome P-450-linked drug-metabolizing systems, in *Concepts in Drug Metabolism Part B* (Testa, B. and Jenner, P., eds.), Marcel Dekker, New York, pp. 53–166.
4. Leinweber, F.-J. (1987) Possible physiological roles of carboxylic ester hydrolases. *Drug Metab. Rev.* **18,** 379–439.
5. Satoh, T. and Hosokawa, M. (1998) The mammalian carboxylesterases: from molecules to functions. *Annu. Rev. Pharmacol. Toxicol.* **38,** 257–288.
6. Bodor, N. (1977) Novel approaches for the design of membrane transport properties of drugs, in *Design of Biopharmaceutical Properties through Prodrugs and Analogs* (Roche, E. B., ed.), Academy of Pharmaceutical Sciences, Washington, DC, pp. 98–135.
7. Bodor, N. (1984) The soft drug approach. *Chemtech* **14 (1),** 28–38.
8. Bodor, N. and Buchwald, P. (2000) Soft drug design: general principles and recent applications. *Med. Res. Rev.* **20,** 58–101.
9. Bodor, N., Kaminski, J. J., and Selk, S. (1980) Soft drugs. 1. Labile quaternary ammonium salts as soft antimicrobials. *J. Med. Chem.* **23,** 469–474.
10. Bodor, N. and Kaminski, J. J. (1980) Soft drugs. 2. Soft alkylating compounds as potential antitumor agents. *J. Med. Chem.* **23,** 566–569.
11. Bodor, N., Woods, R., Raper, C., Kearney, P., and Kaminski, J. (1980) Soft drugs. 3. A new class of anticholinergic agents. *J. Med. Chem.* **23,** 474–480.
12. Erhardt, P. W., Woo, C. M., Anderson, W. G., and Gorczynski, R. J. (1982) Ultra-short-acting β-adrenergic receptor blocking agents. 2. (Aryloxy)propanolamines containing esters on the aryl function. *J. Med. Chem.* **25,** 1408–1412.
13. Erhardt, P. W. (1999) A prodrug and a soft drug., in *Drug Metabolism. Databases and High Throughput Testing During Drug Design and Development* (Erhardt, P. W., ed.), Blackwell Science, Oxford, pp. 62–69.
14. Feldman, P. L., James, M. K., Brackeen, M. F., Bilotta, J. M., Schuster, S. V., Lahey, A. P., et al. (1991) Design, synthesis, and pharmacological evaluation of ultrashort- to long-acting opioid analgetics. *J. Med. Chem.* **34,** 2202–2208.
15. Egan, T. D., Lemmens, H. J. M., Fiset, P., Hermann, D. J., Muir, K. T., Stanski, D. R., and Shafer, S. L. (1993) The pharmacokinetics of the new short-acting

opioid remifentanil (GI87084B) in healthy adult male volunteers. *Anesthesiology* **79,** 881–892.

16. Bodor, N. (1981) Soft steroids having antiinflammatory activity, Belgian patent, BE889,563 (Cl. CO7J).

17. Druzgala, P., Hochhaus, G., and Bodor, N. (1991) Soft drugs. 10. Blanching activity and receptor binding affinity of a new type of glucocorticoid: loteprednol etabonate. *J. Steroid Biochem.* **38,** 149–154.

18. Noble, S. and Goa, K. L. (1998) Loteprednol etabonate. Clinical potential in the management of ocular inflammation. *BioDrugs* **10,** 329–339.

19. Hamilton, T. C. and Chapman, V. (1978) Intrinsic sympathomimetic activity of β-adrenoceptor blocking drugs at cardiac and vascular β-adrenoceptors. *Life Sci.* **23,** 813–820.

20. Machin, P. J., Hurst, D. N., and Osbond, J. M. (1985) β-Adrenoceptor activity of the stereoisomers of the bufuralol alcohol and ketone metabolites. *J. Med. Chem.* **28,** 1648–1651.

21. Francis, R. J., East, P. B., McLaren, S. J., and Larman, J. (1976) Determination of bufuralol and its metabolites in plasma by mass fragmentography and by gas chromatography with electron capture detection. *Biomed. Mass. Spectrom.* **3,** 281–285.

22. Hwang, S.-K., Juhasz, A., Yoon, S.-H., and Bodor, N. (2000) Soft drugs 22. Design, synthesis, and evaluation of soft bufuralol analogues. *J. Med. Chem.* **43,** 1525–1532.

23. Hassall, K. A. (1990) *The Biochemistry and Uses of Pesticides.* 2nd ed. Macmillan, London.

24. Hodgson, E. and Kuhr, R. J. (eds.) (1990) *Safer Insecticides. Development and Use.* Marcel Dekker, New York.

25. Bodor, N., Buchwald, P., and Huang, M.-J. (1999) The role of computational techniques in retrometabolic drug design strategies, in *Computational Molecular Biology* (Leszczynski, J., ed.), Elsevier, Amsterdam, Vol. 8 of series: Theoretical and Computational Chemistry, pp. 569–618.

26. Buchwald, P. and Bodor, N. (1999) Quantitative structure-metabolism relationships: steric and nonsteric effects in the enzymatic hydrolysis of noncongener carboxylic esters. *J. Med. Chem.* **42,** 5160–5168.

33

Statistical Correction of the Area Under the ROC Curve in the Presence of Random Measurement Error and Applications to Biomarkers of Oxidative Stress

Enrique F. Schisterman

1. Introduction

Despite the elegant patho-physiological mechanisms that have been identified linking lipid peroxidation to the development of atherosclerosis, the clinical evidence linking markers of oxidative stress to coronary heart disease is still controversial *(1–4)* in particular for thybarbituric acid reactive substances (TBARS). The reasons for this lack of consistency in research findings may be many and include: 1) the use of small and selected samples; 2) differences in methods used to quantify TBARS; and 3) lack of adjustment for potential confounders. The inconsistency in findings together with issues related to the lack of specificity of TBARS as a measurement of lipid peroxidation have prompted the questioning of TBARS measurement as a marker of lipid peroxidation *(5)*. The use of TBARS as a summarizing value of total circulating oxidative stress in individuals is popular in laboratory research; however, its use is still controversial because of the lack of a uniform methodology between researchers to estimate TBARS levels.

The area under the ROC curve, introduced to the medical literature by Hanley and McNeil *(6)*, is a frequently used criteria to examine the discrimination properties of biomarkers. If X and Y represent the biomarker values on the controls and cases groups, respectively, the area under the ROC curve is $A = P(Y>X)$*(7)*; thus, *A* is a global measure of how well the biomarker distinguishes between cases and controls. Standard statistical methods do not allow for adjustment for measurement error. We introduce a new method to analyze the potential role of markers of oxidative stress as a discriminating

From: *Methods in Molecular Biology, vol. 186: Oxidative Stress Biomarkers and Antioxidant Protocols*
Edited by: D. Armstrong © Humana Press Inc., Totowa, NJ

tool between individuals with and without disease, adjusting for the effect of
random measurement error and to assess the clinical value of this marker.

2. Materials

Statistical software to perform ROC analysis is available upon request at:
schistermane@cshs.org

3. Methods
3.1. Statistical Theory

1. Assume that a biomarker follows the normal distribution $X_i \sim N(\mu_X, \sigma^2_X)$
   ($i = 1,...,n_X$) in the control population and the normal distribution $Y_i \sim N(\mu_Y, \sigma^2_Y)$
   ($i = 1,...,n_Y$) in the cases population. The area under the ROC curve in this
   case is $A = \phi(\delta)$,

   Where $\delta = \dfrac{\mu_Y - \mu_X}{\sqrt{\sigma^2_X + \sigma^2_Y}}$ and ϕ denotes the standard normal cumulative distribution

   function. The almost maximum likelihood estimate of A is $A = \phi \left(\dfrac{\overline{Y} - \overline{X}}{\sqrt{S^2_Y + S^2_X}} \right)$,

   where $\overline{X}, \overline{Y}, S^2_X$, and S^2_Y denote the sample means and variances respectively for
   the two populations (*see* **Note 1**) (controls and cases).

2. In the presence of measurement error the "true" value of the biomarker on the
   cases and controls groups (Y and X) is not available. Instead we assume that
   we actually observe

$$x_i = X_i + \varepsilon^x_i \qquad i = 1,...,n_x \qquad (1)$$
$$y_i = Y_i + \varepsilon^y_i \qquad i = 1,...,n_y$$

   where $\varepsilon^x_i \sim N(0, \sigma^2_\varepsilon)$ and $\varepsilon^y_i \sim N(0, \sigma^2_\varepsilon)$. We further assume that the X,Y, ε^x_i
   and ε^y_i are independent of each other (*see* **Note 2**). Note that the variances of ε^x_i
   and ε^y_i are taken to be equal.

3. In addition we have data available from a reliability study (*see* **Note 3**) designed
   to estimate measurement error. Let w_{ij}, $i = 1,...,n_0$, $j = 1,...,p_i$, denote the jth
   observation on the ith subject, we assume that

$$w_{ij} = W_i + \varepsilon_{ij}$$

   where W_i is the "true" value of the biomarker for the ith subject of the reliability

   study and $\varepsilon_{ij} \sim N(0, \sigma^2_\varepsilon)$. Let $n_f = \sum_{i=1}^{n_0} (p_i - 1)$ and $\overline{w_{i\bullet}} = \sum_{j=1}^{p_i} \dfrac{w_{ij}}{P_i}$. Consequently

$$\hat{\sigma}^2_\varepsilon = \dfrac{\sum_{i=1}^{n_0} \sum_{j=1}^{p_i} (w_{ij} - \overline{w_{i\bullet}})^2}{n_f}$$

is an unbiased estimator of σ^2_ε.

4. Let $\bar{x}, \bar{y}$ denote the sample means for the observed x_i, y_i and set $S_x^2 = \dfrac{\sum\limits_{i=1}^{n_x}(x_i - \bar{x})^2}{n_x - 1}$

and $S_y^2 = \dfrac{\sum\limits_{i=1}^{n_y}(y_i - \bar{y})^2}{n_y - 1}$. Consequently from (1), since $E(S_x^2) = \sigma_X^2 + \sigma_\varepsilon^2$,

an unbiased estimator of σ_X^2 is $\hat{\sigma}_X^2 = S_x^2 - \hat{\sigma}_\varepsilon^2$. Similarly $\hat{\sigma}_Y^2 = S_y^2 - \hat{\sigma}_\varepsilon^2$ is unbiased for σ_Y^2.

Thus, when random measurement error is present, a natural corrected for measurement error, estimate of A is $\hat{A}_c = \phi(\hat{\delta})$ where $\hat{\delta} = \dfrac{\bar{y} - \bar{x}}{\sqrt{(S_x^2 + S_y^2 - 2\sigma_\varepsilon^2)}}$

5. Since ϕ is a monotonically increasing function of $\delta\delta$, finding a confidence interval (*see* **Note 4**) for A is equivalent to finding one for $\delta\delta^8$. Using the Delta method, we estimate the variance of $\hat{\delta}$ by

$$\hat{Var}(\hat{\delta}) = \left[\frac{S_x^2}{n_x} \frac{S_y^2}{n_y}\right] \times (S_x^2 - \hat{\sigma}_\varepsilon^2 + S_y^2 - \hat{\sigma}_\varepsilon^2)^{-1} +$$

$$+ \frac{(\bar{y} - \bar{x})^2}{4(S_x^2 - \hat{\sigma}_\varepsilon^2 + S_y^2 - \hat{\sigma}_\varepsilon^2)^3} \times \left[\frac{2S_x^4}{n_x - 1} + \frac{2S_y^4}{n_y - 1} + \frac{8\hat{\sigma}_\varepsilon^4}{n_f}\right]$$

Therefore an approximate $(1-\alpha\alpha)$ 100% C.I. for δ is given by

$$\hat{\delta} \pm z_{\alpha/2} \sqrt{\hat{Var}(\hat{\delta})}$$

where $z_{\alpha/2}$ is the $1 - \dfrac{\alpha}{2}$ standard normal percentile (*see* **Note 5**), with the corresponding interval for A being

$$\{\phi(\hat{\delta} - z_{\alpha/2}\sqrt{\hat{Var}(\hat{\delta})}), \phi(\hat{\delta} + z_{\alpha/2}\sqrt{\hat{Var}(\hat{\delta})})\} \tag{2}$$

3.2. Application

A population-based sample of randomly selected residents of Erie and Niagara counties, 35–79 yr of age, is the focus of this investigation. The New York State Department of Motor Vehicles drivers' license rolls was utilized as the sampling frame for adults between the ages 35 and 65, while the elderly sample (age 65–79) was randomly selected from the Health Care Financing Administration database.

We defined the cases as individuals with a self-reported history of either prior MI or diagnosis of angina by angiography or stroke. Due to the skewness of the original data, the transformation $(TBARS)^{-1/2}$ was implemented in order to bring the data distribution closer to normality. The transformed data yielded the following results:

$$\bar{x} = 0.604$$

$$\bar{y} = 0.450$$

$$n_x = 928$$

$$n_y = 40$$

$$S_x^2 = 0.0913$$

$$S_y^2 = 0.0886$$

From the reliability study described in the Introduction we obtain the following:

$$\hat{\sigma_\varepsilon}^2 = 0.0567$$

$$n_f = 41$$

$$\hat{R} = 0.59$$

where $\hat{R}$ is obtained by substituting estimates for the parameters in the formula for R. We are assuming the equality of the measurement errors for cases and controls populations.

Not correcting for measurement error results in $\hat{A} = 0.642$ and (0.550–0.725) as the unadjusted 95% confidence interval for A. Applying our methods, we obtain the adjusted area estimate to be $A_C = 0.735$ with the corresponding corrected 95% confidence interval being (0.600–0.888) *(9,10)*.

Correction for measurement error increased the estimate of the area under the ROC curve and shifted the confidence interval to include much higher values. Use of the uncorrected results understates the effectiveness of TBARS as a biomarker capable of discriminating between subjects with and without cardiovascular disease.

4. Notes

1. If the area under the ROC curve is less than 0.5, the results should be reversed.
2. Normality assumptions should be always evaluated using standard statistical procedure like the Q-Q plot.
3. When the reliability or validation study involves only small numbers of individuals and replications, the estimated variance of measurement error will have a large variability leading to uncertainty with respect to corrections.
4. If confidence intervals of the area under the ROC curve include the value 0.5, the marker does not discriminate between cases and controls.

5. If the minimum detectable levels of the marker of interest are high, this procedure might be biased.

References

1. Duthie, G. G., Beattie, J. A., Arther, J. R., Franklin, M., Morrice, P. C., and Jams, W. P. (1994) Blood antioxidants and indices of lipid peroxidation in subjects with angina pectoris. *Nutrition* **10,** 313–316.
2. Karmansky, I., Shnaider, H., Palant, A., and Gruener, N. (1996) Plasma lipid oxidation and susceptibility of low-density lipoproteins to oxidation in male patients with stable coronary artery disease. *Clin. Biochem.* **29(6),** 573–579.
3. Jayakumari, N., Ambikakumari, V., Balakrishnan, K. G., and Subramonia Iyer, K. (1992) Antioxidant status in relation to free radical production during stable and unstable angina syndromes. *Atherosclerosis* **94,** 183–190.
4. Miwa, K., Miyagi, U., and Fujita, M. (1995) Susceptibility of plasma low density lipoprotein to cupric ion-induced peroxidation in patients with variant angina. *J. Am. Coll. Cardiol.* **26,** 632–638.
5. Esterbauer, H., Gebicki, J., Puhl, H., and Jurgens, G. (1992) The role of lipid peroxidation and antioxidants in oxidative modification of LDL. *Free Radical Biol. Med.* **13,** 341–390.
6. Hanley, J. A., and McNeil, B. J. (1982) The meaning and use of the area under a receiver characteristic curve (ROC). *Radiology* **13,** 129–133.
7. Bamber, D. (1975) The area above the ordinal dominance graph and the area bellow the receiver operating characteristic graph. *J. Mathemat. Psychol.* **12,** 387–415.
8. Schisterman, E. F., Faraggi, D., Reiser, B., and Trevisan, M. (2001) Statistical inference for the area under the ROC curve in the presence of random measurement error. *Am. J. Epidemiol.* **154,** 174–179.
9. Schisterman, E. F., Faraggi, D., Browne, R., Freudenheim, J., Dorn, J., Muti, P., et al. (2001) TBARS and cardiovascular disease in a population-based sample. *J. Cardiovasc. Risk.* **8,** 219–225.

Index

1-hydroxyethyl radical, 93, 95
  PBN trapping efficiency, 98
Hydroxyl radical
  GC-MS analysis, 91
  generation by Fenton reaction, 91
  PBN spin trap, 92, 96

I

Immunoblotting, 50, 268
Isoelectric focusing, 143
Isoprostanes
  extraction, 62
  GC/negative ion chemical
    ionization/MS analysis, 64
  handling and storage, 60
  plasma levels, 64
  purification, 63
  regioisomers, 58

L

Lipid hydroperoxides, 13, 21
  carboxyl, aldehyde analysis, 25
  cis, trans analysis, 25
  HPLC photodiode assay
    detection, 15, 17
  infrared spectroscopy, 22
  molar extinction coefficients, 16
  nomenclature, 16
  serum values, 19
  standards, 15, 18, 38
Lipid hydroperoxide-derived protein
  modification, 40
ELISA, 39
  lipid-lysine adduct, 38, 42
  SDS-PAGE, 39
Lipid hydroxides
  simultaneous separation from
    lipid hydroperoxides, 19
  standards, 15

M

Metallothionein
  fluorescence derivatization, 277
  HPLC procedure
    chromatographic profile,
      278, 279
    copper content of MT, 281
    fluorescence detection, 277
    standard curve, 280
  metal chelation, 273
  sample preparation, 276
  standard, 275
  structure, 274
  thiol oxidation, 285
    colorimetric assay, 287
    gel chromatography, 289
    oxidation conditions, 287
    recovery, 290
    sample preparations, 286

N–P

8-nitroguanine
  HPLC electrochemical detection, 81
  DNA content, 78
8-oxoguanine, 78
  HPLC electrochemical detection, 81
Peroxynitrite, 29
  ADP-ribose synthetase
    activation, 83
  DNA injury, 77
  8-nitroguanine, 78
  8-oxoguanine, 82
Phospholipase A_2
  immunoblotting, 49
  metabolic pathway, 46
  secretory and cytosolic forms, 45
Polyunsaturated fatty acids
  HPLC procedure, 17